OCEANIC WHITECAPS

and Their Role in Air–Sea Exchange Processes

Quilted patchwork wall-hanging depicting sea foam (oceanic whitecaps) as perceived in various cultures. Executed by Helen Hardesty and Elizabeth Monahan.

Oceanic Whitecaps

and Their Role in Air–Sea Exchange Processes

Edited by

EDWARD C. MONAHAN

Marine Sciences Institute, University of Connecticut, Avery Point, U.S.A.

and

GEARÓID MAC NIOCAILL

University College, Galway, Ireland

D. Reidel Publishing Company

A MEMBER OF THE KLUWER ACADEMIC PUBLISHERS GROUP

Dordrecht / Boston / Lancaster / Tokyo

Published in association with the GALWAY UNIVERSITY PRESS

Library of Congress Cataloging in Publication Data

CIP

Oceanic whitecaps and their role in air-sea exchange processes.

(Oceanic sciences library)
"Published in association with the Galway University Press."
"Proceedings of the 1983 Galway Whitecap Workshop"–CIP foreword.
Bibliography: p.
Includes index.
1. Ocean waves–Congresses. 2. Ocean-atmosphere interaction–Congresses. I. Monahan, Edward C., 1936- II. Mac Niocaill, Gearóid. III. Galway Whitecap Workshop (1983) IV. Series.
GC206.027 1986 551.47′02 86-6521
ISNB 90-277-2251-X

Published by D. Reidel Publishing Company
P.O. Box 17, 3300 AA Dordrecht, Holland

Sold and distributed in the U.S.A. and Canada
by Kluwer Academic Publishers,
190 Old Derby Street, Hingham, MA 02043, U.S.A.

In all other countries, sold and distributed
by Kluwer Academic Publishers Group,
P.O. Box 322, 3300 AH Dordrecht, Holland

This work is related to Department of the Navy Grant N00014-83-G-0110
issued by the Office of Naval Research.

Printed in The Netherlands

CONTENTS

FOREWORD

While various volumes have previously been devoted to such topics as droplets and bubbles, it is our conceit that this is the first volume dedicated to the description of the phenomenon of oceanic whitecapping, and to a consideration of the role these whitecaps play in satellite marine remote sensing, in sea-salt aerosol generation, and in a broad range of other sea surface processes. This observation, reflecting in part the relatively modest attention paid until recently by the scientific community to whitecaps, is noteworthy when one considers that collectively whitecaps are to the general public one of the most striking features of the seascape. Whitecaps feature prominently in many paintings of nautical subjects, and in numerous poems, both ancient and modern. No novelist in describing a stormy sea, fails to mention these transient foam patches. In the vernacular of various cultures there are diverse, often animate, terms, such as white horses, hares, or geese,* used to describe these commonly occuring features on the sea surface. (Perhaps the ultimate measure of their impact on the public consciousness is that at least one professional sports team, i.e. the Vancouver, British Columbia, soccer side, has adopted them as their symbol!).

Why until lately has the scientific community shown scant interest in whitecaps, when they represent so obvious a feature of the sea surface? One possible, and perhaps the most probable, answer to this question lies in the observation that while whitecaps are some of the most apparent features associated with high sea states, they have also proved to be some of the most difficult objects to measure and describe quantitatively, and while scientists as a group may like to tackle difficult problems, we should not be accused of undue modesty when we observe that as a group we also have a finite tolerance for frustration and a human, perhaps aesthetic, prejudice in favour of natural phenomena that are amenable to detailed description. It is appropriate to note that Professor Woodcock, to whom this volume is dedicated, approached the task of describing the bubbles that comprise oceanic foam patches from the perspective of an observer, keen of eye and unencumbered by the biases of the academic scientist.

That a considerable research effort, involving theoretical and laboratory approaches as well as work at sea, has recently been committed to obtaining a description and understanding of the whitecapping phenomenon and related processes, will, we hope, be manifest to anyone who reads the various contributions that comprise this volume. Resources have been made available for the investigation of whitecaps now that their important role in various sea-air exchange processes has become apparent. The recent attention given to the physics of the marine atmospheric boundary layer has helped to stimulate the study of whitecaps. With the coming-of-age of satellite remote sensing, we find that whitecaps may prove to be more rea-

*See the workshop banner in the group photograph for a depiction of a number of these representations.

dily detectable and amenable to measurement, using various passive microwave systems, than are near-surface wind speeds, and that indeed whitecap measurements may be used as a practical basis of estimating these wind speeds, an idea that would not come as a surprise to Admiral Beaufort if he were alive in retirement in his native Co. Meath today.

This volume, and the 1983 Galway Whitecap Workshop whose proceedings it represents, has been made possible by a grant from the Office of Naval Research of the United States Navy to University College, Galway, of the National University of Ireland. It is the fervent wish of the organisers of this workshop that research on this topic will henceforth progress at such a rate that another such workshop, and a similar proceedings volume, will within a few years be deemed desirable and indeed necessary. We would hope that such a second meeting on whitecaps could be held before the end of the current decade. We trust that this present volume will serve as a source book for those now turning to the study of whitecaps. To this end, we have included as an appendix a supplementary bibliography, in which are listed pertinent articles not cited by the authors of the various chapters.

As the convenor of the 1983 Whitecap Workshop, I wish to express my gratitude to all of the authors who have contributed to this volume, to my colleagues at University College, Galway, who have helped with the various aspects of the workshop and with the preparation of the volume, and to the Office of Naval Research, Washington, which has provided the necessary funding, and given me, and a number of the other participants in this workshop, the encouragement to pursue our individual investigations of whitecap-related phenomena.

Edward C. Monahan
Aboard the FS *Polarstern*
81°31′N, 5°39′E, July 1983.

AFTERWORD

Some time has elapsed since the Whitecap Workshop was held in University College, Galway. Despite the best of intentions of the editors and the conscientious adherence to deadlines on the part of all the contributors, publication of this volume has suffered an unanticipated delay which was outside the control of the publishers and editors. Nevertheless, the material presented in this volume still represents an up-to-date authoritative state-of-the-art review in this area of physical oceanography.

I would like to thank my co-editor Gearóid Mac Niocaill, the staff of Officina Typographica, the Galway University Press, and Mrs. Françoise Yates, Mrs. Elizabeth Monahan, Miss Ann Duffy, and Miss Nancy Monahan, for their editorial and administrative efforts. I would also like to thank D. Reidel Publishing Company for efficiently organising publication after the final camera-ready typescript became available. Lastly, our thanks go to the U.S. Office of Naval Research and University College, Galway, for making the Workshop, and this resulting volume, possible.

Edward C. Monahan
Galway
January 1986

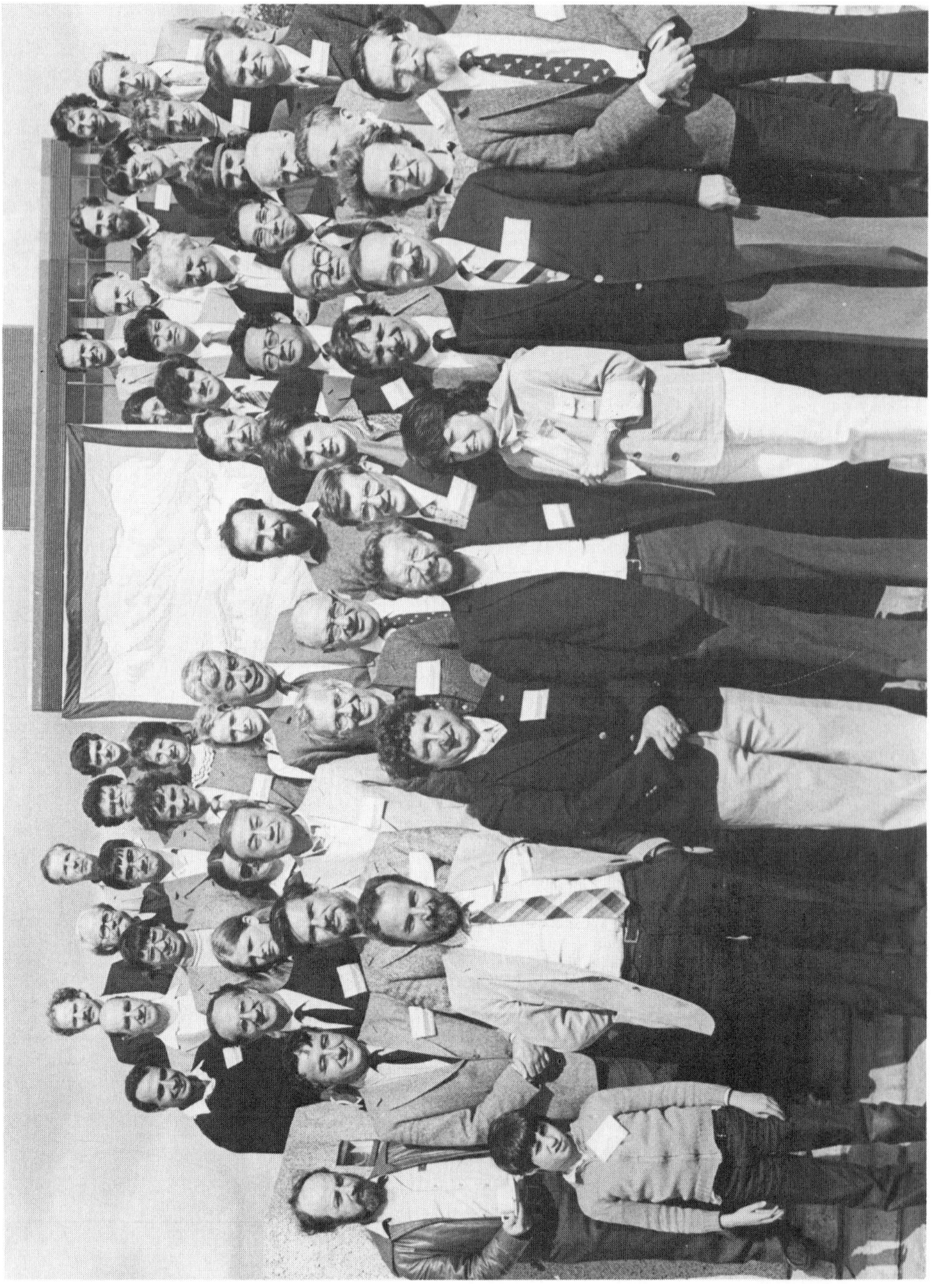

Key to Group Photo

1. J. C. Scott, Portland
2. L. D. Syzdek, Albany
3. H. Gucinski, Anne Arundel
4. L. Merlivat, Gif-sur-Yvette
5. S. Colgan, Washington
6. E. C. Monahan, Galway
7. M. Spillane, Galway
8. B. Farrell, Haulbowline
9. A. W. Hogan, Albany
10. D. C. Blanchard, Albany
11. R. Abbey, Washington
12. K. Davidson, Monterey
13. D. M. Doyle, Galway
14. M. Griggs, San Diego
15. J. Wu, Lewes
16. P. Twitchell, Washington
17. H. J. Exton, Manchester
18. W. Martin, Washington
19. B. Vonnegut, Albany
20. P. J. Stabeno, Galway
21. P. Park, Manchester
22. A. J. Croft, Coventry
23. E. G. Smith, Coventry
24. M. Y. Su, Bay St. Louis
25. C. McKinney, London
26. L. Cavaleri, Venice
27. S. Burk, Monterey
28. S. J. Hogan, Cambridge
29. J. P. Ralls, Bristol
30. L. Hasse, Kiel
31. S. G. Jennings, Galway
32. J. Latham, Manchester
33. Y. Toba, Sendai
34. Q. Espey, Southampton
35. S. Gathman, Washington
36. M. S. Longuet-Higgins, Cambridge
37. M. H. Smith, Manchester
38. A. Duffy, Secretariat
39. N. E. Monahan, Secretariat
40. M. Koga, Sapporo
41. T. I. Bern, Trondheim
42. P. Koepke, Munich
43. T. O'Connor, Galway
44. B. D. Johnson, Halifax
45. L. H. Holthuysen, Delft
46. S. Thorpe, Wormley
47. ? , Venice
48. A. Roddy, Galway
49. F. Resch, Toulon
50. F. MacIntyre, Narragansett
51. B. Kerman, Downsview
52. R. Cipriano, Albany

P. A. Bowyer, Galway
M. Briquet, Arcueil
I. G. Ó Muircheartaigh, Galway
D. E. Spiel, Monterey

Dr. Alfred H. Woodcock
to whom this volume is dedicated

THE LIFE AND SCIENCE OF ALFRED H. WOODCOCK

DUNCAN C. BLANCHARD
Atmospheric Sciences Research Center
State University of New York at Albany

The stock market crash of 1929 plunged the United States into the greatest crisis since the Civil War, and by the spring of 1931 the country had entered the long dark night of an economic depression. But in the small seaside town of Daytona, Florida, Mrs. Woodcock didn't have time to worry about the state of the country. She had something else to worry about. Her youngest son, Alfred, had just quit his job on a farm in Massachusetts. Leaving one job for something better is perhaps an admirable thing to do, but this was his 25th job since dropping out of high school about 10 years before. And what was his 26th job to be? A sailor! Woodcock decided that he wanted to go to sea. His mother, of course, had no way of knowing this was to be his last job, one that would lead him into science. To her it was just the latest in a long line of jobs that seemed to go on and on and on. She was really worried.

Alfred H. Woodcock was born in Atlanta, Georgia, on 7 September 1905. His parents had come from England, his father from Manchester and his mother from Kent. His father didn't think very highly of formal education. He had only three years of schooling, but that's all he needed to learn how to 'read, write, and do sums.' At the age of eight he had a job in the mills in Manchester and at age 12, after coming to America, did a man's work in running four looms in a mill in Lawrence, Massachusetts. Little wonder that he felt his young son was wasting time in school.

After dropping out of high school, Woodcock drifted into and out of jobs. He left some because they ceased to exist, being only seasonal in the tourist economy of Florida. He left others because of boredom, and some, well, he had no choice. He was fired. Woodcock once told me, 'I was fired from the front desk at the Clarendon Hotel because I did not remember the guests' names and from the Western Union Telegraph Company file-clerk job because, while daydreaming, I once filed an unsent telegram with those already transmitted. That was a no-no of the first magnitude.'

Woodcock had long dreamed of becoming a farmer, but decided he'd better learn something about the farming trade. He drove north, entered the Massachusetts Agricultural College, took the two-year course in practical agriculture, and went to work for a Massachusetts farmer. But no ordinary farmer was his employer. He had graduated from Harvard, was comfortably fixed in life, and owned a yacht, a 35-foot yawl. In the late spring of 1930, after spraying all the orchards, Woodcock, along with others, was aboard the yawl enjoying a leisurely sail southward along the Massachusetts coast to Cape Cod. They dropped anchor at the sleepy fishing village of Woods Hole. It was here that serendipity was to point the way to a life in science.

Over the only grocery store in Woods Hole was the barber shop, in the front room of the barber's cramped quarters. Needing a haircut, Woodcock went up and, while awaiting his turn in the chair, asked the gentleman next to him, whom he later found out to be captain of a United States Fisheries vessel, what that large

E. C. Monahan and G. Mac Niocaill (eds.), Oceanic Whitecaps, 1–14.

brick structure being built on the waterfront directly across from the barbershop was to be. The captain told him it would be an oceanographic institution, that the director was to be Dr. Henry Bigelow of Harvard University, and that they were looking around for a crew to go to Copenhagen to bring back the *Atlantis*, a new research vessel being built there. At that moment Woodcock decided that he needed more seagoing experience than he had gotten aboard the yawl on its cruise into Woods Hole. A job aboard the *Atlantis* would be just dandy.

Weeks later, after having found out that Columbus Iselin, a 25-year-old protégé of Dr. Bigelow, was to be the captain of the new *Atlantis*, Woodcock approached him for a job. His interview took place during a coffee break on the back steps of Harvard's Museum of Comparative Zoology. He was hired on the spot. Years later Iselin, who succeeded Dr. Bigelow as director of the Woods Hole Oceanographic Institution, said this: 'I liked him at once, and signed him on as an ordinary seaman at $45 a month, feeling that he would be a steadying influence among what looked like a rather turbulent crew... From the outset it was obvious that Woodcock was much more than a young sailor... He has been scientifically...productive... In fact, he is a remarkable person.'

In the spring of 1931, Woodcock and others of the fledgling crew went to Copenhagen to claim the *Atlantis*. She was a beautiful ship. With a length of 142 feet and rigged as a ketch with masts 100 and 140 feet high, she was the largest such vessel in the world. They took the *Atlantis* to Plymouth, England, to pick up the scientific staff. There, on 16 July, 1931, with all aboard, including the scientific paraphernalia they hoped to use on that first voyage and, oh yes, numerous cases of Carlsberg beer, they weighed anchor and headed out to sea. With the jib and mizzen sheeted in the *Atlantis* began beating to windward. As they headed into the open Atlantic the real education of Alfred Woodcock was about to begin.

During the next 10 years he would make many voyages on the *Atlantis*, far more than anyone else at the Woods Hole Oceanographic Institution. And far more than anyone else, he would let nature teach him the subtleties of how she worked. He would learn that nature knows no boundaries, and that the sea and the air, and the animals that live therein, are all part of one grand fabric that has evolved to be in exquisite ecological balance. Nature taught him that it wasn't quite fair to call one part biology, another part chemistry, and still another physics. Such compartmentalization in thinking about nature's mysteries was man's doing, not nature's. But best of all, nature taught Woodcock to look carefully at what he saw, and then to look again and again. He learned how to frame hypotheses and how to test them with simple experiments. He learned that while it might not be nice to fool Mother Nature, the reverse was not necessarily true. Mother Nature will gleefully trip you up whenever she can. Her secrets are not easily uncovered.

During that first crossing of the Atlantic the work of science went on. Hydrographic stations were made, the penetration of light into the sea was noted, the vertical migration of plankton was studied, and some meteorological data were collected. By the time the *Atlantis* arrived in Woods Hole, Woodcock had had his initiation into oceanographic science. What are his memories of that time? He recently told me '... I had little or no inkling of the real meaning of science before I arrived in Woods Hole on the *Atlantis*. However, during this period I was learning that my interest in farming was really an interest in nature, all nature, whether airborne, water borne, or ashore. From the beginning I did not like school, unfortunately. Why, I am not sure. For years after leaving school a recurring nightmare was to find myself back in the classroom again.'

Although Woodcock had been hired as a sailor-technician, it became apparent to Iselin and others that he could be far more than that. Most of the *Atlantis*' cruises were in the summer, as that was the time when scientists from universities came to Woods Hole to augment the tiny year-round staff. But in 1935 Iselin, who no longer was Captain but Chief Scientist, decided to take the *Atlantis* on its first winter cruise into the stormy North Atlantic. Other than the ship's crew there would be only two persons aboard, himself and Woodcock. Just before sailing in February, Iselin's wife became ill and he had to let Woodcock go to sea to do the science alone. Iselin wrote to Woodcock, 'You will undoubtedly have trouble at this time of year. There will be storms, fog, rain, sleet. If field ice is reported along the south-west edge of the Grand Banks, the shallow stations may be omitted.' When he read that, Woodcock probably wished he had stayed on the farm.

But away to sea he went and a splendid job he did as a one-man scientific staff. Schlee (1978), in her book about the *Atlantis*, has this to say: 'Under either condition, mild or miserable, Woodcock appeared in the deck lab for every station. In the highly unusual role of single scientist he felt himself responsible for all data gathered. As if his scientific responsibilities were not enough, he chose to spend hours standing watch. Huddled in the bow or clinging halfway up the mast, he watched all that moved through sea and air.'

Five of Woodcock's first six papers were about birds at sea. In between and after his daily tasks on the *Atlantis* were done, he observed and patiently recorded bird behavior. Since birds have been flying the marine airways for hundreds of thousands of years, they must have evolved to take maximum advantage of the ups and downs of the unseen rivers of air. Thus, Woodcock reasoned, the birds have a lot to tell us about air motion over the sea. All we have to do is watch them. And watch them he did, and after seven years his first paper on birds at sea appeared in the May, 1938, *Atlantic Monthly*. His first writing is not the cold, harsh logic of science, but reveals his respect for the sea and his admiration for the birds who must survive there. Listen:

> 'The constant westerly gale has built up long high waves, so large that there are green plateaus on some of them, topped by snowy peaks... The fulmar petrels fly among these shifting peaks and crags, swooping down into the valley eddies, constantly battling the mighty river of air which would sweep them ever backward... Tired of climbing the everlasting meteorological hill, they come to rest on the sea surface and sit jauntily hove-to under slow speed ahead, occasionally lifting their wings to jump a breaking crest.'

In his third paper, published in 1940 in the *Journal of Marine Research*, he presented his now-classic observations of herring-gull soaring at sea. With nothing more than a thermometer, an anemometer, and a sharp pair of eyes, he deduced from the soaring routines of the gulls the existence of two types of organized convection in the subcloud layer at sea. When the air was warmer than the water, or stable, there was no convection, and he never saw free soaring among the gulls. No dummies, these gulls. When the air was stable they spent a lot of time sitting in the water. In the air they had to work mightily with wind-flapping flight to stay aloft. However, when Woodcock found the air colder than the sea, or unstable, the gulls took to the air and soared easily for minutes on end.

But more importantly, Woodcock observed two types of soaring. For wind speeds less than about 7m s^{-1}, the gulls soared in lazy circles in the sky and slowly moved upward. He called this circle soaring and deduced that the gulls were in organized updrafts that could be the roots of clouds. At wind speeds approaching 7m s^{-1} the chimney-like thermals began to tilt precariously, as if to dump the gulls soaring at different levels. When the wind exceeded 7m s^{-1} the soaring routine of the gulls changed dramatically. Circle soaring gave way to what Woodcock called linear soaring, a type of soaring '... as nearly two-dimensional as it is possible for flight to be.' Here the birds, distributed in a thin vertical sheet, soared directly into the wind, rising as they went and finally disappearing from sight far upwind. Woodcock deduced that the gulls were taking advantage of the updrafts in longitudinal roll convection cells, the type responsible for cloud streets. Above wind speeds of about 13m s^{-1} the gulls did not soar and Woodcock concluded that organized convection breaks down. These classic observations, made nearly half a century ago, have long since been confirmed, but the end is not in sight. A few years ago, Woodcock (1975) returned to this work and argued that under some conditions the gulls, with a mighty community effort of wind flapping, can start their own thermals.

If we can learn from the birds about organized circulations in the air, then perhaps the animals living in the sea have something to tell us about organized circulations in the sea. Woodcock had long been puzzled about the sailing before the wind of the Portuguese Man-of-War, a jellyfish of the genus *Physalia* with a large balloon-like float that caught the wind. It was not that it sailed with the wind that puzzled Woodcock, but it always sailed about forty-five degrees to the left of the wind. Why? He reasoned that this was an evolutionary adaptation that increased the survival of the animal. But the question still persisted; why is it safer for a *Physalia* to sail left of the wind rather than right or directly downwind? Maybe the answer has to do with organized circulation in the surface waters. Woodcock had heard of the work of Langmuir (1938), who had shown on Lake George in New York that large counter-rotating pairs of cells, with horizontal axes parallel to the wind, are what cause surface slicks and debris to be lined up with the wind. Something about these Langmuir circulations may provide a clue to the answer. No one had ever measured Langmuir circulations at sea, so Woodcock went out and did it. After many experiments in which ballasted bottles were tossed into the sea from the *Atlantis*' dinghy that he was rowing crosswind, he found that the two cells in each counter-rotating pair were *unequal* in size. The one to the left of the slick as you looked downwind was much smaller. With this finding he presented a clever explanation of why a *Physalia* sailing left of the wind would have the best chance of escaping entanglement and death in the sargassum weed and other materials caught in the convergence zone of the Langmuir circulations (Woodcock, 1944).

This was not the end of the story. Woodcock postulated that the asymmetry in the Langmuir cells was caused somehow by the Coriolis force. If that was true, then in the Southern Hemisphere the asymmetry should be reversed and the *Physalia* there should have evolved to sail to the right of the wind. He found some pickled *Physalia* that had been collected in the Southern Hemisphere and which were preserved in a museum. To his delight 86 percent of these *Physalia* were mirror images of their Northern Hemisphere neighbors. In life they must have sailed to the right of the wind. And there the story ends. Although numerous investigations of Langmuir circulations have been made since, not a one has attempted to extend, much less confirm, Woodcock's puzzling finding of the asymmetry of these circulations.* And no one has been south of the equator to see if, as Alice saw in her Looking-Glass House, '...things go the other way.'

* * *

In 1940, as the winds of war swept ever closer to the United States, Woodcock and others at Woods Hole became more and more involved in classified work for the Navy. He was chief scientist on a cruise of the *Atlantis* early in 1941. The object of the cruise? To learn more about a problem that plagued the Navy sonars. Sometimes they could not detect underwater submarines. Sound refraction by the temperature structure in the water was the cause. The way to predict it was to know the temperature structure. The bathythermograph had just been developed at Woods Hole and the Woods Hole scientists were busy teaching Navy personnel how to use it.

During the war Woodcock worked briefly with a group that was trying to defoam the wakes of the Navy's amphibious landing craft. These brilliant white wakes were highly visible from the air, sometimes even at night. He tells of the many hours he spent on the fantail of the *Atlantis* helping to pour gallon after gallon of various so-called foam breakers, including peanut oil, in an attempt to destroy the tell-tale wake of the *Atlantis*.

Because he had learned so much about convection over the sea from watching the birds,

* Not quite true. In a theoretical analysis of the problem, Munk (*Annals N. Y. Acad. Sci.*, **48**, 815–820, 1947) offered a possible explanation in which '... the asymmetry of the convection cells is brought about not by a direct action of the earth's rotation upon the convective motion, but by the effect of the Coriolis force upon the surface current, which in turn distorts the convection cells into the observed asymmetrical pattern.' And Savilov (*Trudy Inst. Okeanol. Akad. Nauk SSSR*, **45**, 223–239, 1961) did observe that a majority of the *Physalia* in the Southern Hemisphere sailed to the right of the wind. However, he argued that this was due to the strength and position of regions of high pressure, but he admitted that he could not rule out Woodcock's hypothesis. In conclusion, the jury is still out.

Woodcock was asked to work with the smoke screen project. The laying down of dense layers of smoke to hug the sea and conceal ships and landing craft from enemy operations was a top priority for the Navy. Such work was not without its lighter moments. One day, while working in the Gulf of Mexico with the small armada of 22 ships and planes, including a blimp, that had been assigned to the smoke screen project. Woodcock was asked to report aboard one of the ships to do some trouble shooting. The ship was the U.S.S. *Woodcock*. When he identified himself as Woodcock from Woods Hole his credibility took a sudden plunge.

What was the trouble aboard the ship? The officers told him that something was wrong with the Langmuir-Schaefer fog-oil generators (Langmuir, 1948a). Sometimes they worked just dandy, making heavy smoke that hugged the sea and hid all the ships, but on other days they weren't worth a damn. They made smoke so light that it went straight up into the air! Woodcock had to teach the officers what the birds had learned long ago. Air and smoke moves upward only when the sea is warmer than the air. Don't blame the smoke generators. Blame the atmospheric stability.

On the more serious side, Woodcock's analysis of many experiments on the dissipation of long lines of smoke laid down in unstable air by fast-moving ships indicated the existence of hexagonal convection cells at low wind speeds and longitudinal roll vortices at higher wind speeds (Woodcock and Wyman, 1947). Of course, by then he was not surprised.

At the end of the war the Navy, still interested in the movement of air over the sea, released several ships and planes for researches in the trade wind areas of the North Atlantic. The coleaders of the expedition were not meteorologists. They were Jeffries Wyman, a physical chemist from Harvard, and Woodcock, the self-taught man for all seasons. For the first time an extensive series of soundings of wet and dry bulb temperature and the vertical component of acceleration were made in the subcloud layer, the cloud layer, and within the clouds themselves. Now up until this time the conventional wisdom of the day proclaimed that the temperature lapse rate in a cloud was wet adiabatic. But that was not what Wyman and Woodcock found. They found it to be in between the wet lapse rate and the lapse rate in the clear air outside the cloud. In 1947, Henry Stommel of Woods Hole explained this in terms of cloud mixing with the environment. Today, students and teachers alike accept this without question, few if any knowing that the toppling of conventional wisdom was done with the help of the Wyman-Woodcock data.

The impact of the Wyman-Woodcock Report was felt in many places, and more than one eminent meteorologist was deeply influenced by its findings. Listen to what Joanne Simpson has to say:

'... it is to Al and the infectious enthusiasm and originality of his work to which I owe my choice of research area. In 1947, I was one of about a dozen graduate students at the University of Chicago to enroll in Herbert Riehl's first course in Tropical Meteorology. Also in this course were T. C. Yeh, H. L. Kuo, Seymour Hess, Werner Baum and Noel LaSeur. The lectures began with convection in the tropics. Riehl had just received a copy of the later famous Wyman-Woodcock Report and spent two weeks discussing it, emphasizing tropical cumuli, their unexpectedly non-wet-adiabatic cores and apparent lack of 'roots' in the subcloud layer.

'Almost immediately a bolt a lightning struck me and I said to myself and my colleagues 'This is it - tropical cumuli are what I want to work on.' Al's marvelous paper on the soaring of herring gulls and what it said about air motions was the subject of my term paper. I immediately wrote to Woods Hole applying for a summer job and started working on the Wyman-Woodcock cumulus data, which became the subject of my Ph.D. dissertation... Since then, in the dozen or more times I have taught tropical meteorology, the course always begins with the Wyman-Woodcock expedition.'

* * *

The stage was now set for the entry of Woodcock into the study of sea-salt particles and their role in the formation of rain. The curtain rose on this work to find him puzzling over a bizarre finding of the expedition. There appeared to be a small but significant source of sensible heat in the trade wind air between the sea and the clouds. If this was real, it might explain another problem he had pondered for a long time. He had seen frigate birds soaring in the trade winds when there was little or no difference in temperature between the sea and the air. His herring gulls in more northern latitudes couldn't soar in such conditions of neutral stability, so how could the frigate birds do it? Were they more efficient soaring machines or was this apparent heat source in the air producing thermals?

What was producing the heat? Could it be the latent heat from the condensation of water vapor on the sea-salt particles in the air? That might take a lot of salt particles. Were there enough in the air? Woodcock looked into the literature for an answer, but found none. No one had measured the size distributions of sea-salt particles. He'd have to do it himself. Thus he began a long series of experiments to perfect a method for the measurement of airborne salt particles. A year or more went by and he began to get some answers. Under normal conditions there were not enough salt particles in the air to produce much of a temperature difference and, besides, it appeared that the heat source may not have been real, just an artifact of instrument error. But by that time Woodcock had become excited about another possible role of the salt particles. Could they be responsible for the ease with which marine clouds produce rain? During the war meteorologists began to realize that the Bergeron ice-crystal mechanism (Bergeron, 1935; Blanchard, 1978) was not the only way in which rain could be produced. Warm rain was a reality and the hygroscopic sea-salt particles might trigger it off.

Before the war a number of eminent meteorologists had down-played the role of salt particles in atmospheric processes. Sir George Simpson (1939) of England did not think they were important for visibility and Findeisen (1937) of Germany didn't think they were important for much of anything. He flatly stated, without any evidence, that droplets smaller than 10 μm radius could not be produced by the sea. What should one do in such a situation? Exactly what Woodcock did. Make the measurements. He collected salt particles on small glass slides that were treated to make a 90-degree contact angle with water. Since for a given wind speed the collection efficiency varied not only with particle size but with the width of the glass slide, he exposed a series of different-sized slides to get a salt-particle spectrum that ranged over seven orders of magnitude by weight. The small slides (in later years only 60 μm wide!) collected primarily small particles, the large slides the large particles. Utilizing the isopiestic method, and by applying the newly-obtained Langmuir calculations of collection efficiency (Langmuir, 1948b), Woodcock calculated the salt-particle distribution captured by each slide. By piecing it all together he got one continuous distribution. Always suspicious of theoretical calculations, especially those he didn't really understand, Woodcock used two different experimental methods that convinced him that the Langmuir calculations were reasonably correct.

Over the next four years, starting in 1949, he published a series of seven pioneering papers that constitute a tour de force in the measurement of atmospheric sea salt (Woodcock and Gifford, 1949; Woodcock, 1950a, 1950b, 1950c, 1951, 1952, 1953a). His first measurements were made at the end of the Woods Hole dock in air blowing in from the sea, but with mastery of his glass slides he quickly took to an aircraft. At various altitudes and wind speeds over the sea, he made numerous measurements of salt particle spectra in the trade wind areas of both the Atlantic and the Pacific Oceans.

He even chased hurricanes and finally caught up with one. Huddled in the top of a Florida lighthouse that was shaking from the blasts of hurricane winds and from the battering of tons of water from the wind-ravaged surf, Woodcock spent a long night in September, 1948, collecting salt particles on his slides (Woodcock, 1950b). To this day, in spite of major advances in instrumentation, no one has repeated these measurements (I wonder why!). Aware of this, Woodcock would like to get more data. In 1958 he published his ideas about the role of salt particles in the release of latent heat in hurricanes, but now has further ideas he'd like to test. Two years ago, while telling me about these ideas, and getting more excited as he talked, Woodcock suddenly stopped and asked 'How do you suppose Jim Hughes would react if I were to tell him I'd like to make more salt particle measurements in hurricanes?' Al Woodcock was quite serious!

His early measurements showed that with winds normally found over the sea, the concentrations of sea-salt particles carried up to cloud base altitudes easily exceeded 10^5 m^{-3}. Most of them were less than 10 μm radius as drops of seawater, the ones Findeisen said couldn't be produced by the sea! With this evidence Woodcock, in 1952, published his first paper on the role of sea-salt particles in the formation of warm rain. His hypothesis predicted a one-to-one correspondence between the giant salt particles that enter the cloud and the raindrops that are produced, with the largest salt particles producing the largest raindrops. Field tests of this hypothesis, which involved comparing the salinity of raindrops to the mass of salt particles, suggested that the salt particles did indeed play a role in the formation of raindrops (Woodcock and Blanchard, 1955).

Woodcock's conviction of the importance of sea salt in the formation of cloud and rain, and the relative ease in which studies could be carried out in the trade winds of Hawaii, was firmly expressed in 1953 in his paper 'Hawaii as a Cloud Physics Laboratory.' In referring to the orographic rain produced on the windward slopes of the 14,000 foot volcano of Mauna Kea, he said: 'geophysical events of the past and present seem to have fortuitously conspired to produce a setting in which simple natural experiments in rain-making are an almost daily occurrence.' Thus, primarily inspired by Woodcock, Project Shower, the first large-scale, international attack on the physics and chemistry of warm rain took place in Hilo, Hawaii, in the fall of 1954. Those from outside the United States included Christian Junge from Germany, Erik Eriksson from Sweden, and from Australia Patrick Squires, Sean Twomey, and Jack Warner. The huge mass of data collected during Project Shower and some of the results were published in a special issue of *Tellus* in 1957.

After Project Shower Woodcock continued his studies on the role of sea-salt particles in atmospheric processes. No idea was too wild nor was any experiment too difficult to do. If he couldn't do it himself, he'd not hesitate to get advice and help from others. With Ted Spencer, his colleague and a superb instrument designer, he built and used in 1957 the first airborne flame photometer for the measurement of salt particles, and in 1964 they developed a photographic raindrop size recorder that could operate on batteries for several days in the rain forests of Hawaii. When the analytical chemists began to use neutron activation analysis in the 1960s to measure aerosol trace elements, Woodcock worked with Robert Duce and Jarvis Moyers. Duce and Moyers were among his new colleagues at the University of Hawaii. He had moved to Hawaii in 1963, partly to be where '...simple natural experiments in rain-making are an almost daily occurrence.'

In 1967 they determined the variation of ion ratios with size among sea-salt particles in the Hawaiian trade winds and related this to fractionation processes occurring during bubble bursting at the surface of the sea. In 1971 they succeeded in measuring the iodine-to-chloride ratio as a function of size on both salt particles and raindrops. To Woodcock's surprise, an analysis of the data revealed that his giant sea-salt particles probably were not the culprits responsible for the formation of the raindrops. The evidence pointed to, in Junge's terminology, the large salt particles, those from 10^{-12} to 10^{-14} g (between 0.1 and 0.5 μm radius as a dry salt particle). Though he had believed in his one-to-one hypothesis for about 20 years, Woodcock was never one to fly in the face of the facts. With Duce in 1972 he gave center stage to the large salt particles.

* * *

I first met Woodcock in March, 1951, in Woods Hole. Two months before, he had written me at Penn State where I was finishing a Master's degree in physics. He needed someone to help him measure raindrop sizes and the chlorinity of shower rains. Vincent Schaefer had told him I might be interested in the job. Would I like to visit him at Woods Hole and talk about it? Would I! As the train passed the last salt pond and cranberry bog and creaked into Woods Hole, I wondered what sort of person I was about to meet. I soon found out. The single, colonial brick building that comprised the Woods Hole Oceanographic Institution was clearly visible from the station, perched at the very edge of the harbor with its white cupola reflecting the morning sun. Far above, several herring gulls were soaring. Woodcock met me at the door. He was a middle-aged man of moderate build whose trim physique indicated a lot of exercise. A small fringe of hair surrounded a prematurely bald head and rimless glasses perched high on his nose.

In his office, with a soft-spoken, almost diffident manner one does not associate with those who have spent years at sea, Woodcock told me about his ideas on sea-salt particles and the formation of raindrops. He outlined some field experiments that could test his hypothesis. The measurement of raindrop-size distributions was crucial. I told him I had played around with the Bentley flour method of getting drop size, and that I had developed a screen technique that held promise, especially for large drops. I said that some people had used filter papers, though I had never tried them.

At lunch at a nearby restaurant he pointed out the window to some soaring gulls and told me how you could learn about convection in the air by watching them. Soon he was talking about an organized circulation in the ocean and how it can be revealed by the sailing of jellyfish and alignment of *Sargassum* plants on the surface of the sea. Several napkins and many sketches later he had told me about his measurements of gas pressure in the bladders of *Sargassum*. He said that these plants are often dragged far beneath the surface in downwelling water, but he had a suspicion that they could counter the increased hydrostatic pressure by pumping more gas into their bladders to maintain their positive buoyancy and perhaps make it back to the surface (Woodcock, 1950d). We parted company soon after this, but not before I told him I'd be delighted to accept his job offer and start work right after my classes ended at Penn State. No small part of my delight was to hear Woodcock say that in September we would be going to Hawaii to begin a 10-month long field trip.

The days and months in Hawaii sped by. We spent countless hours using filter papers to measure raindrop-size distributions high in the rain forests of Maui or at cloud base on the windward side of the volcano of Mauna Kea. On many days Woodcock made flights over the windward sea to get the salt-particle distribution. We watched the dramatic and rapid formation of a stationary cloud at the convergence of the sea breeze and the trade winds on the lee side of the Big Island of Hawaii. These clouds formed in an area where the rainfall was less than 10 inches per year, a sharp change from the more than 300 inches per year on the windward side of the island. Since these clouds seldom rained, Woodcock wondered if salt seeding might work here. We wandered along the shores of Hawaii, watching the waves breaking and seeing in the beam of a flashlight at night great clouds of salt droplets being carried upward. We gazed in awe at an eruption of the volcano of Kilauea and saw an aerosol of another kind being forcibly ejected into the atmosphere.

But through all of this Woodcock kept returning to one problem, that of the formation of the sea-salt aerosol. He thought that upon our return to Woods Hole we should make a major effort to understand just what happens when air bubbles burst at the surface of the sea. He had done some work on that a few years before and felt that the bubbles held a lot of secrets we should know.

A year or so later, back at Woods Hole and after the preparation of several papers on the Hawaii trip, we turned to the bubble-bursting problem. We knew we'd have to use a high-speed camera running at several thousand frames per second to catch the details of bubble collapse. With some advice from Papa Flash, Harold Edgerton of M. I. T., and lots of trial and error, we succeeded in photographing the jet of water that rose rapidly from the collapsing bubble cavity, and we saw how the jet drops were pinched off to continue their upward flight (Woodcock et al., 1953). In 1957 (Blanchard and Woodcock) we published all we had learned about how bubbles are produced in the sea. In addition to some measurements of bubble spectra in breaking waves, we had found that rain and snow falling into the sea were prolific sources of bubbles.

It was no accident that Woodcock zeroed in on the bubble problem so quickly. He knew all about bubble-produced aerosols. In the summer of 1947 a rapid growth of phytoplankton in the coastal waters near Venice, Florida, discolored the water and produced what is called the Red Tide. When the winds were onshore, the people in Venice experienced a respiratory irritation and a burning sensation in the eyes, nose, and throat. At the request of the United States Fish and Wildlife Service Woodcock went to Venice to see what he could learn about the cause of these problems. It was not long before he observed that the problems were present only when the winds were strong enough to producee breaking waves or when raindrops fell into the sea. In either case, bubbles were produced.

In the lab he breathed in air that was passed over buckets of the red water and experienced no problem, but the instant he produced some air bubbles in the water and breathed in some of the aerosol from the bursting of the bubbles, he started coughing and felt the burning sensation in his nose and throat. But was it really caused by the droplets produced by the bubbles? Perhaps the irritant was in gaseous form in the bubbles. He quickly settled the question by breathing through some absorbent cotton. That filtered out the droplets and the irritation disappeared. It was clear that the irritant was being carried by the droplets. In 1948 Woodcock published this work in a pioneer paper that foreshadowed the present day findings on the enrichment of toxins, virus, and bacteria in jet and film drops (Baylor et al., 1977; Wendt et al., 1980; Blanchard, 1983).

* * *

Today, at age 78, Woodcock happily and doggedly still travels along the research trail he started to blaze over 50 years ago on the maiden cruise of the *Atlantis*. Although honors have come in the form of an honorary Doctor of Science degree from C. W. Post College in New York and election as a Fellow of the American Meteorological Society, they have not changed his advance along the trail. Only serendipitous happenings can do that, and usually they lead to new trails. Woodcock has made over 350 trips up the 14,000 ft. volcano of Mauna Kea, the most recent in May of 1983. He has published papers on mountain breathing (Woodcock and Friedman, 1979), on permafrost (Woodcock, 1974), and on the changes in lake level and the temperature gradient beneath Lake Waiau, one of the highest alpine lakes in the United States (Woodcock and Groves, 1969; Woodcock, 1980).

For the past two or three years he and his wife, Harriet, have spent June or July on Cape Cod, Massachusetts, where he has been busy trying to understand the often rapid formation of fog in the Cape Cod Canal and Buzzards Bay. He's anxious to test a fog forecasting scheme he's devised (Woodcock, 1983). In a letter to me last spring he said, 'It looks as if Harriet and I will be on the Cape again this summer, probably in July. I am excited by the prospect of trying to forecast the Canal fogs for the Corps of Engineers, and of learning more about the possible role of condensational heating of the canal surface film waters [Woodcock, 1982]... The canal fogs sometimes do not form when the Taylor diagram says that they should.'

Al Woodcock may never solve this mystery, but it won't be from lack of trying. And he won't accept the easy answers any more than he did on that flight we made from Honolulu to Hilo over 30 years ago. The planes were slow in those days and they flew low, weaving their way around the trade wind clouds. It was early in the morning and there were plenty of seats. Wanting to carefully observe the clouds, Al Woodcock took a window seat directly in front of me. Peering intently out the window, moving his head this way and that to get a better view, he was in marked contrast to the other passengers, most of whom were asleep. Noticing this, the flight attendant, a young woman, approached him and asked,

'Is there anything wrong, sir? Can I help you?'

'No, thank you,' Woodcock replied, 'My friend and I have come all the way from Massachusetts just to study your clouds. We'd like to understand how they make rain so easily.'

A puzzled look on the young woman's face turned solemn and she said, 'I know what makes the clouds rain.'

'You do,' said Woodcock. 'What?'

'God makes it rain.'

'Oh no,' replied Woodcock, with the slightest trace of a wry grin on his face, 'that answer's too easy. But if God does make the rain, we want to know exactly how he does it.'

ACKNOWLEDGEMENTS

I could not have written this biographical sketch without Woodcock's help in the form of many letters and conversations about his early years. His scientific and other adventures abroad the *Atlantis* are like a chapter out of Dana's *Two Years Before the Mast*. Some of this can be found in Susan Schlee's fine book *On Almost Any Wind*. I am grateful for her permission to use both the remarks of Columbus Iselin about Woodcock and her description of the activities of Woodcock on the first winter cruise of the *Atlantis*. The quote from Joanne Simpson, the 1983 recipient of the Carl-Gustaf Rossby Research Medal from the American Meteorological Society, is from a letter of hers about Woodcock's work. Mentioned only once in this sketch, but without whom it is doubtful that Woodcock could have accomplished all that he has, is James Hughes of the Office of Naval Research. As astute an observer of the scientific scene and disperser of government funds for research has seldom appeared on the American scene. Hughes, a one-man band who plays like a symphony in a huge bureaucracy, recognized early the talents of Woodcock and has supported his work since 1951.

REFERENCES

Baylor, E. R., Baylor, M. B., Blanchard, D. C., Syzdek, L. D., and C. Appel. Virus transfer from surf to wind. *Science*, 198, 575-580, 1977.

Bergeron, T. On the physics of cloud and precipitation. *Proceedings of the 5th General Assembly of the UGGI*, Lisbon, 2, 156-175, 1935.

Blanchard, D. C. The life and science of Tor Bergeron. *Bull. Amer. Meteor. Soc.*, 59, 389-392, 1978.

— The production, distribution, and bacterial enrichment of the sea-salt aerosol, in: *Air-Sea Exchange of Gases and Particles* (eds. P. S. Liss and W. G. N. Slinn). D. Reidel Pub. Co., 1983.

— and A. H. Woodcock. Bubble formation and modification in the sea and its meteorological significance. *Tellus*, 9, 145-158, 1957.

Duce, R. A., Woodcock, A. H., and J. L. Moyers. Variation of ion ratios with size among particles in tropical oceanic air. *Tellus*, 19, 369-379, 1967.

Findeisen, W. Entstehen die Kondensationskerne an der Meeresoberfläche? *Met. Zeit.*, 54, 377-379, 1937.

Langmuir, I. Surface motion of water induced by wind. *Science*, 87, 119-123, 1938.

— The growth of particles in smokes and clouds and the production of snow from supercooled clouds. *Proc. Amer. Philosophical Soc.*, 92, 167-185, 1948a.

— The production of rain by a chain reaction in cumulus clouds at temperatures above freezing. *J. Meteor.*, 5, 175-192, 1948b.

Schlee, S. On almost any wind, the saga of the oceanographic research vessel 'Atlantis.' Cornell University Press, 301 pages, 1978.

Simpson, G. C. Sea-salt and condensation nuclei. *Quart. J. Roy. Met. Soc.*, 65, 553-554, 1939.

Spencer, A. T. and A. H. Woodcock. A photographic rain recorder for studying showers in marine air. *J. Appld. Meteor.*, 3, 105-109, 1964.

Stommel, H. Entrainment of air into a cumulus cloud. *J. Meteor.*, 4, 91-94, 1947.

Wendt, S. L., George, K. L., Parker, B. C., Gruft, H., and J. O. Falkinham, III. Epidemiology of infection by nontuberculous mycobacteria III. Isolation of potentially pathogenic mycobacteria from aerosols. *Am. Rev. Respir. Dis.*, 122, 259-263, 1980.

Woodcock, A. H. Birds at sea. *Atlantic Monthly*, May, 678-681, 1938.

— Convection and soaring over the open ocean. *J. Marine Res.*, 3, 248-253, 1940.

— A theory of surface water motion deduced from the wind-induced motion of the *Physalia*. *J. Marine Res.*, 5, 196-205, 1944.

— Note concerning human respiratory irritation associated with high concentrations of plankton and mass mortality of marine organisms. *J. Marine Res.*, 7, 56-62, 1948.

— Condensation nuclei and precipitation. *J. Meteor.*, 7, 161-162, 1950a.

— Sea salt in a tropical storm. *J. Meteor.*, 7, 397-401, 1950b.

— Impact deposition of atmospheric sea salts on a test plate. *Amer. Soc. Testing Materials*, 50, 1151-1166, 1950c.

— Subsurface pelagic *Sargassum*. *J. Mar. Res.*, 9, 77-92, 1950d.

— Remarks on 'Sea salt in a tropical storm.' *J. Meteor.*, 8, 362-363, 1951.
— Atmospheric salt particles and raindrops. *J. Meteor.*, 9, 200-212, 1952.
— Salt nuclei in marine air as a function of altitude and wind force. *J. Meteor.*, 10, 362-371, 1953a.
— Hawaii as a cloud physics laboratory, *Pacific Science*, 7, 522-524, 1953b.
— Atmospheric sea-salt nuclei data for Project Shower. *Tellus*, 9, 521-527. Other papers by various authors are on pages 471-590, 1957.
— The release of latent heat in tropical storms due to the fall-out of sea-salt particles. *Tellus*, 10, 355-371, 1958.
— Permafrost and climatology of a Hawaii volcano crater. *Arctic and Alpine Research*, 6, 49-62, 1974.
— Thermals over the sea and gull flight behavior. *Boundary Layer Meteorology*, 9, 63-68, 1975.
— Hawaiian alpine lake level, rainfall trends, and spring flow. *Pacific Science*, 34, 195-209, 1980.
— Fog and tidal current connection at Cape Cod Canal—early recognition and recent measurements. *Bull. Amer. Meteor. Soc.*, 63, 161-166, 1982.
— Winds, upwelling and fog at Cape Cod Canal, Massachusetts. *J. Clim. Appld. Meteor.* **23**, 611-616, 1984.
— and D. C. Blanchard. Tests of the salt-nuclei hypothesis of rain formation. *Tellus*, 7, 437-448, 1955.
— and R. A. Duce. The 'large' salt nuclei hypothesis of raindrop growth in Hawaii: further measurements and discussion. *J. Rech. Atmos.*, 6, 639-649, 1972.
—, Duce, R. A. and J. L. Moyers. Salt particles and raindrops in Hawaii. *J. Atmos. Sci.*, 28, 1252-1257, 1971.
— and I. Friedman. Mountain breathing—preliminary studies of air-land interaction on Mauna Kea, Hawaii. U. S. Geological Survey Professional Paper 1123-A, A1-A8, 1979.
— and M. M. Gifford. Sampling atmospheric sea-salt nuclei over the ocean. *J. Mar. Res.*, 8, 177-197, 1949.
— and G. W. Groves. Negative thermal gradient under alpine lake in Hawaii. *Deep-Sea Res.*, 16 (supplement), 393-405, 1969.
—, Kientzler, C. F., Arons, A. B. and D. C. Blanchard. Giant condensation nuclei from bursting bubbles. *Nature*, 172, 1144, 1953.
— and A. T. Spencer. An airborne flame photometer and its use in the scanning of marine atmospheres for sea-salt particles. *J. Meteor.*, 14, 437-447, 1957.
— and J. Wyman. Convective motion in air over the sea. *Annals N. Y. Acad. Sci.*, 48, 749-776, 1947.

SCIENTIFIC PAPERS OF DR. ALFRED H. WOODCOCK

The 76 papers of Alfred H. Woodcock (as of March, 1985) can be loosely divided into five categories. In the first we find the 14 papers on the observations of animals and plants that revealed organized motion in the air and the sea. A number of Woodcock's early papers were on this subject. The second category contains 11 papers on convection and heat exchange. They cover work that ranges from the temperature in the surface microlayer of a fresh water pond to the release of latent heat in hurricanes. The third category, 41 papers, contains his work on sea salt, bubbles, rain, and fog. These papers constitute the body of work for which he is best known. In the fourth category, 8 papers, we find his researches on ice, heat flow, and alpine lakes. Although the first paper was published in 1947, Woodcock's interest in the subject was rekindled by his many trips to the summit of Mauna Kea, Hawaii. The fifth category contains only two papers that might have been listed in the third category. They are listed separately because they were seminal papers on the origin of biological aerosols, a subject of much interest today.

A. *Animals and Plants as Indicators of Organized Motion in the Air and in the Sea.*

1. Birds at sea. *Atlantic Monthly*, 678-681, May 1938.
2. Observations on herring gull soaring. *Auk*, 57, 219-224, 1940.
3. Convection and soaring over the open ocean. *J. Mar. Res.*, 3, 248-253, 1940.
4. Soaring over the open sea. *Sci. Monthly*, 55, 226-232, 1942.
5. A theory of surface water motion deduced from the wind-induced motion of the *Physalia. J. Mar. Res.*, 5, 196-205, 1944.
6. The swimming of dolphins. *Nature*, **161**, 602, 1948.
7. Subsurface Pelagic *Sargassum. J. Mar. Res.*, 9, 77-92, 1950.
8. Wave-riding dolphins. *J. Exp. Biology*, **28**, 215-217, 1951 (A. H. Woodcock and A. F. McBride).
9. Dimorphism in the Portuguese Man-of-War. *Nature*, **178**, 253-255, 1956.
10. Sustained swimming in dolphins. *Science*, **133**, 952, 1961.
11. The spacing of windrows of *Sargassum* in the ocean. *J. Mar. Res.*, 22, 22-29, 1964 (A. J. Faller and A. H. Woodcock).
12. Note concerning *Physalia* behavior at sea. *Limnol. Oceanog.*, **16**, 551-552, 1971.
13. Thermals over the sea and gull flight behavior. *Boundary-Layer Meteor.*, 9, 63-68, 1975.
14. Discussion of 'Thermals over the sea and gull flight behavior.' *Boundary-Layer Meteor.*, **10**, 247-248, 1976 (A. H. Woodcock and J. W. Deardorff).

B. *Convection and Heat Exchange.*

1. Surface cooling and streaming in shallow fresh and salt waters. *J. Mar. Res.*, 4, 153-161, 1941.
2. Vertical motion and exchange of heat and water between the air and the sea in the region of the Trades. Report of studies conducted during the spring of 1946 in the Panama and Caribbean Sea frontiers by the Woods Hole Oceanographic Institution for the Bureau of Ships, U.S. Navy Department; Contract NObs-2083 (J. Wyman and A. H. Woodcock).
3. Convective motion in air over the sea. *Annals N. Y. Acad. Sci.*, **48**, 749-776, 1947 (A. H. Woodcock and J. Wyman).
4. Temperatures observed near the surface of a fresh-water pond at night. *J. Meteor.*, **4**, 102-103, 1947 (A. H. Woodcock and H. Stommel).
5. Diurnal heating of the surface of the Gulf of Mexico in the spring of 1942. *Trans. Amer. Geophys. Union*, **32**, 565-571, 1951 (H. Stommel and A. H. Woodcock).
6. The release of latent heat in tropical storms due to the fall-out of sea-salt particles. *Tellus*, **10**, 355-371, 1958.
7. The origin of trade-wind orographic shower rains. *Tellus* **12**, 315-326, 1960.
8. A February storm and limnology. *Oceanus*, 7, 14-15, 1961.
9. Salt-induced convection and clouds. *J. Atmos. Sci.*, **20**, 159-169, 1963 (A. H. Woodcock, D. C. Blanchard and C. G. H. Rooth).
10. Latent heat released experimentally by adding sodium chloride particles to the atmosphere. *J. Appld. Meteor.*, **6**, 95-101, 1967 (A. H. Woodcock and A. T. Spencer).
11. Comments concerning 'convection patterns in a pond'. *Bull. Amer. Meteor. Soc.*, **64**, 274-277, 1983 (A. H. Woodcock and R. B. Lukas).

C. *Sea-salt, Rain and Fog.*

1. Sampling atmospheric sea-salt nuclei over the ocean. *J. Mar. Res.*, **8**, 177-197, 1949 (A. H. Woodcock and M. M. Gifford).

2. Condensation nuclei and precipitation. *J. Meteor.*, 7, 161-162, 1949.
3. Impact deposition of atmospheric sea salts on a test plate. *Proc. Amer. Soc. Testing Materials*, **50**, 1151-1166, 1950.
4. Sea salt in a tropical storm. *J. Meteor.*, 7, 397-401, 1950.
5. Remarks on 'Sea salt in a tropical storm.' *J. Meteor.*, **8**, 362-363, 1951 (A. H. Woodcock and J. E. McDonald).
6. Atmospheric salt particles and raindrops. *J. Meteor.*, 9, 200-212, 1952.
7. Salt nuclei in marine air as a function of altitude and wind force. *J. Meteor.*, **10**, 362-371, 1953.
8. Hawaii as a cloud physics laboratory. *Pacific Science*, 7, 522-524, 1953.
9. Giant condensation nuclei and bursting bubbles. *Nature*, **172**, 1144-1146, 1953 (A. H. Woodcock, C. F. Kientzler, A. B. Arons and D. C. Blanchard).
10. Remarks on 'Atmospheric salt particles and raindrops.' *J. Meteor.*, **10**, 165-166, 1953 (A. H. Woodcock and J. E. McDonald).
11. Photographic investigation of the projection of droplets by bubbles bursting at a water surface. *Tellus*, **6**, 1-7, 1954 (C. F. Kientzler, A. B. Arons, D. C. Blanchard and A. H. Woodcock).
12. Salt nuclei, wind and daily rainfall in Hawaii. *Tellus*, **7**, 291-300, 1955 (A. H. Woodcock and W. A. Mordy).
13. Tests of the salt-nuclei hypothesis of rain formation. *Tellus*, 7, 437-448, 1955 (A. H. Woodcock and D. C. Blanchard).
14. Salt dust from the sea. *Naval Research Reviews*, pp. 5-7, August 1955.
15. Bubbles, salt dust and raindrops. *Oceanus*, **4** 17-20, 1956.
16. Salt and rain. *Sci. American*, **197** (October), 42-47, 1957.
17. An airborne flame photometer and its use in the scanning of marine atmospheres for sea-salt particles. *J. Meteor.*, **14**, 437-447, 1957 (A. H. Woodcock and A. T. Spencer).
18. Project shower—an investigation on warm rainfall in Hawaii. *Tellus*, **9**, 471-590, 1957 (A. H. Woodcock and many other authors).
19. Atmospheric salt in nuclei and in raindrops. *in* Artificial Stimulation of Rain, Pergamon Press, New York, pp. 202-206, 1957.
20. Atmospheric sea-salt nuclei data for Project Shower. *Tellus*, **9** 521-524, 1957.
21. Bubble formation and modification in the sea and its meteorological significance. *Tellus*, **9**, 145-158, 1957 (D. C. Blanchard and A. H. Woodcock).
22. Determination of deuterium-hydrogen ratios in Hawaiian waters. *Tellus*, **9** 553-556, 1957 (I. Friedman and A. H. Woodcock).
23. Lava-sea-air contact areas as sources of sea-salt particles in the atmosphere. *J. Geophys. Res.*, **66** 2873-2887, 1961 (A. H. Woodcock and A. T. Spencer).
24. Interchange of properties between sea and air (solubles). *The Sea*, I, 305-312, 1962, John Wiley and Sons, New York.
25. The deuterium content of raindrops. *J. Geophys. Res.*, **68**, 4477-4483, 1963 (A. H. Woodcock and I. Friedman).
26. A portable flame photometer for analysis of sodium in individual raindrops. *J. Atmos. Sci.*, **20**, 343-347, 1963 (A. T. Spencer and A. H. Woodcock).
27. A photographic rain recorder for studying showers in marine air. *J. Appld. Meteor.* **3**, 105-109, 1964 (A. T. Spencer and A. H. Woodcock).
28. Variation of ion ratios with size among particles in tropical oceanic air. *Tellus*, **19**, 369-379, 1967 (R. A. Duce, A. H. Woodcock and J. L. Moyers).
29. Sodium-to-chloride ratio in Hawaiian rains as a function of distance inland and of elevation. *J. Geophys. Res.*, **74** 1101-1103, 1969 (Y. B. Seto, R. A. Duce and A. H. Woodcock).
30. Rainfall trends in Hawaii. *J. Appld. Meteor.*, **9**, 690-696, 1970 (A. H. Woodcock and R. H. Jones).
31. Salt particles and raindrops in Hawaii. *J. Atmos. Sci.*, **28**, 1252-1257, 1971 (A. H. Woodcock, R. A. Duce and J. L. Moyers).
32. Difference in chemical composition of atmospheric sea salt particles produced in the surf zone and on the open sea in Hawaii. *Tellus*, **23**, 427-435, 1971 (R. A. Duce and A. H. Woodcock).
33. The 'large' salt nuclei hypothesis of raindrop growth in Hawaii: further measurements and discussion. *J. Rech. Atmos.*, **6**, 639-649, 1972 (A. H. Woodcock and R. A. Duce).
34. Smaller salt particles in oceanic air and bubble behavior in the sea. *J. Geophys. Res.*, 77 5316-5321, 1972.
35. Anomalous orographic rains of Hawaii. *Mon. Wea. Rev.*, **103**, 334-343, 1975.
36. Marine fog droplets and salt nuclei - part I. *J. Atmos. Sci.*, **35**, 657-664, 1978.
37. The production, concentration, and vertical distribution of the sea-salt aerosol. *Ann. N. Y. Acad. Sci.*, **338** 330-347, 1980 (D. C. Blanchard and A. H. Woodcock).
38. Marine fog droplets and salt nuclei - part II. *J. Atmos. Sci.*, **38**, 129-140, 1981 (A. H. Woodcock, D. C. Blanchard and J. E. Jiusto).
39. Fog and tidal current connection at Cape Cod Canal —early recognition and recent measurements. *Bull. Amer. Meteor. Soc.*, **63**, 161-166, 1982.
40. Fog inception in strong salt-laden marine winds. *Tellus*, **34**, 127-134, 1982.
41. Winds, upwelling and fog at Cape Cod Canal, Massachusetts. *Tellus*, in press.

D. *Ice and Alpine Lakes*

1. Patterns in pond ice. *J. Meteor.*, **4**, 100–101, 1947 (A. H. Woodcock and G. A. Riley).
2. Melt patterns in ice over shallow waters. *Limnol. Oceanog.*, **10** (supplement), R290-R297, 1965.
3. Deep layer of sediments in alpine lake in the tropical mid-Pacific. *Science*, **154**, 647–648, 1966 (A. H. Woodcock, M. Rubin and R. A. Duce).
4. Negative thermal gradient under alpine lake in Hawaii. *Deep-Sea Research*, **16** (supplement), 393–405, 1969 (A. H. Woodcock and G. W. Groves).
5. Fossil ice in Hawaii? *Nature*, **226**, 873, 1970 (A. H. Woodcock, A. S. Furumoto and G. P. Woollard).
6. Permafrost and climatology of a Hawaii volcano crater. *Arctic and Alpine Research*, **6**, 49–62, 1974.
7. Mountain breathing-preliminary studies of air-land interaction on Mauna Kea, Hawaii. U. S. G. S. Professional Paper 1123A, pages A1–A8, 1979. This paper was published, along with three other papers, in a single document by the U. S. Government Printing Office, Washington, D. C. (A. H. Woodcock and I. Friedman).
8. Hawaiian alpine lake level, rainfall trends, and spring flow. *Pacific Science*, **34**, 195–209, 1980.

E. *Biological Aerosols.*

1. Note concerning human respiratory irritation associated with high concentrations of plankton and mass mortality of marine organisms. *J. Mar. Res.*, 7, 56–62, 1948.
2. Bursting bubbles and air pollution. *J. Sewage Industrial Wastes*, 27, 1189–1192, 1955.

WAVE GROUP STATISTICS

M. S. LONGUET-HIGGINS
Department of Applied Mathematics and Theoretical Physics, University of Cambridge, Silver Street, Cambridge, England *and* Institute of Oceanographic Sciences, Wormley, Surrey, England.

Because of the dispersive property of surface waves in deep water, whitecaps are intermittent, and their lifetime is controlled partly by the lengths of wave groups, or of runs of successive high waves. This paper discusses and unifies two different approaches to the problem of wave grouping: (a) treating the sea state as a gaussian process, with group properties given by the wave envelope function, and (b) treating the sequence of wave-heights as a one-step Markov process. The latter is here related, with examples, to the spectral density function $E(\sigma)$, and it is shown that the spectral width parameter ν plays an important role in both (a) and (b). It is pointed out, however, that any group analysis implicitly requires a prefiltering of the data, and an appropriate band-pass filter is determined. The Markov predictions that the distribution of high runs is a negative exponential, and that the total group length is the difference between two exponentials, are both confirmed by comparison with numerical data.

1. Introduction

The intermittent behaviour of whitecaps, and wave breaking, in a wind-driven sea is closely related to the tendency of high waves to occur in groups. The observation that each whitecap, in deep water, appears only for a limited time, and is succeeded periodically by a whitecap on the wave next following it (Donelan, Longuet-Higgins and Turner 1972) can be accounted for by the dispersive property of waves in deep water; the phase velocity being greater than the group velocity, individual waves tend to move forwards relative to the group, and to break only for a limited time during which their steepness exceeds a certain threshold. Hence the length of a group of dominant waves, along with the average wave steepness, is a controlling factor in determining the lifetime of a typical whitecap.

Similar considerations will apply to the breaking of short waves riding on the backs of dominant longer waves, for the short waves will steepen and break preferentially on the steeper members of the dominant waves.

Accordingly, we see that the statistics of wave groups and the statistical distribution of group lengths will be basic to the study of whitecaps.

There have been two distinct approaches to the analysis of wave groups. The first, adopted by Longuet-Higgins (1957, 1962), Nolte and Hsu (1972) and Ewing (1973) treats the sea surface as a narrow-band gaussian noise $\zeta(t)$ and derives the group properties through the envelope function $\rho(t)$ of ζ (cf. Rice 1944, 1958). The group properties are then given in terms of the lower-order moments of the spectral density $E(\sigma)$.

A second approach, due to Sawnhey (1962), Goda (1970, 1983) and Kimura (1980) treats the sequence of wave heights as a Markov process, with certain transition probabilities between one wave height and the next. Though quite successful, this approach has less apparent connection with the wave spectrum, or with any physical model of the sea surface.

The purpose of the present paper is to test each theory against wave data, and to show the theoretical connection between the two approaches. On the one hand it is pointed out that

E. C. Monahan and G. Mac Niocaill (eds.), Oceanic Whitecaps, 15–35.

the concept of a wave group, or even of a single dominant wave, requires that we consider only a narrow, or band-passed, spectrum, so that the first approach (via the wave envelope) is more applicable to sea waves than is sometimes thought. On the other hand, in the second (Markov) approach, it is shown how the transition probabilities can be calculated directly from the wave spectrum, and may in some cases be expressed explicitly in terms of the lower spectral moments.

Though the correlation between successive waves may be high, we are able to show from theory that the correlation between *alternate* waves may simultaneously be quite low.

2. *Definitions: The wave envelope*

We assume that the surface elevation ζ may be represented as a stationary random function of the time t, with correlation function

$$\psi(\tau) = \overline{f(t)\, f(t+\tau)} \tag{2.1}$$

(a bar denoting the mean value with respect to t). The energy spectrum $E(\sigma)$ is related to $\psi(\tau)$ by

$$E(\sigma) = \frac{1}{\pi}\int_0^\infty \psi(\tau)\cos\sigma\tau\, d\tau \tag{2.2}$$

and so

$$\psi(\tau) = \int_0^\infty E(\sigma)\cos\sigma\tau\, d\sigma \tag{2.3}$$

We assume also that over some finite time-interval (-T/2, T/2) the function ζ may be represented as a Fourier sum:

$$\zeta = \sum_{n=0}^{\infty} c_n \cos(\sigma_n t + \epsilon_n) \tag{2.4}$$

where $\sigma_n = 2n\pi/T$, the phases ϵ_n are distributed uniformly over $(0, 2\pi)$ and the amplitudes c_n are such that

$$\lim_{T\to\infty} \frac{\sum \tfrac{1}{2} c_n^2}{d\sigma} = E(\sigma)\, d\sigma \tag{2.5}$$

to order $d\sigma$, the summation being over any small but fixed frequency range $(\sigma, \sigma + d\sigma)$.

The spectral moments m_r are defined by

$$m_r = \int_0^\infty \sigma^r E(\sigma)\, d\sigma \tag{2.6}$$

so that by (2.1)

$$m_0 = \int_0^\infty E(\sigma) d\sigma = \psi(0) = \overline{\zeta^2} \tag{2.7}$$

represents the mean-square surface displacement and

$$\bar{\sigma} = m_1/m_0 \tag{2.8}$$

may be defined as the 'mean frequency'. If μ_r denotes the rth moment of $E(\sigma)$ about the mean, i.e.

$$\mu_r = \int_0^\infty (\sigma - \bar{\sigma})^r E(\sigma) d\sigma \tag{2.9}$$

then clearly

$$\mu_0 = m_0 \quad \mu_1 = 0, \quad \mu_2 = m_2 - m_1^2/m_0 \tag{2.10}$$

and we may define the spectral width parameter ν by

$$\nu^2 = \frac{\mu_2}{\mu_0 \bar{\sigma}^2} = \frac{m_2 m_0}{m_1^2} - 1 \tag{2.11}$$

When $\nu^2 \ll 1$ we say that the spectrum is narrow.

Even when the spectrum is not necessarily narrow, it is useful to define the complex envelope function A(t) by writing (2.4) in the form

$$\zeta = Re\, A(t) e^{i\bar{\sigma}t} \tag{2.12}$$

where, for instance,

$$A = \sum_n c_n e^{i[(\sigma_n - \bar{\sigma})t + \epsilon_n]} = \rho e^{i\phi} \tag{2.13}$$

$\rho(t)$ may be called the real envelope function, or wave-amplitude, and $\phi(t)$ the phase. The real and imaginary parts of A are given by

$$\begin{aligned} \rho\cos\phi &= \sum_n c_n \cos[(\sigma_n - \bar{\sigma})t + \epsilon_n] \\ \rho\sin\phi &= \sum_n c_n \sin[(\sigma_n - \sigma)t + \epsilon_n] \end{aligned} \tag{2.14}$$

and it can be shown that these may always be computed, given only the initial function $\zeta(t)$.

From (2.13) it will be seen that the time-derivative $\mathring{A}_t = dA/dt$ contains, under the summation on the right, the factor $(\sigma_n - \sigma)$, so that

$$|\overline{\mathring{A}_t}|^2 = \mu_2 \tag{2.15}$$

and when the spectrum is narrow A varies slowly, on average, compared with the carrier wave $e^{i\sigma t}$. Thus, the wave record is practically sinusoidal, and the local, crest-to-trough, wave height is given by 2ρ, very nearly. A closer inspection suggests that the assumption is correct at least to order ν. Some terms of order ν^2 will nevertheless be carried along in the analysis.

In practice, the envelope $\rho(t)$ of a function $\zeta(t)$ given over some finite length of time T may be calculated by analysing ζ into its Fourier components, then shifting the phase of each component by 90° so as to obtain the Hilbert transform $\eta(t)$. The envelope $\rho(t)$ is then given by

$$\rho = (\eta^2 + \zeta^2)^{1/2} \tag{2.16}$$

3. The mean group-length $\bar{G}$ and high run length $\bar{H}$

Under general conditions the probability density of ζ is gaussian:

$$p(\zeta) = \frac{1}{(2\pi\, m_0)^{1/2}} e^{-\zeta^2/2m_0} \tag{3.1}$$

(see Rice 1944) and the mean number of up-crossings of ζ per unit time is given by

$$N(\zeta) = \frac{1}{2\pi}\left(\frac{m_2}{m_0}\right)^{1/2} e^{-\zeta^2/2m_0} \tag{3.2}$$

Similarly the probability density of the envelope $\rho(t)$ can be shown to be Rayleigh:

$$p(\rho) = \frac{\rho}{\mu_0} e^{-\rho^2/2\mu_0} \tag{3.3}$$

and the mean number of up-crossings of ρ per unit time is given by

$$N(\rho) = \left(\frac{\mu_2}{2\pi\mu_0}\right)^{1/2} \xi\, e^{-\frac{1}{2}\xi^2}, \quad \xi = \rho/\mu_0^{1/2} \tag{3.4}$$

We note that $N(\rho)$ is a maximum when $\xi = 1$ or $\rho = \mu_0^{1/2}$.

We may define the length l of a wave group as the time interval between two successive up-crossings of an arbitrary level ρ, so that the mean group-length is

$$\bar{l} = 1/N(\rho) \tag{3.5}$$

The mean number of waves $\bar{G}$ in a group may then be estimated as $\bar{l}N(0)$ where $N(0)$ is the number of zero-upcrossings of the mean level $\zeta = 0$. From equation (3.2) we have

$$\bar{G} = \left(\frac{m_2}{2\pi\mu_2}\right)^{1/2} \xi^{-1}\, e^{\frac{1}{2}\xi^2} \tag{3.6}$$

which can also be written

$$\bar{G} = \frac{1}{(2\pi)^{1/2}} \frac{(1+\nu^2)^{1/2}}{\nu} \xi^{-1}\, e^{\frac{1}{2}\xi^2} \tag{3.7}$$

where ν is the spectral width parameter defined in (2.14). When ν is small, clearly $\bar{G}$ varies as ν^{-1}, for any given constant level ξ or ρ.

Three levels are of particular interest: the level $\xi=1$, where G is a minimum; the mean level $\xi=(\pi/2)^{1/2}$; and the level $\xi=2$, which corresponds nearly to the 'significant' wave amplitude $\rho=2.003\,\mu_0^{1/2}$. If we neglect ν^2 compared with 1, then equation (3.7) gives the values of the mean group-length in these three cases as

$$\bar{G} = \begin{cases} 0.6577\ \nu^{-1}, & \xi = 1, \\ 0.6981\ \nu^{-1}, & \xi = (\pi/2)^{1/2}, \\ 0.14739\,\nu^{-1}, & \xi = 2 \end{cases} \tag{3.8}$$

A *high run* of waves H may be defined as a sequence of waves each of which exceeds a given reference level ρ (Rice 1958). The mean length $\bar{H}$ of high runs can be estimated by multiplying $\bar{G}$ by the proportion of time q that the envelope is above this level. From (3.3) this proportion is

$$q = \int_\rho^\infty p(\rho)\, d\rho = e^{-\frac{1}{2}\xi^2} \tag{3.9}$$

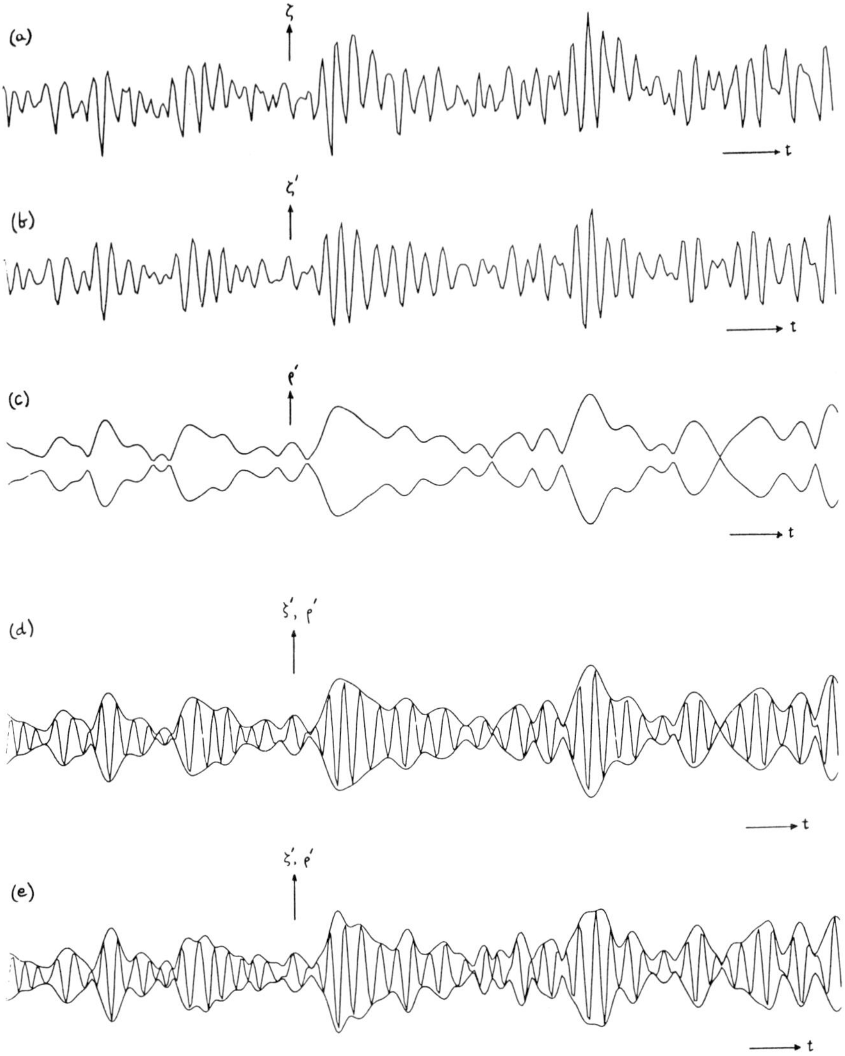

Figure 1. Part of a record of sea-surface elevation made with a shipborne wave recorder in the North Sea, at 55°04N, 7° 32E, on 22 September 1973. Digitisation: 1 c/s.

(a) Original record.
(b) Filtered record: $n'/n_p = 0.5, n''/n_p = 1.5$.
(c) Envelope function (2.14) of (b).
(d) Superposition of (b) and (c).
(e) As (d), but with $n'/n_p = 0.25, n''/n_p = 1.75$.

Thus the average number of waves $\bar{H}$ in a high run is given by

$$\bar{H} = \frac{1}{(2\pi)^{1/2}} \frac{(1+\nu^2)^{1/2}}{\nu} \xi^{-1} \tag{3.10}$$

from (3.7). We see that $\bar{H}$ is inversely proportional to ξ, and also to ν, when $\nu^2 \ll 1$.

4. Comparison with observation

Figure 1a shows a typical section of a record of surface elevation taken in the southern North Sea by a ship-borne wave recorder. The record is digitised at time-intervals of 1 sec.

The spectrum of a stretch of the record lasting 19½ minutes is shown in Figure 2. Each ordinate represents $\Sigma \frac{1}{2} c_n^2$ summed over 10 successive harmonics. The vertical scale has been normalised so that

$$\bar{\zeta} = c_0 = 0 \tag{4.1}$$

and

$$\overline{\zeta^2} = \sum_n \tfrac{1}{2} c_n^2 = 1 \tag{4.2}$$

where $c_n^2 = a_n^2 + b_n^2$. It can be seen that, apart from the slight rise in energy at very low frequencies (which may be partly due to the method of measurement) there is a single dominant peak in the spectrum at about $n = n_p = 165$ (corresponding to a frequency 0.141 c/s). On the high-frequency side, the energy falls away rapidly.

Now in studying group properties of the dominant waves we are not interested in very short waves; we do not include them in our count. In fact we are not even interested in waves as short as half the dominant wavelength. In effect, therefore, we apply a high-frequency cut-off to the wave spectrum at some frequency less than $2n_p$.

Similarly, we are not interested in very low frequencies, or in a slowly varying mean level, but only in the local crest-to-trough wave heights. So in effect we apply a lower cut-off to the spectrum as well.

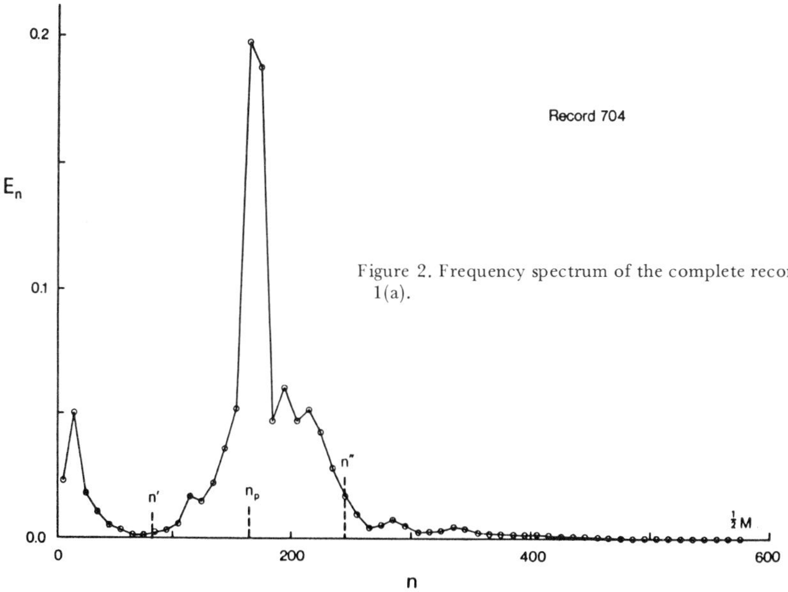

Figure 2. Frequency spectrum of the complete record 1(a).

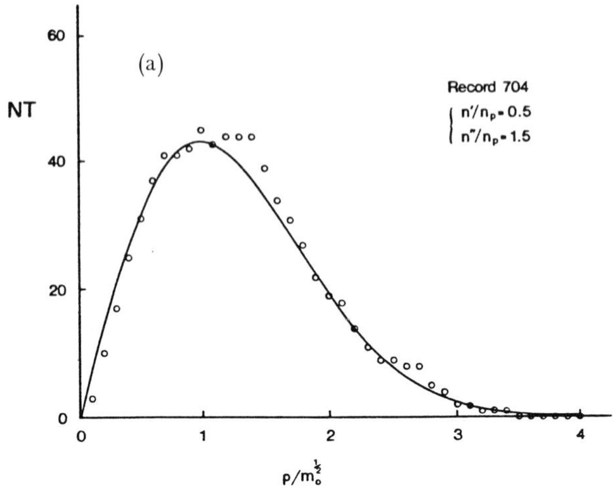

Figure 3. Number of level crossings of the wave envelope in the complete length of record shown partly in Figure 1, as a function of the critical level.
(a) when $n'/n_p = 0.5$, $n''/n_p = 1.5$,
(b) when $n'/n_p = 0.25$, $n''/n_p = 1.75$.

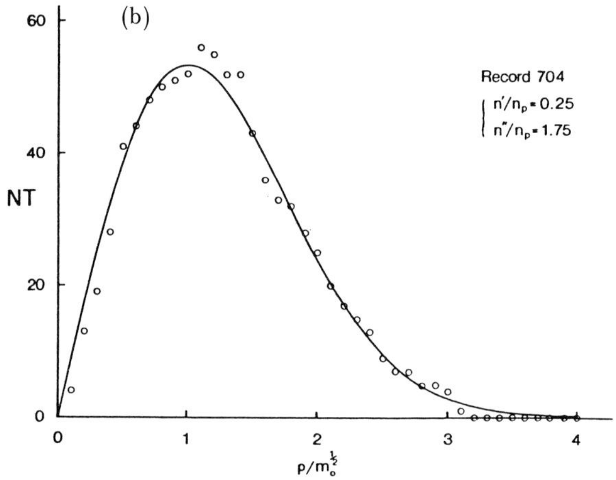

Suppose we take the lower and upper cut-off frequencies at $n'=0.5n_p$ and $n''=1.5n_p$, for example. Figure 1b shows the resulting filtered record $\zeta'(t)$. Corresponding crests and troughs of the dominant waves between Figures 1a and 1b can easily be identified. The envelope function $\rho(t)$ is shown by itself in Figure 1c and it will be noticed at once that there are a surprising number of points where ρ seems to approach zero, so that the positive and negative branches cross over. The function $\zeta'(t)$ and its envelope are shown superimposed in Figure 1d.

Figure 3a shows the total number of up-crossings of a given level ρ by the envelope function throughout the record (with the same choice of n', n''). The solid curve represents equation (3.4). The fit appears reasonable; statistical fluctuations might be expected to produce differences of order $(NT)^{1/2}$. It will be noticed that the maximum theoretical value, $NT = 43$ is quite close to the value $NT = 45$ which is obtained if one constructs a visual envelope of the original record ζ by drawing straight lines between successive crests.

Figure 1e shows the effect of taking different cut-off frequencies, so that now $n' = 0.25n_p$ and $n'' = 1.75n_p$. The envelope has many more fluctuations (maxima and minima) which seem to be irrelevant to the fluctuations in the height of the dominant waves. The corresponding number of level-crossings is shown in Figure 3b. Again, the empirical points agree reasonably with the theoretical curve, but the maximum value of NT is now 56, or somewhat greater than the visual value.

Table 3 summarises the results for various values of n'/n_p and n''/n_p. It will be seen that a change in n'/n_p from 0.5 to 0.25 has relatively little effect, but as n''/n_p is varied from 1.5 to 2.5, so NT departs more and more from the visual value.

Figure 4 shows the average number $\bar{G}$ of waves in a group corresponding to Figure 3a, that is to say when $n'/n_p = 0.5$ and $n''/n_p = 1.5$. The full curve represents the theory, equation (3.7). Except for the cases corresponding to very low levels of ρ, there is fair agreement. The minimum value of $\bar{G}$ at $\rho/\mu_0^{1/2} \cong 1$ is about 4.1, and at the significant wave amplitude ($\rho/\mu_0^{1/2} \cong 2$), $\bar{G}$ is about 9.2.

Figure 5 shows corresponding results for the mean number of waves $\bar{H}$ in a high run, given by equation (3.10). Though the two curves for $\bar{G}$ and $\bar{H}$ are quite different, the agreement is of course similar in Figures 4 and 5.

From this and other examples we may conclude that typical wind-wave spectra are effectively filtered by a 'group analysis', and that the cutoff frequencies $n' = 0.5n_p$, $n'' = 1.5n_p$ are appropriate. As seen from Table 3, this filtering of the record reduces slightly the total energy m_0 in the record. For a satisfactory analysis we may specify that m_0 shall not be changed significantly by the filtering.

5. The Distribution of Group-Lengths

The length l of a group was defined in Section 3 as the interval between two successive up-crossins of $\rho(t)$. The statistical distribution of l, apart from its mean $\bar{l}$, is difficult to determine

Table 1. Summary of the effect of varying the cut-off frequencies n' and n'' on the analysis of the record in Figure 1.

				N'			
n'/n_p	$n''n_p$	ν	$m_0^{1/2}$	$\rho_0 = m_0^{1/2}$ theor	obs	$\rho_0 = 2m_0^{1/2}$ theor	obs
0.50	1.50	0.160	15.5	43.2	45	19.3	19
0.50	1.75	.172	15.8	48.9	52	21.8	24
0.25	1.75	.196	15.9	53.4	56	23.8	25
0.25	2.00	.213	16.1	58.8	59	26.2	25
0.25	2.25	.237	16.2	66.1	69	29.5	26
0.25	2.50	.250	16.2	70.5	72	31.5	30
visual					45		18.5

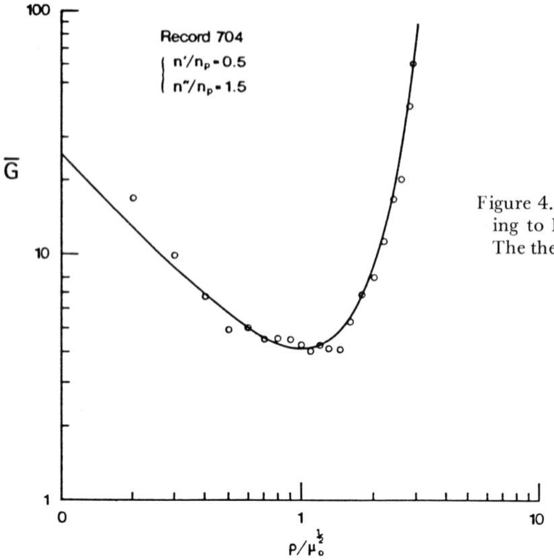

Figure 4. Plot of the mean group length $\bar{G}$ corresponding to Figure 1(d), as a function of the critical level. The theoretical curve represents equation (3.7).

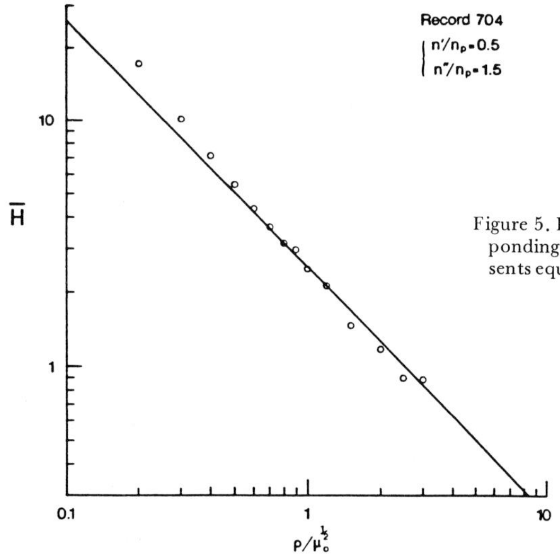

Figure 5. Plot of the mean length of high runs $\bar{H}$ corresponding to the record of Figure 1 (d). The line represents equation (3.10).

in general (see Rice 1958). However, for narrow spectra an approximation may be derived from the notion that since the spectrum of ρ is predominantly low-pass, we expect successive up-crossings to be uncorrelated, at least when l is sufficiently large. Hence the distribution of l will be asymptotically the same as in a 'shot-effect', where the time-axis is peppered randomly with points at a mean rate

(5.1) $$\lambda = 1/\bar{l}$$

per unit time. The density $p(l)$ for this process is known to be simply

(5.2) $$p(l) = \lambda e^{-\lambda l}$$

that is a negative exponential (see Rice 1954, Section 3.4). Rice gives a proof involving an infinite series of terms. A more direct proof is as follows. Divide a given interval $(t, t + l)$ into a large number of equal parts. The probability that ρ have no level crossing in any of these subintervals is $(1-\lambda l/m)^m$, and in the limit as $m \to \infty$ this tends to

(5.3) $$p(l) = e^{-\lambda l}$$

The density (5.2) then follows on applying to (5.3) the general formula

(5.4) $$p(l) = \frac{1}{\lambda}\frac{d^2 p}{d l^2}$$

where λ is the mean number of up-crossings per unit time (see Longuet-Higgins 1958, Section 2).

Incidentally it may be noted that for the low-pass spectrum $E(\sigma) = (1 + \sigma^2)^{-2}$, the distribution of zero-crossing intervals of ζ is almost (but not quite) negative exponential; see Favreau, Low and Pfeffer (1956); Longuet-Higgins (1962).

Assuming (5.2) to be valid, we have simply

(5.5) $$p(l) = \bar{l} e^{-c/\bar{l}}$$

and so for the number of waves G in a complete group

(5.6) $$p(G) = \bar{G} e^{-G/\bar{G}}$$

where $\bar{G}$ is given by (3.7). Some comparisons with observation will be given below.

To estimate the statistical density $p(H)$ of high runs, we may assume as an approximation that each high run H is, on the whole, in proportion to the corresponding group-length G, so $H = qG$, where q is given by (3.9). It follows that the distribution of H, like that of G, is also a negative exponential:

$$p(H) \cong \bar{H} e^{-H/\bar{H}} \tag{5.7}$$

Does this fit existing observations? Most data are given for integer values of the group-length G or run-length H. We may reasonably assume that the probability H_j of H for an integer value $j > 0$ is related to the continuous probability density $p(H)$ by

$$H_j \propto \int_{j-\frac{1}{2}}^{j+\frac{1}{2}} p(H)\, dH \tag{5.8}$$

that is to say, the probability density of a run of length H contributes to the probability of the discrete run having the nearest integer value. The densities for runs $H < \frac{1}{2}$ contribute only to runs of zero length, that is, they are ignored.

If $p(H)$ is a negative exponential, then the assumption (5.8) has two simple consequences: (1) the probability H_j is also negative exponential, that is

$$H_j \propto e^{-j/H} \tag{5.9}$$

and (2) because of the effective truncation of the distribution at $H = \frac{1}{2}$, the mean value $\bar{H}_j$ is increased by approximately the same amount, i.e.

$$\bar{H}_j \cong \bar{H} + 0.5 \tag{5.10}$$

Sufficiently long wave records are quite rare, but the numerically simulated data of Kimura (1980), reproduced in part in Figure 12, show conclusively that the distribution of H_j is indeed negative exponential, over practically the whole range of j. Figure 13 shows that the distribution of G_j is almost exponential, particularly when j is large, as expected, but for small values of j there are systematic differences.

For H_j it is therefore worth testing the second

conclusion (5.10) just mentioned. The 'target spectra' used by Kimura (1980) and shown in his Figure 8 appear to be of the form

$$(5.11)\quad S(f) = f^{-n}\, e^{(n/\gamma)(1-f^{-\gamma})}, \quad (f = 2\pi\sigma)$$

where $\gamma = 4$ and n runs from 4 to 8 in Kimura's Cases 1 to 5, respectively. It will be seen that the analytic form (5.11) has a peak at $f = f_p = 1$, where $f = f_{max} = 1$, as required.

With cut-off frequencies at $f = 0.5$ and 1.5, we calculate the values of ν, $\bar{H}$ and $\bar{H}_j$ seen in Table 2, both for $\rho = \rho_{mean}$ $(\rho/\mu_0^{1/2} = \sqrt{2/\pi})$ and for $\rho = \rho_{1/3}$ $(\rho/\mu_0^{1/2} = 2)$. Comparison with the data, taken from Table 1 of Kimura (1980) shows good agreement when $\rho = \rho_{mean}$, though less so when $\rho = \rho_{1/3}$.

We have seen then that the 'envelope approach' gives results for the mean values $\bar{G}$ and $\bar{H}$ in fair agreement with observation, but that it is unable, as yet, to predict the complete distributions p(G) and p(H) or their discrete counterparts G_j and H_j. In the following Sections we shall outline a different approach to finding G_j and H_j based partly on Kimura's (1980) work, but in a way that shows how it can be related theoretically to the previous method.

6. *Correlation between successive wave heights*

Consider the joint density of $\rho_1 = \rho(t_1)$ and $\rho_2 = \rho(t_2)$ at two points separated by a constant time interval $\tau = t_2 - t_1$. This is known exactly from the work of Uhlenbeck (1943) and Rice (1944, 1968). The general result may be written

$$p(\rho_1, \rho_2) = \frac{\rho_1 \rho_2}{\mu_0^2 (1-k^2)} e^{-(\rho_1^2 + \rho_2^2)/2\mu_0(1-k^2)} \times I_0\left(\frac{k}{1-k^2}\frac{\rho_1 \rho_2}{\mu_0}\right) \qquad (6.1)$$

where

$$X = \int_0^\infty E(\sigma) \cos(\sigma - \bar{\sigma})\tau \, d\sigma, \qquad Y = \int_0^\infty E(\sigma) \sin(\sigma - \bar{\sigma})\tau \, d\sigma, \qquad (6.2)$$

$$k = (X^2 + Y^2)^{1/2}/\mu_0 \qquad (6.3)$$

and I_0 denotes the modified Bessel function of order zero:

$$I_0(z) = \frac{1}{2\pi}\int_0^{2\pi} e^{z\cos\theta}\, d\theta \qquad (6.4)$$

Equation (6.1) has been called the 'two-dimensional Rayleigh distribution' (see Kimura 1980).

We shall assume that when the separation τ equals $2\pi/\bar{\sigma}$, then ρ_1 and ρ_2 approximate the amplitudes of two successive waves.

The correlation coefficient γ, defined as $M_{11}/(M_{20} M_{02})^{1/2}$ where

$$M_{pq} = \int_0^\infty \int_0^\infty (\rho_1 - \bar{\rho})^p (\rho_2 - \bar{\rho})^q \, p(\rho_1, \rho_2)\, d\rho_1\, d\rho_2, \qquad (6.5)$$

has been evaluated by Uhlenbeck (1943); see also Middleton (1960), as

Table 2. Comparison of theoretical and observed values of the mean probability $\bar{H}_1$ of high runs, in the data of Kimura (1980).

			$\rho = \rho$ mean			$\rho = \rho_{1/3}$		
Case	n	ν'	$\bar{H}$	$\bar{H}_j$	data	$\bar{H}$	$\bar{H}_j$	data
1	4	.1879	1.72	2.22	2.20	1.08	1.58	1.28
2	5	.1805	1.82	2.32	2.29	1.12	1.62	1.29
3	6	.1742	1.85	2.35	2.34	1.16	1.66	1.29
4	7	.1686	1.92	2.42	2.42	1.20	1.70	1.37
5	8	.1635	1.96	2.46	2.45	1.23	1.73	1.53

(6.6) $\gamma = [E - \frac{1}{2}(1-k^2)K - \frac{1}{4}\pi]/(1-\frac{1}{4}\pi)$

where E and K are complete elliptic integrals:

(6.7) $E(k) = \int_0^{\pi/2} (1-k^2 \sin^2 \theta)^{1/2} d\theta$

and

(6.8) $K(k) = \int_0^{\pi/2} (1-k^2 \sin^2 \theta)^{-1/2} d\theta$

γ is shown as a function of k^2 in Figure 11 (cf. Kimura 1980, Figure 1, where γ is shown as a function of k).

For values of k very close to 1 it may be shown that

$$\gamma \sim 1 - \frac{1-k^2}{4-\pi} \qquad (6.9)$$

and this is represented by the tangent at k = 1 to the curve in Figure 6. However, it can immediately be seen that for values of k^2 less than 0.6 a closer approximation to γ is given by the simple expression

$$\gamma \simeq k^2 \qquad (6.10)$$

represented by the straight diagonal in Figure 6. This holds good to within a few percent over the whole range of k.

Consider the interpretation of these results for a narrow spectrum. From (2.10) and (2.7) we have

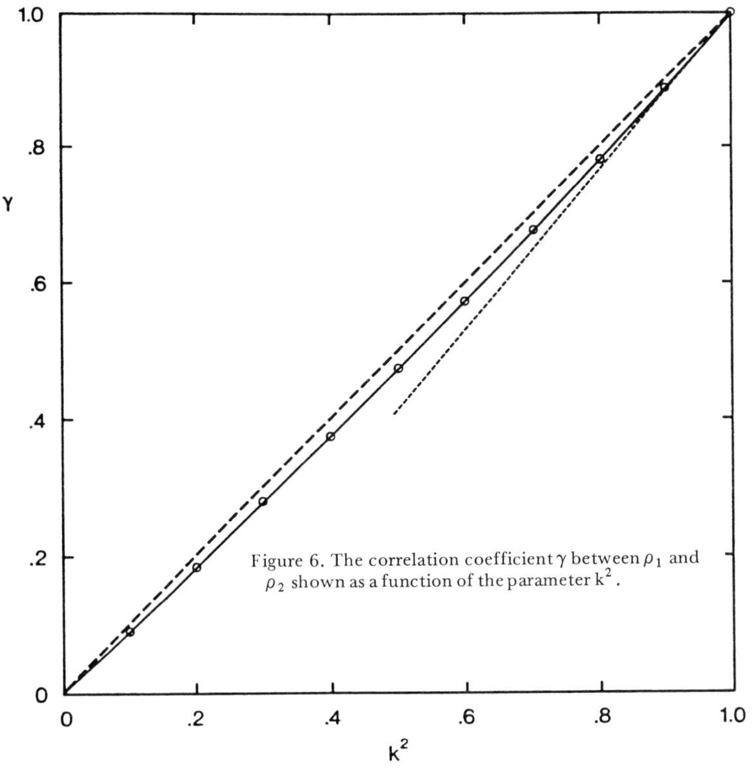

Figure 6. The correlation coefficient γ between ρ_1 and ρ_2 shown as a function of the parameter k^2.

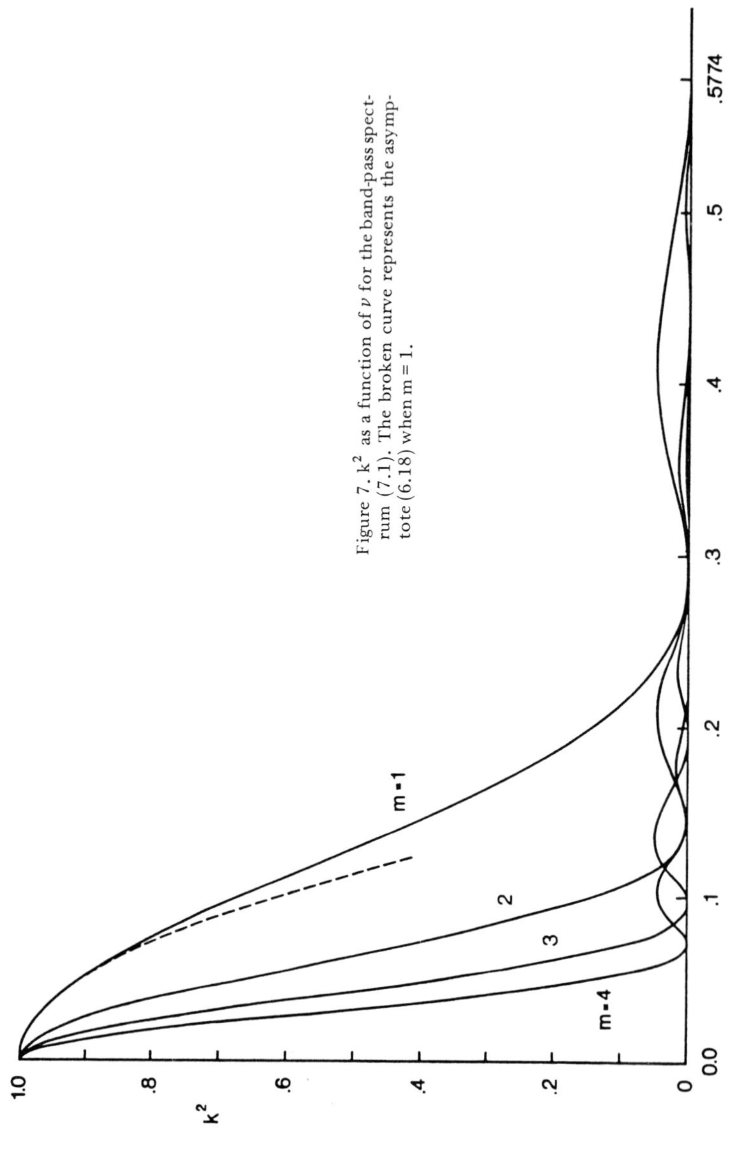

Figure 7. k^2 as a function of ν for the band-pass spectrum (7.1). The broken curve represents the asymptote (6.18) when m = 1.

$$\mu_0{}^2 = \int_0^\infty\int_0^\infty E(\sigma)\,E(\sigma')\,d\sigma\,d\sigma' \tag{6.11}$$

and similarly from (6.2)

$$X^2 + Y^2 = \int_0^\infty\int_0^\infty E(\sigma)E(\sigma')\cos(\sigma-\sigma')\tau\,d\sigma\,d\sigma' \tag{6.12}$$

So by (6.3)

$$\mu_0{}^2(1-k^2) = 2\int_0^\infty\int_0^\infty E(\sigma)E(\sigma')\sin^2\frac{(\sigma-\sigma')\tau}{2}\,d\sigma\,d\sigma' \tag{6.13}$$

For a narrow spectrum let us formally replace the trigonometric term in (6.12) by the first term in its power series, that is set

$$\sin^2\frac{(\sigma-\sigma')\tau}{2} \cong \tfrac{1}{4}(\sigma-\sigma')^2\tau^2 \tag{6.14}$$

$$= \tfrac{1}{4}\left[(\sigma-\bar\sigma)^2-(\sigma'-\bar\sigma)^2\right]\tau^2 \tag{6.15}$$

Then we obtain

$$\mu_0{}^2(1-k^2) \cong \tfrac{1}{2}(\mu_2\mu_0 - 2\mu_1{}^2 + \mu_0\mu_2) = \mu_0\mu_2\tau^2 \tag{6.16}$$

since $\mu = 0$. Hence writing

$$\tau = 2\pi/\bar\sigma \tag{6.17}$$

we see from (6.16) that, to lowest order,

$$1-k^2 = \frac{\mu_2}{\mu_0}\tau^2 = 4\pi^2\nu^2 \tag{6.18}$$

So from (6.9)

$$\gamma \sim 1 - 45.99\,\nu^2 \tag{6.19}$$

Because of the large coefficient multiplying ν^2, this formula for γ can be expected to be adequate only when $\nu \leqslant 0.1$, say.

If we wished to calculate the correlation γ_2 between *alternate* wave heights, we would have to substitute $\tau = 4\pi/\bar\sigma$ in (6.16), thus doubling τ and restricting the range of validity of the linear theory to $\nu \leqslant 0.025$ at most. Nevertheless the linearised theory does suggest qualitatively the very dramatic decrease in γ to be expected as ν and τ are increased.

For larger values of ν or τ we may use the accurate expressions for k^2 provided by equations (6.1) and (6.3), together with the relation between γ and k^2 indicated by the solid curve in Figure 6, or its approximation, equation (6.11).

7. *The correlation coefficient: examples*

To illustrate the dependence of γ on ν for typical spectra, consider the band-pass spectrum

$$E(\sigma) = \left.\begin{matrix}\alpha, & |\sigma-\bar\sigma| < \delta\bar\sigma \\ 0, & |\sigma-\bar\sigma| > \delta\bar\sigma\end{matrix}\right\} \tag{7.1}$$

for which $\nu = 3^{-1/2}\,\delta$. From equation (6.1) we have immediately

$$X = m_0\frac{\sin\delta\bar\sigma\tau}{\delta\bar\sigma\tau}, \quad Y = 0 \tag{7.2}$$

Hence

$$k^2 = \left(\frac{\sin\delta\bar\sigma\tau}{\delta\bar\sigma\tau}\right)^2 \tag{7.3}$$

To find $\gamma = \gamma_1$, write $\tau = 2\pi/\bar\sigma$, so

$$k^2 = \left(\frac{\sin 2\pi\delta}{2\pi\delta}\right)^2 \tag{7.4}$$

As $\delta \to 0$ we have $k^2 \sim 1-4\pi^2\delta^2/3$, in agreement with (6.18). As δ increases from 0, at first k^2 decreases monotonically to 0 at $\delta = 0.5$ ($\nu = 0.289$, see Figure 7). However as δ increases further, k^2 rises again to a maximum value 0.047 before falling finally to zero at $\delta = 1$ ($\nu = 0.577$).

In Figure 7 we also show γ_m, the correlation coefficient corresponding to $\tau = 2m\pi/\bar\sigma$. This shows that $\gamma_2 < \gamma_1$ always, but as m increases, it is not always true that $\gamma_{m+1} < \gamma_m$. For instance when $\nu = 0.14$, γ_3 may exceed γ_2.

This non-monotonic behaviour may be associated with the sharp cut-off in a band-pass spectrum. An example when the cut-off is

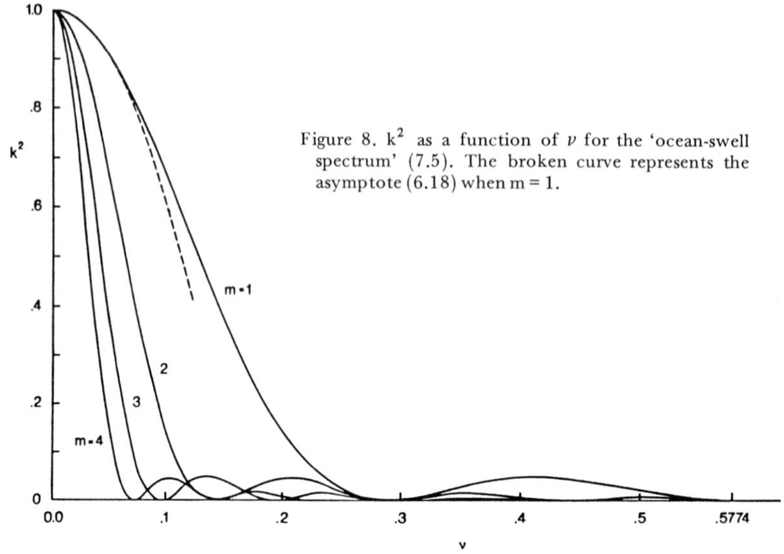

Figure 8. k^2 as a function of ν for the 'ocean-swell spectrum' (7.5). The broken curve represents the asymptote (6.18) when m = 1.

smooth, but still decisive, is provided by the 'ocean-swell' spectrum

$$E(\sigma) = \alpha\,\sigma^{-1/2}\,e^{-n[\beta\sigma + (\beta\sigma)^{-1}]/2} \qquad (7.5)$$

where α, β and n are constants. For such a spectrum it may be shown that

$$\nu = \frac{(n+2)^{1/2}}{n+1} \qquad (7.6)$$

and

$$k^2 = (1+r^2)^{-1/2}\, e^{-2n\,[\frac{1}{2}(1+r^2)^{1/2} + \frac{1}{2}]^{1/2}} - 1 \qquad (7.7)$$

where

$$r = 4m\pi/(n+1) \qquad (7.8)$$

The expression (7.7) is plotted against ν in Figure 8. Each curve is now monotonic in both ν and m, over the ranges shown, and when $\nu >$ 0.15, γ_2, γ_3 and γ_4 are all very small.

When ν = 0.082, for example, the sequence of values of k^2 for m = 1, 2, 3 and 4 is 0.76, 0.34, 0.10 and 0.02, giving γ_m = 0.71, 0.30, 0.09 and 0.02. This compares with Goda's (1983) values for swell of γ_m = 0.65, 0.35, 0.18 and 0.07.

For wind-waves, however, very different results are to be expected. Figure 9 shows k^2 plotted against ν for the Pierson-Moskowitz spectrum

$$E(\sigma) = \alpha\,\sigma^{-5}\,e^{-(\beta/\sigma)^{\gamma}}$$

for which

$$\nu^2 = \frac{\Gamma(2/\gamma)\,\Gamma(4/\gamma)}{\Gamma(3/\gamma)^2} - 1 \qquad (7.9)$$

(Here $\Gamma(z)$ is the standard gamma function.) The integrals X and Y of Section 6 were found by numerical integration. The behaviour of k^2, shown in Figure 8, differs from that in Figure 7. For one thing, the value of ν for the Pierson-Moskowitz spectrum is never less than 0.3536. Also the maximum value of k^2 is always less than 0.34. It is clear that for this spectrum the narrow-band expression (6.18) never applies.

Table 3a. Parameters of the Kimura spectrum (5.11).

					h_{mean}		$h_{1/3}$	
Case	n	ν	k^2	γ	p_+	p_-	p_+	p_-
1	4	.8319	.1253	.116	.490	.574	.194	.874
2	5	.6980	.1513	.140	.498	.581	.207	.876
3	6	.6118	.1820	.169	.507	.589	.225	.879
4	7	.5507	.2152	.200	.517	.598	.242	.881
5	8	.5047	.2493	.232	.528	.606	.260	.883

Table 3b. Parameters of the truncated Kimura spectrum.

					h_{mean}		$h_{1/3}$	
Case	n	ν	k^2	γ	p_+	p_-	p_+	p_-
1	4	.5857	.2071	.192	.516	.595	.237	.880
2	5	.5453	.2412	.224	.526	.604	.256	.883
3	6	.5113	.2723	.254	.535	.613	.275	.885
4	7	.4820	.3011	.281	.545	.620	.292	.888
5	8	.4565	.3284	.307	.555	.628	.308	.890

Table 3c. Mean values of H_j for the truncated Kimura spectrum

	h_{mean}		$h_{1/3}$	
Case	(8.2)	obs.	(8.2)	obs.
1	1.96	2.20	1.16	1.28
2	1.99	2.29	1.26	1.29
3	2.03	2.34	1.29	1.29
4	2.07	2.42	1.32	1.37
5	2.12	2.45	1.35	1.53

Table 3d. Mean values of G_j for the truncated Kimura spectrum

	h_{mean}		$h_{1/3}$	
Case	(8.5)	obs.	(8.5)	obs.
1	4.31	4.66	9.18	9.33
2	4.38	4.67	9.33	9.47
3	4.46	4.94	9.55	10.00
4	4.56	5.17	9.72	9.95
5	4.66	5.32	9.90	10.71

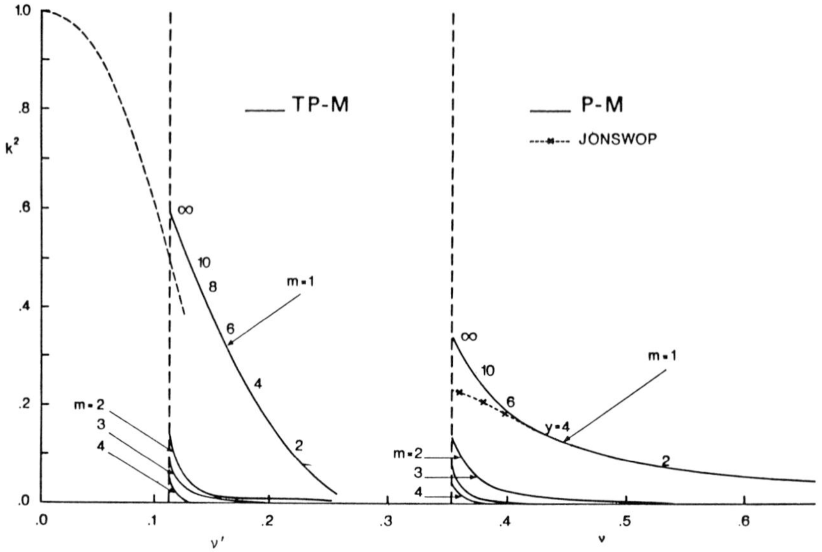

Figure 9. k^2 as a function of ν for the Pierson-Moskowitz spectrum (P–M), and JONSWAP spectrum. These are on the right. The corresponding curves for the truncated Pierson-Moskowitz spectrum (TP–M) are shown on the left.

However, as we saw earlier, it is implicit in the definition of a wave group that we use a filtered form of the record, with both a high-frequency and a low-frequency cut-off. The result after applying cut-offs at 1.5 and 0.5 times the peak frequency respectively, is shown in the left of Figure 9.

The lower bound for ν is now reduced to 0.113 and k^2 can be as great as 0.595 (compared with the narrow-band approximation 0.500). The sequence of values for $\gamma_1, \gamma_2, \gamma_3, \gamma_4$ is then 0.595, 0.139, 0.080 and 0.052. However, only a slight shift to the right, to say $\nu' = 0.16$ reduces γ_1 to about 0.33, which is typical of wind-waves. Further, γ_2, γ_3 and γ_4 are each reduced to less than 0.01, which can be considered insignificant.

8. *Distribution of G_j and H_j: Markov theory*

Kimura (1980) has given a rough but simple theory for the distribution of group lengths and of high runs, treating the sequence of waveheights as a Markov chain, as first suggested by Sawnhey (1962). Kimura's theory can be presented in an even simpler way, without the use of matrices, as follows.

Choose a critical waveheight h^* as in Figure 10. Given that a certain waveheight h_1 exceeds h^*, let p_+ denote the probability that the next waveheight h_2 also exceeds h^*. To determine the probability of a high run of length j we know already that the first waveheight exceeds h^*; the next (j-1) waveheights must then exceed h^* and the one after must *not* exceed h^* (see Figure 10a). The probabilities being assumed independent, the combined probability is

$$p(H_j) = p_+^{j-1}(1-p_+) \tag{8.1}$$

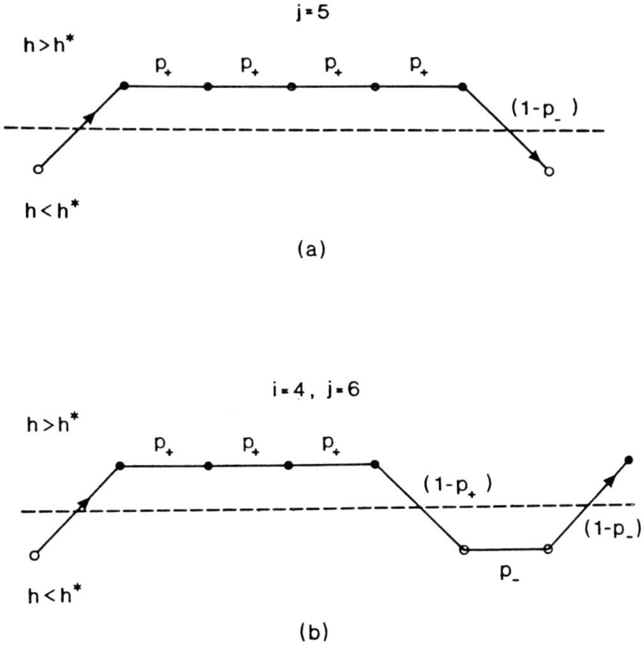

Figure 10. Diagram showing the basis for (a) the probability of a high run of j waves, equation (8.1), and (b) the probability of a wave group of j waves, equations (8.3) and (8.4).

The mean length of high runs is then given by

$$\bar{H} = \sum_{1}^{\infty} j\, p\,(H_j) = \frac{1}{1 - p_+} \tag{8.2}$$

To find the distribution of total runs we may reason as follows. In a total run of length j the first i waves, say, will be a high run of length i and the remaining (j-i) waves will be a low run of length (j-i) (see Figure 10b). The probability of such an event is clearly

$$p_+^{i-1}\,(1-p_+)\,p_-^{j-i-1}(1-p_-) \tag{8.3}$$

where p_- denotes the probability that $h_2 < h^*$ given that $h_1 < h^*$. Summing the above expression from i = 1 to i = j-1 we obtain

$$p(G_j) = (1-p_+)\,(1-p_-)\,\frac{p_+^{j-1}-p_-^{j-1}}{p_+-p_-} \tag{8.4}$$

when $j \geqslant 2$. The mean length of a total run is then

$$\bar{G} = \sum_{2}^{\infty} j\, p\,(G_j) = \frac{1}{1-p_+} + \frac{1}{1-p_-} \tag{8.5}$$

The only question then is to detemine p_+ and p_- for a given wave record.

Kimura (1980) proposed that $p\,(h_1, h_2)$ be approximated by a two-dimensional Rayleigh distribution of the form (6.1), which is reasonable if we assume that h_1 and h_2 can be approximated by $2\rho_1$ and $2\rho_2$ respectively (though Kimura does not explicitly make this assumption). Then the conditional probabilities p_+

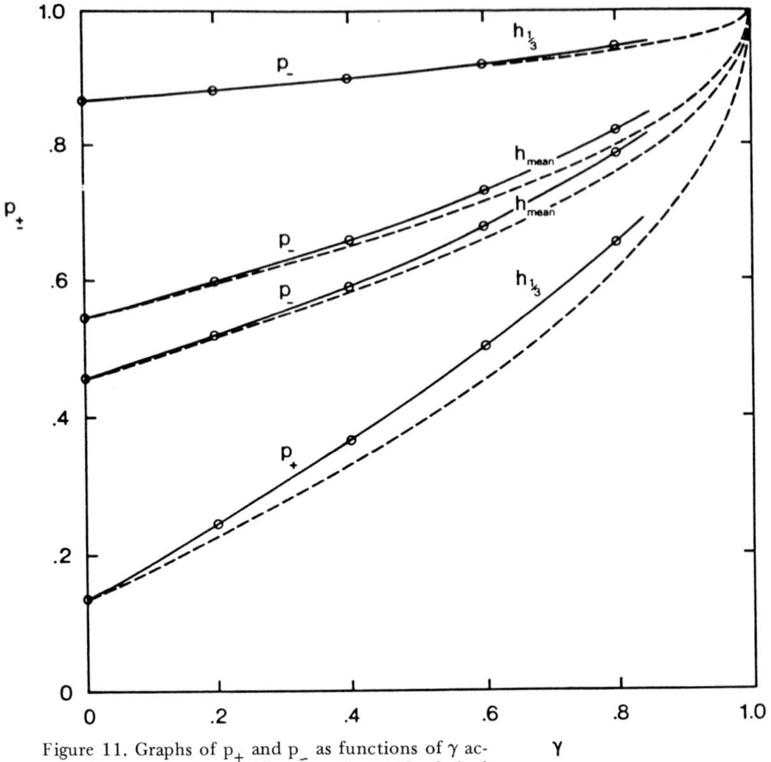

Figure 11. Graphs of p_+ and p_- as functions of γ according to equations (8.6) and Section 6. The dashed curves represent the parabolic approximations (8.8).

and p_- can be calculated directly from

$$p_+ = \frac{\int_{\rho^*}^{\infty}\int_{\rho^*}^{\infty} p(\rho_1,\rho_2)\,d\rho_1\,d\rho_2}{\int_0^{\infty}\int_{\rho^*}^{\infty} p(\rho_1,\rho_2)\,d\rho_1\,d\rho_2}$$

(8.6)

$$p_- = \frac{\int_0^{\rho^*}\int_0^{\rho^*} p(\rho_1,\rho_2)\,d\rho_1\,d\rho_2}{\int_0^{\infty}\int_0^{\rho^*} p(\rho_1,\rho_2)\,d\rho_1\,d\rho_2}$$

where $\rho^* = \tfrac{1}{2}h^*$. Such probabilities are then a function only of γ, as shown in Figure 11, and hence of k^2 by Figure 6.

Assuming that Kimura's five 'target spectra' are given by equation (5.11), we have calculated (see Table 3a) the corresponding values of k^2 and hence γ; also the values of p_+ and p_- using Figure 11. The corresponding values for the truncated spectra are given in Table 3b. It will be seen that while the truncation changes the values of ν, k^2 and γ very considerably, the values of p_+ and p_- are much less affected.*

*The values of γ used by Kimura (1980) to calculate p_+ and p_- were determined empirically, and not calculated from the frequency spectra as here.

The distributions of G_j and H_j corresponding to a typical spectrum (case 5) are seen in Figures 12 and 13. Also shown are Kimura's observations. (See also Tables 3c and 3d). From these results we may conclude

(1) that truncation of the spectra has a small but appreciable effect upon the theoretical distributions,

(2) that the observations agree fairly well with either set of curves, but distinctly better with those for the truncated spectra (solid lines).

Based on Figure 11, we may also give some rough analytic expressions for $\bar{H}$ and $\bar{G}$. For the values of p_+ and p_- on the left-hand axis ($\gamma = 0$) are known:

$$p_+ = e^{-\xi^2/2}, \quad p_- = 1 - e^{-\xi^2/2} \tag{8.7}$$

where $\xi = \rho/\mu_0^{1/2}$. If we approximate the curves in Figure 11 by parabolas through the point (1, 1) with horizontal axes we must have in general

$$\left.\begin{aligned} 1 - p_+ &= (1-e^{-\xi^2/2})(1-\gamma)^{1/2} \\ 1 - p_- &= e^{-\xi^2/2}(1-\gamma)^{1/2} \end{aligned}\right\} \tag{8.8}$$

Now by (6.11), we have $(1-\gamma)^{1/2} \cong (1-k^2)^{1/2} = k'$ and so from (6.18)

$$\left.\begin{aligned} 1 - p_+ &= 2\pi\nu\,(1-e^{-\xi^2/2}) \\ 1 - p_- &= 2\pi\nu\, e^{-\xi^2/2} \end{aligned}\right\} \tag{8.9}$$

Now substituting in (8.2) we get

$$\bar{H}_j = \frac{1}{2\pi\nu}\,\frac{e^{\xi^2/2}}{e^{\xi^2/2}-1} \tag{8.10}$$

and similarly from (8.5)

$$\bar{G}_j = \frac{1}{2\pi\nu\, e^{\xi^2/2}-1} \tag{8.11}$$

These equations indicate that $\bar{H}$ and $\bar{G}$ are both inversely proportional to ν, as was also found in Section 3. In fact if ν^2 is negligible, equations (3.7) and (3.10) can be written

$$\bar{G} = \frac{1}{(2\pi)^{1/2}}\,\frac{e^{\xi^2/2}}{\nu\xi} \tag{8.12}$$

and

$$\bar{H} = \frac{1}{(2\pi)^{1/2}}\,\frac{1}{\nu\xi} \tag{8.13}$$

respectively.

The functional dependence on ξ in equations (8.10) and (8.11) is quite different from that in the two last equations. However, a numerical comparison is interesting. Table 4 shows the functions of ξ evaluated at three different levels: $h^* = h_{mode}$, h_{mean} and $h_{1/3}$ ($\xi = 1$, $\sqrt{\pi/2}$ and 2). In every case the pairs of formulae, though analytically different, agree to within ten percent. Hence over a certain range of ν and of ξ the two theories give quite similar results.*

Table 4. A comparison of theoretical values of $\nu\bar{H}$ and $\nu\bar{G}$.

		$\nu\bar{H}_j$	$\nu\bar{H}$	$\nu\bar{G}_j$	$\nu\bar{G}$
h^*	ξ	(8.10)	(8.13)	(8.11)	(8.12)
h_{mode}	1	.404	.399	.667	.658
h_{mean}	$\sqrt{\pi/2}$	.293	.318	.642	.698
$h_{1/3}$	2	.184	.199	1.360	1.474

9. *Discussion and Conclusions*

We have seen how two different approaches to the analysis of wave grouping can lead to almost identical results. Of these approaches, the first or Gaussian noise theory is valid asymptotically only as $\nu \to 0$. The second, or Markov theory, is valid for narrow, but not too narrow spectra, since when $\nu < 0.1$ the coefficient of correlation between non-adjacent pairs of wave heights becomes significant (cf. Goda 1983). Thus Markov theory, and also the approximations (8.10) and (8.11), are valid in only an intermediate range of ν, which seems to include, however, most wind-wave spectra.

*In fact, according to equation (5.10) we would expect the corresponding values of $\nu\bar{H}_j$ and $\nu\bar{H}$ to differ by a small amount of order 0.5ν.

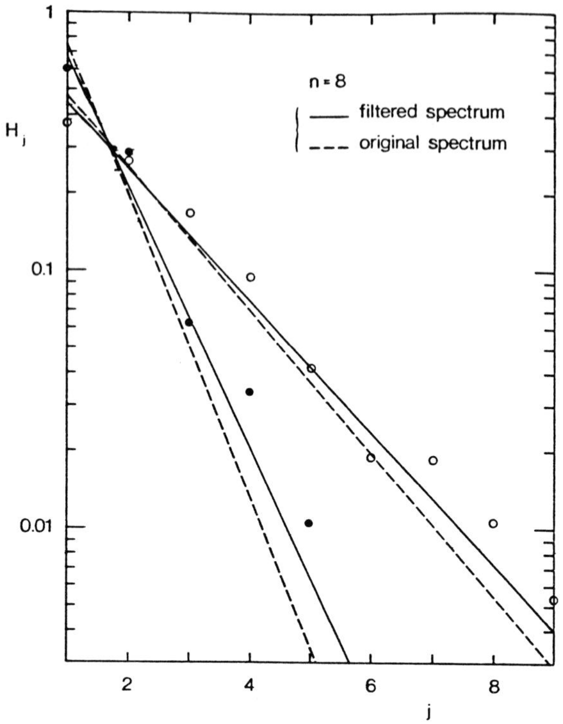

Figure 12. The probability H_j of a high run, as a function of j for the Kimura spectrum (5.11) when n = 8. The curves represent equation (8.1).
---- original spectrum
— truncated spectrum
Plotted points are data from Kimura (1980), Figure 9(a).

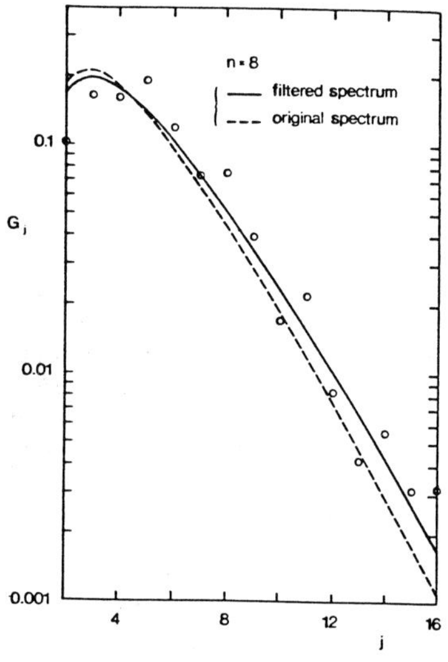

Figure 13. The probability G_j of a group of total length j for the Kimura spectrum (5.11) when n = 8. The curves represent equation (8.4):
--- original spectrum
— truncated spectrum
Plotted points are data from Kimura (1980), Figure 10(a).

As against this the Gaussian theory is applicable strictly only to linear surface waves. When the waves become steep the harmonic components in a wave record are not independent, and the surface must become non-Gaussian. Markov theory, however, can still be applied, though its physical basis is not yet secure.

Whatever the relative merits of the two approaches, it appears that neither can be applied in a sensible way except to sufficiently narrow-band processes, or to data that have been filtered so as to eliminate both high and low frequencies. The same conclusion was also reached by Nolte and Hsu (1972, 1979) though the arguments for the tapered filter suggested in their 1979 paper do not appear to be conclusive. We have recommended a surface 'square-topped' filter with limits $0.5f_p$ and $1.5f_p$ which has two advantages:

(1) it leaves the peak frequency f_p unchanged;

(2) two successive applications of the filter have the same effect as only one.

Moreover, the chosen limits have been shown to give answers in agreement with a visual assessment of the group properties of the record.

This paper has confined attention to the essentially linear properties of wave groups. Some nonlinear statistical properties of wave groups deserve further attention, and studies directed towards this aspect are under way.

ACKNOWLEDGEMENTS

This paper was begun while the author was visiting the Cal. Tech. Jet Propulsion Laboratory, Pasadena, in July 1983. He is indebted to Dr O. H. Shemdin, Dr V. Hsiao and Mr J. A. Ewing for kindly supplying the wave data discussed in Section 4. Useful discussions have been held with Mr D. J. T. Carter and Mr P. G. Challenor at I.O.S., Wormley.

REFERENCES

Donelan, M., Longuet-Higgins, M. S. and Turner, J. S. 1972. Periodicity in whitecaps. *Nature, Lond.* **239**, 255-261.

Ewing, J. A. 1973. Mean length of runs of high waves. *J. Geophys. Res.* **78**, 1933-1936.

Goda, Y. 1970. Numerical experiments on wave statistics with spectral simulation. *Rep. Port and Harbour Res. Inst.* **9**, 3-75.

Goda, Y. 1976. On wave groups. *Proc. Conf. on Behaviour of Offshore Structures*, Trondheim, Norway, Vol. 1, pp. 1-14.

Goda, Y. 1983. Analysis of wave grouping and spectra of long-travelled swell. *Rep. Port and Harbour Res. Inst.* **22**, 3-41.

Kimura, A. 1980. Statistical properties of random wave groups. *Proc. 17th Int. Conf. on Coastal Eng.*, pp. 2955-2973.

Longuet-Higgins, M. S. 1957. The statistical analysis of a random moving surface. *Phil. Trans. R. Soc. Lond.* A **249**, 321-387.

Longuet-Higgins, M. S. 1958. On the intervals between zeros of a random function. *Proc. R. Soc. Lond.* A **246**, 99-118.

Longuet-Higgins, M. S. 1962. The distribution of intervals between a stationary random function. *Phil. Trans. R. Soc. Lond.* A **154**, 557-599.

Middleton, D. 1960. *Statistical Communication Theory*. New York, McGraw-Hill, 1140 pp.

Nolte, K. G. and Hsu, F. H. 1972. Statistics of ocean wave groups. *Proc. 4th Offshore Tech. Conf.*, Dallas, Texas, Preprint No. 1688, pp. 139-146.

Rice, S. O. 1944-1945. The mathematical analysis of random noise. *Bell Syst. Tech. J.* **23**, 282-332 and **24**, 46-156.

Rice, S. O. 1958. Distribution of the duration of fades in radio transmission-gaussian noise model. *Bell Syst. Tech. J.* **37**, 581-635.

Sawnhey, M. D. 1962. A study of ocean wave amplitudes in terms of the theory of runs and a Markov train process. *N. York Univ. Tech. Rep.* 29pp.

Siefert, W. 1974. Wave investigation in shallow water. *Proc. 14th Int. Conf. on Coastal Eng.* pp. 151-178.

Uhlenbeck, G. E. 1943. Theory of random process. *M. I. T. Radiation Lab. Rep.* 454, October 1943.

Watson, G. N. 1958. *Theory of Bessel Functions*. 2nd ed. Cambridge University Press, 804 pp.

Wilson, J. R. and Baird, W. F. 1972. A discussion of some measured wave data. *Proc. 13th Int. Conf. on Coastal Eng.* pp. 113-130.

A PARAMETER DESCRIBING OVERALL CONDITIONS OF WAVE BREAKING, WHITECAPPING, SEA-SPRAY PRODUCTION AND WIND STRESS

YOSHIAKI TOBA
Department of Geophysics, Faculty of Science,
Tohoku University, Sendai 980 Japan

MOMOKI KOGA
Department of Geophysics, Faculty of Science,
Hokkaido University, Sapporo 060 Japan

The dimensionless parameter $u_*^2/\nu\sigma$ can be used widely as a parameter to describe the overall conditions of air-sea boundary processes, where u_* is the friction velocity of air, ν is the kinematic viscosity of air and σ is the spectral peak frequency of the wind waves. A critical value of this parameter for the appreciable commencement of breaking of wind waves is 10^3. Beyond this value, the percentage of waves passing a fixed point that are breaking, α, and the percentage of whitecap coverage, P, are both approximately proportional to this parameter. The number concentration of sea-salt particles containing salt in the vicinity of 10^{-10} g at the 6-m level is also proportional to this parameter. The dimensionless roughness length associated with the air flow over water, $u_* z_0/\nu$, also correlates better with this parameter than with a parameter which does not contain the spectral peak frequency. This gives an approximate relation of $z_0\sigma/u_* = 0.025$, and a corresponding formula for the drag coefficient is proposed. The dimensional and physical interpretation of the parameter $u_*^2/\nu\sigma$ is presented.

1. *Introduction*

The overall condition of air-sea boundary processes should be determined by the local state of the wind and wind waves. Since world-wide distributions of wind and wind waves are now going to be collected by satellite sensors, the development of quantitative expressions for related phenomena as functions of a parameter which includes both the wind and the wind waves should be useful.

In this article, problems related to the thermal stratification are excluded, and we confine ourselves to dynamical processes under neutral conditions.

Many efforts to obtain reliable expressions for air-sea boundary processes such as wind stress or whitecap coverage as functions of the wind speed at the 10-m level have been made. For example, comprehensive reviews were given by Garratt (1977) and Wu (1980) on wind stress, and by Wu (1979), Monahan and Ó Muircheartaigh (1980) and Monahan et al. (1983) on whitecap coverage. Marine aerosols also have been studied as functions of the wind since Woodcock (1953), as can be seen from the review of Blanchard and Woodcock (1980). Approaches that include information on waves have also been undertaken. For example Kitaigorodsky (1968), and Kitaigorodskii and Zaslavsky (1974), proposed to incorporate wave age into parameters influencing the roughness length z_0, and Melville (1977) tried to use quantities derived in the study of wave breaking conditions, including the surface vortical layer, as discussed by Banner and Phillips (1974), and Phillips and Banner (1974), in developing an expression for z_0.

Recent studies of wave breaking by Longuet-Higgins (1978a, b), McLean (1982), Melville (1982) and Su et al. (1983) clearly suggest that instabilities of strongly nonlinear waves should be important elements in the processes related to wind waves. Experimental studies of wind waves in our laboratory have demonstrated that the wind wave phenomena are strongly related to the existence of turbulence above and below the water surface as will be mentioned in the next section.

The phenomena associated with wind waves, including wave breaking and the production of

E. C. Monahan and G. Mac Niocaill (eds.), Oceanic Whitecaps, 37–47.

droplets, are thus strongly nonlinear ones including turbulence, and a dynamically rigorous approach to formulate proper expressions seems beyond our present state of achievement. In this situation, the most effective approach would be a combination of some dimensional considerations, which take into account the actual physical processes, together with the analysis of measurement data. The present study follows this line of approach.

2. Physical processes of wind waves, similarity structure and a characteristic quantity

It is already known that the generation of initial wavelets by wind at a still water surface is by the instability of a two-layer shear flow of viscous fluids, i.e. of air and water (Kawai, 1979). However, in a time scale of the order of 10s, the wavelets change to irregular waves that include three-dimensional turbulence, and following this transition the growth of the wind waves commences (Toba et al., 1975; Kawai, 1979).

The local tangential wind stress is very large at the crest and the windward face of individual waves (Okuda et al., 1976), and a separation bubble in the air flow is formed at the leeward face of a considerable percentage of young laboratory wind waves (Kawai, 1982; Kawamura et al., 1981), triggering ordered motions in the boundary layers (Toba et al., 1984). In the water, there is a special region of very high vorticity, of the order of $10^2\ s^{-1}$, below the crest of individual waves of about 1 cm wave height and 0.24s wave period (Okuda, 1982). Particle speed at the crest is often larger than the phase speed, and in this case there is a recirculation of water below the crest, corresponding to the air-flow separation behind the crest. When this recirculation is intense, visible wave breaking or bubble entrainment occurs (Koga, 1981a; Toba et al., 1975).

These properties of wave breaking or air flow separation are associated with individual waves (not with the Fourier component waves), and the characteristics of the individual waves are affected by the nature of nonlinear water waves (e.g., Hatori and Toba, 1983) as was mentioned in Section 1. Thus the mechanics of wind waves are characterized by a combination of the turbulent aspect with the nonlinear water waves.

Presumably as a consequence of these strong nonlinearies, there is a rather simple similarity in wind waves. For example, there is a simple relation between the nondimensional significant wave height $H^* \equiv gH/u_*^2$ and the nondimensional period $T^* \equiv gT/u_*$ expressed by

$$H^* = BT^{*3/2}, \quad B = 0.062 \tag{1}$$

where g is the acceleration of gravity (Toba, 1972, 1978). This relation is also applicable statistically to individual waves in the main frequency range (Tokuda and Toba, 1981). There is also a similarity structure in the energy spectra of growing pure wind waves (e.g., Kawai et al., 1977), and these overall similarity situations result in a simple power law of the form,

$$E^* = B_\sigma \sigma^{*-3}, \quad B_\sigma = 0.051 \tag{2}$$

where $E^* \equiv g^2 E/u_*^4$ is the nondimensional total energy and $\sigma^* \equiv u_*\sigma/g$ is the nondimensional angular frequency at the spectral peak (Toba, 1978). Consequently, the state of wind waves can be expressed by the single nondimensional parameter σ^*. This parameter is effectively used in wave prediction models (The SWAMP group, 1983), and it also has a nice application to the present problem.

3. Dimensional considerations

There are several dimensional variables or parameters which are involved in the phenomena concerned. The fetch χ and the duration t are independent variables, u_* and the peak frequency σ are quantities characterising the wind and wind waves; the density of air ρ_a, and of water ρ_w, the viscosity coefficient of air μ_a, and of water μ_w, g and the surface tension S are physical constants. Among these, χ and t are not concerned with the local physical processes. Since ρ_a and ρ_w are constants with a constant value of their ratio, we may use either one, e.g. ρ_a, and the situation is the same for μ_a and μ_w and we use only μ_a. Further, we will use $\nu = \mu_a/\rho_a$, the kinematic viscosity of air. The surface tension may be included in the extended gravity $g = g + Sk^2/\rho_w$, where k is the wave number (e.g., Toba, 1978), and for the conditions where σ is in the gravity wave region, S is no longer necessary. The application of the π theorem by Vaschy and Backingham results in only two nondimensional variables determining the local physical processes in these situations, and we can use $u_*^* \equiv u_*^3/g\nu$, and σ^*, as the nondimensional parameters of state of the wind and wind waves.

Consequently, the percentage of the waves which pass a fixed point which are breaking, α, the percentage of whitecap coverage, P, or the nondimensional roughness parameter representing wind stress, $z_0^* \equiv u_* z_0/\nu$, would all be expressed as functions of u_*^* and σ^*.

4. The parameter $u_*^2/\nu\sigma$ and expressions for its use

If we are concerned solely with the air-sea boundary on the earth, g has a constant value. Consequently, measured data can be expressed by a parameter which does not include g. Proceeding from this expectation, we eliminate g from u_*^* and σ^*, and obtain a new nondimensional parameter $u_*^2/\nu\sigma$. The use of this parameter was proposed initially by Toba (1979), but in the present paper we present a more comprehensive description.

Toba (1961; a table of values was given in Toba, 1972) and Toba et al. (1971) measured α in a wind-wave tunnel, and at an oceanographic tower station in a bay, respectively. The definition of α involved the occurrence of air entrainment. The plot of α against the 10-m wind speed u is shown in Fig. 1. The same data plotted against $u_*^2/\nu\sigma$ is given in Fig. 2, and in this case the entire combined data set lies along a 45° line, indicating that α is approximately proportional to $u_*^2/\nu\sigma$. The critical value for the commencement of breaking is about 10^3 in $u_*^2/\nu\sigma$. Beyond this critical value, the formula

$$\alpha(\%) = 4.3\times10^{-3}u_*^2/\nu\sigma,\quad u_*^2/\nu\sigma \gtrsim 10^3 \tag{3}$$

is proposed.

The plot of the available values of the percentage of whitecap coverage is given in Fig. 3. It was reported by Wu (1979), Monahan and Ó Muircheartaigh (1980) and Monahan et al. (1983) that data by Toba and Chaen (1973) were in remarkable agreement with data by Monahan (1971), and by Monahan et al. (1983). However, only the former data were plotted since the latter data do not include the information of waves. Eq. (1) was used to estimate σ from the data on significant wave height and wind found by Nordberg et al. (1971), and Ross and Cardone (1974). In Fig. 3 P is approximately proportional to $U_*^2/\nu\sigma$, leading to the proposed formula

$$P(\%) = 8.9\times10^{-5}u_*^2/\nu\sigma,\quad u_*^2/\nu\sigma \gtrsim 10^3 \tag{4}$$

The relation between P and α is thus given by

$$P(\%) = 0.021\alpha(\%) \tag{5}$$

This relation is interpreted to mean that if 100% of the significant wave crests are breaking, the whitecap coverage is 2.1%.

The production of sea water droplets also seems to be proportional to this parameter. Chaen (1973), and Toba and Chaen (1973), observed the concentration of sea salt particles, the significant wave period and other marine meteorological variables on board R. V. *Hakuho-Maru*. From their data, the concentration of sea salt particles θ (m^{-3}) in the range of log m = 1.25 to 2.25 (m: the mass of salt contained in a particle in units of 10^{-12} g) at a 6-m level, collected by use of an impactor, has been determined and plotted in Fig. 4. We considered this range of particles because, in this range, their fall velocity is small enough for the sedimentation diffusion equilibrium to hold (Koga and Toba, 1981), and the vertical distribution in the equilibrium state in this range should give almost the same value of θ between the 6-m level and the water surface, as Chaen's data proved.

It is seen that these points are also distributed along a 45° line, except for the closed circles which are data obtained in Osaka Bay which is enclosed by land. Since θ reflects the history of the particles in the air to some extent, the critical value for $u_*^2/\nu\sigma$ of 10^3 is not clear in θ. The proposed relation is

$$\theta(\log m = 1.25\text{-}2.25;\ m^{-3}) = 65\,u_*^2/\nu\sigma \tag{6}$$

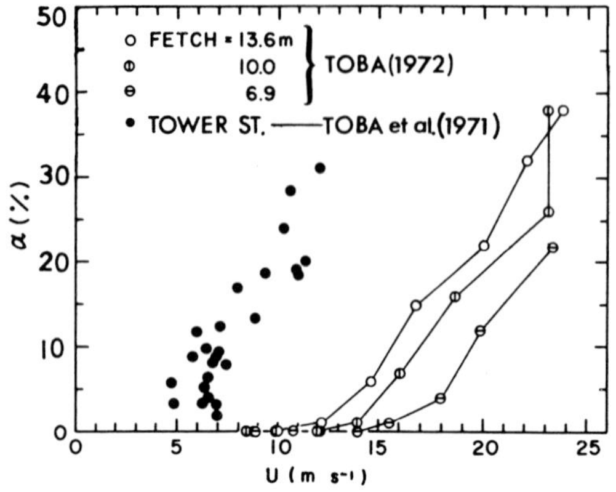

Fig. 1. Rate of breaking crests among individual waves travelling through a fixed point, α, plotted against the 10-m wind speed, u.

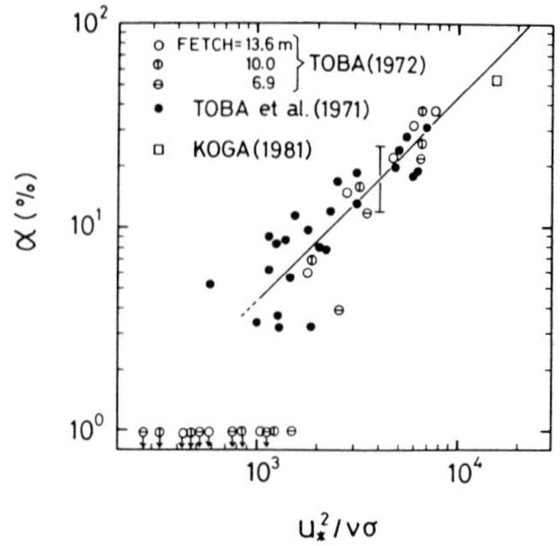

Fig. 2. Rate of breaking crests among individual waves travelling through a fixed point, α, plotted against $u_*^2/\nu\sigma$. The data is the same as that of Fig. 1.

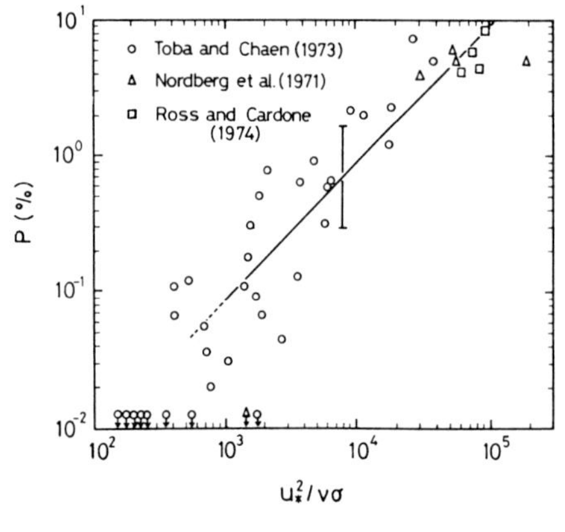

Fig. 3. Whitecap coverage, P, plotted against $u_*^2/\nu\sigma$. For data by Nordberg et al. (1971) and Ross and Cardone (1974), Eq. (1) was used to estimate σ from significant wave height.

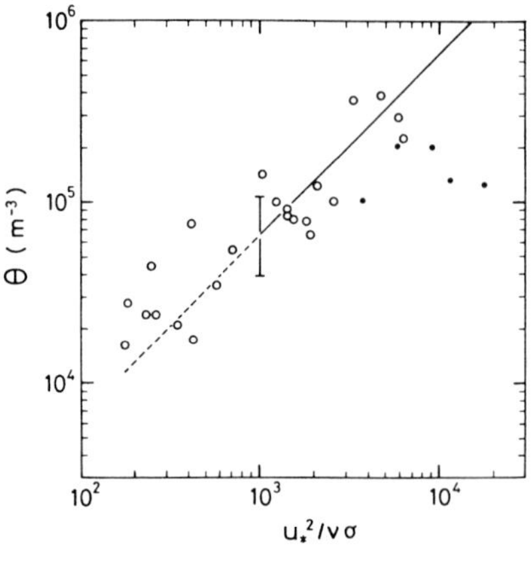

Fig. 4. Concentration of sea-salt particles, θ, of the range of mass of salt containing: log m = 1.25 to 2.25 at 6-m level plotted against $u_*^2/\nu\sigma$. The data is from Chaen (1973) and Toba and Chaen (1973).

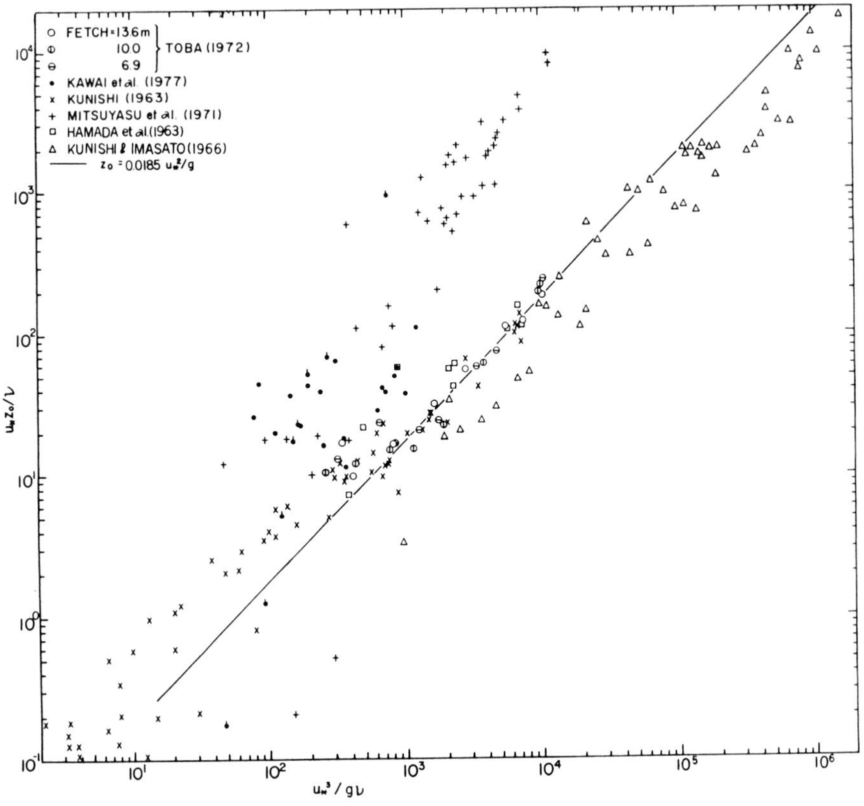

Fig. 5. $u_* z_0/\nu$ plotted against $u_* \bar{*}$. The straight line corresponds to Charnock's formula of Eq. (8) with a constant proposed by Wu (1980). Closed circles with a stick at the top indicate data in which swells were included.

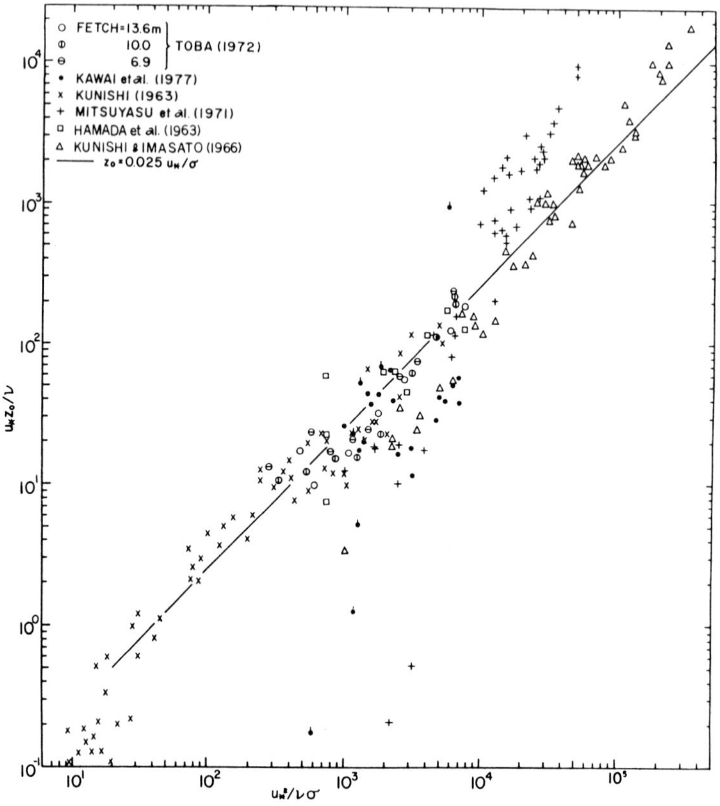

Fig. 6. $u_* z_0/\nu$ plotted against $u_*^2/\nu\sigma$. The straight line corresponds to Eq. (9). Closed circles with a stick at the top indicate data in which swells were included. The data is the same as that of Fig. 7.

Table 1. Average values and the standard deviation of γ in Eq. (9) and of β in Eq. (7), obtained from six data sources, with the means of these average values and the standard deviations of these mean values. Magnitudes of the standard deviations relative to the average values are shown in percentages in parentheses.

Data source and remark *Wind wave tunnel data †Tower station data	Condition of C/u	γ	β
Hamada (1963)*	<0.08	0.030 ± 0.020 (67%)	0.035 ± 0.022 (63%)
Kunishi (1963)* (for points $u_* z_0/\nu > 0.5$)	<0.08	0.028 b 0.010 (36%)	0.031 ± 0.017 (55%)
Kunishi and Imasato (1966)*	<0.06	0.026 ± 0.013 (50%)	0.011 ± 0.005 (45%)
Toba (1972)*	<0.06	0.024 ± 0.009 (38%)	0.022 ± 0.010 (45%)
Mitsuyasu et al. (1971)† (for points $u_* z_0/\nu > 0.6$)	<0.94	0.062 ± 0.045 (73%)	0.408 ± 0.241 (60%)
Kawai et al. (1977)† (run nos. 9–26)	<0.6	0.015 ± 0.011 (73%)	0.142 ± 0.126 (89%)
For the above six mean values		0.031 ± 0.015 (48%)	0.109 ± 0.141 (129%)
For the above four laboratory-work mean values		0.027 ± 0.002 (7%)	0.025 ± 0.009 (36%)

The standard deviations in the logarithmic scales are shown in these figures by an error bar.

It should be noted that although θ carries the units of m^{-3}, it is physically a kind of nondimensional variable since we may consider it as a mixing ratio with the molecules of air.

With respect to z_0, Charnock's (1955) formula

$$z_0 = \beta u_*^2/g \tag{7}$$

has been widely accepted. Wu (1980) proposed an optimal value of β of 0.0185. This corresponds to a relation between $z_0{}^*$ and $u_*{}^*$:

$$z_0{}^* = 0.0185 u_*{}^* \tag{8}$$

in which σ^* is disregarded. We can compare the plot of $z_0{}^*$ against $u_*{}^*$ in Fig. 5 with another plot of $z_0{}^*$ against $u_*{}^2/\nu\sigma$ using the same data in Fig. 6. The number of values of z_0 available together with associated σ values were not many, and all were Japanese data. Kunishi (1963), Hamada (1963), Kunishi and Imasato (1966) and Toba (1972; original data in Toba, 1961) contain data from wind-wave tunnels, while Mitsuyasu et al. (1971) and Kawai et al. (1977) presented data from tower stations in the sea. The data in Toba (1972) and Kunishi (1963) were the same as those used in Toba and Kunishi (1970). The 45° line in Fig. 5 is given by Eq. (8). But the scattering of points is clearly narrower in Fig. 6. The straight line in Fig. 6 corresponds to

$$z_0 \sigma/u_* = \gamma, \quad \gamma = 0.025 \tag{9}$$

a form in which ν has been eliminated for the same reason that g was eliminated earlier.

Average values and standard deviations of γ

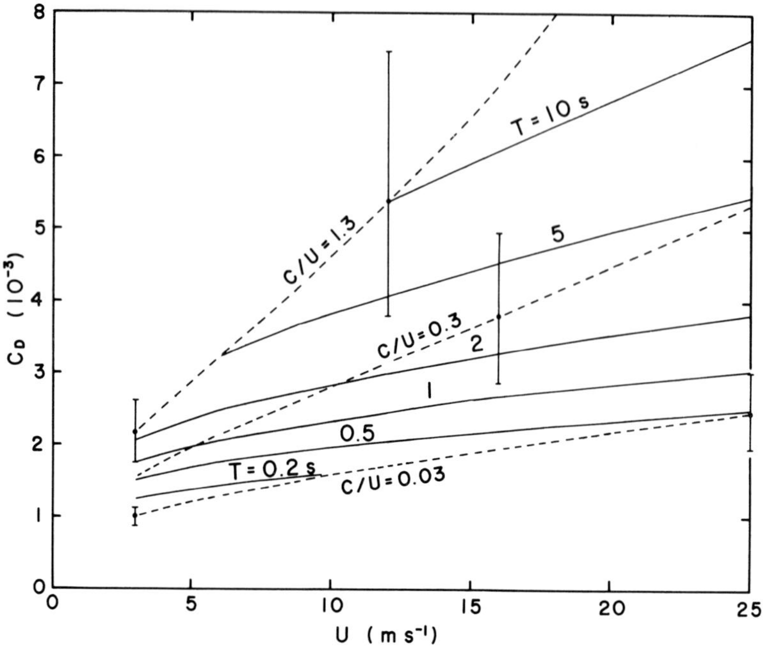

Fig. 7. Calculated value of C_D by use of Eq. (11). Vertical segments indicate the range of variations of C_D corresponding to an extreme range of γ from 0.01 to 0.05.

for the six data sources are shown in Table 1. As to the Kawai et al. (1977) data, only the points for run nos. 9-26 were used in the calculation, since the data of run nos. 1-8 were obtained under the presence of some swells with a resultant deviation of the points from Eq. (9). Points of run nos. 1-8 carry a small stick at the top of closed circles in Figs. 5 and 6.

Corresponding values for β in Eq. (7), calculated from the same data, are also entered in Table 1 for comparison. It is seen that the relative standard deviation is generally smaller in γ than in β.

5. The drag coefficient corresponding to eq. (9)

If we assume Eq. (9), the definition of the drag coefficient C_D for 10-m wind speed $C_D = u_*^2/u^2$, and the logarithmic law for the wind profile $u/u_* = (1/\kappa)\ln(z_{10}/z_0)$, then it follows that

$$C_D = \kappa^2 [\ln(z_{10}\sigma/\gamma u_*)]^{-2} \qquad (10)$$

where z_{10} is 10-m, and κ is the von Kármán constant, 0.41. If we express the right hand side of Eq. (10) in terms of the conventional parameter u and the significant wave period of the wind waves T, it further follows that

$$C_D = \kappa^2 [\ln(2\pi z_{10}/\gamma) - \ln u - \ln T - 0.5 \ln C_D]^{-2} \qquad (11)$$

The drag coefficient in the right hand side of Eq. (11) appears because of the use of u instead of u_*, and this C_D may easily be computed through a few iterations.

In Fig. 7 the calculated values of C_D from (11) are shown as a function of u ($>3\,\mathrm{m\,s^{-1}}$) for specified values of the parameter T. It is noted that the value of C_D is much more sensitive to T than to u. Consequently, there is a possibility that the scattering of the observed C_D data plotted against u is due to this reason, although stability effects in the lowest atmosphere and errors in the measurements may also be amongst the causes. Equation (9) predicts very large values of C_D under conditions of large wind waves. However, measurements to check this prediction under such conditions are not available, so its substantiation is left for the future. It should be noted that the value of C_D is not very sensitive to a change in γ. In fact, for an extreme range of γ from 0.01 to 0.05, the range of calculated C_D values is as shown in Fig. 7.

Since the ratio of the phase speed of representative wind waves, C, to u_* may be expressed by

$$C/u_* = C/(C_D)^{1/2}u = g/\sigma u_* \quad (12)$$

Eq. (9) may be rewritten as

$$z_0 = \beta' u_*^2/g \ , \quad \beta' = \gamma C/u_* \quad (13)$$

In this form, Eq. (9) may be interpreted as an extension of Charnock's formula to include information on wind waves by the incorporation of the wave age C/u_*. This corresponds to the ideas expressed by Kitaigorodsky (1968) and Melville (1977), but is here couched in simpler form. In fact the mean values of β in Table 1 really seem to depend on the magnitude of C/u.

In Fig. 7 Eq. (13) is shown by broken lines for three values of wave age, C/u = 0.03, 0.3 and 1.3, where the last value predicts the maximum values of C_D, corresponding to the saturation condition of wind waves under the respective winds.

It should be noted that the waves considered here are purely wind waves. Consequently, in cases where large swells are present, we should use σ or C values that relate to the wind wave components.

6. Physical interpretation

Since Munk (1955), it has been considered that the elements effective in determining the stress characteristics of air flowing over water would be the high-frequency components of the waves. Following along the lines of this idea the relationship of z_0 with a roughness Reynolds number was examined where the wave height of the high-frequency components of the waves was used as the amplitude of the roughness elements (e.g., Kitaigorodsky, 1969; Kondo et al., 1973). However, the energy spectrum of wind waves has a continuous structure and it is difficult to define the frequency range of the waves which work as the roughness elements. If we consider the recirculation of water particles under the crest as was mentioned in section 2, it is apparent that the elements effective in wind stress are not restricted to the high-frequency component of the waves, but that the waves near the peak frequency might be more effective.

The inverse of the peak frequency, σ^{-1}, represents π^{-1} times the wind interval (T/2), the time required for a fixed point to reach the top of a crest from having been at the bottom of a trough, or vice versa. If we consider a length scale $L_u = u_*(T/2\pi)$, which is the product of this time interval divided by π, and u_* (the velocity representing the wind stress acting on the surface of individual waves) and which is a length scale relating to the recirculation, then $u_*^2/\nu\sigma$ may be interpreted as a kind of Reynolds number which is constructed from u_*, L_u and ν, or $u_* L_u/\nu$.

Also, since $u_*^2 = \nu \partial u/\partial z$ by its definition, $u_*^2/\nu\sigma$ represents the product of $T/2\pi$ and $\partial u/\partial z$, where $\partial u/\partial z$ is directly related to the horizontal vorticity at the surface of individual waves. It seems very natural that this Reynolds number, which corresponds to the product of the time scale $T/2\pi$ and the vorticity at the surface, becomes the parameter describing the commencement and intensity of the breaking of wind waves, and other characteristics of the air-sea boundary processes.

ACKNOWLEDGEMENTS

The authors express their thanks to the late Dr. Sanshiro Kawai for his valuable discussion, to Mrs. Fumiko Ishii and Miss Yoko Inohana for their assistance, and to Mrs. Cynthia Jones for correcting their English sentences. This study was partially supported by a Grant-in Aid for Scientific Research by the Japanese Ministry of Education, Science and Culture, Project No. 58101004.

REFERENCES

Banner, M. L. & O. M. Phillips (1974): On the incipient breaking of small-scale waves. *J. Fluid Mech.*, **65**, 647-656.

Blanchard, D. C. & A. H. Woodcock (1980): The production, concentration and vertical distribution of the sea-salt aerosol. *Annals, New York Acad. Sci.*, **338**, 330-347.

Chaen, M. (1973): Studies on the production of sea-salt particles on the sea surface. *Mem. Fac. Fish., Kagoshima Univ.*, **22**, 49-107.

Charnock, H. (1955): Wind stress on a water surface. *Quart. J. Roy. Meteor. Soc.*, **81**, 639-640.

Garratt, J. R. (1977): Review of drag coefficients over oceans and continents. *Mon. Wea. Rev.*, **105**, 915-929.

Hamada, T. (1963): An experimental study of development of wind waves. *Rep. Port and Harbour Tech. Res. Inst.*, No. 2, 1-41.

Hatori, M. & Y. Toba (1983): Transition of mechanically generated regular waves to wind waves under the action of wind. *J. Fluid Mech.*, **130**, 397-409.

Kawai, S. (1979): Generation of initial wavelets by instability of a coupled shear flow and their evolution to wind waves. *J. Fluid Mech.*, **93**, 661-703.

Kawai, S. (1982): Structure of the air flow over wind wave crests revealed by flow visualization techniques. *Boundary-Layer Met.*, **23**, 503-521.

Kawai, S., K. Okada & Y. Toba (1977): Field data support of three-seconds power law and $gu_*\sigma^{-4}$-spectral form for growing wind waves. *J. Oceanogr. Soc. Japan*, **33**, 137-150.

Kawamura, H., K. Okuda, S. Kawai & Y. Toba (1981): Structure of turbulent boundary layer over wind waves in a wind wave tunnel. *Tokohu Geophys. Journ.* (Sci. Rep. Tohoku Univ. Ser. 5), **28**, 69-86.

Kitaigorodsky, S. A. (1968): On the calculation of aerodynamic roughness of the sea surface. Izv. Akad. Sci. USSR, *Atmos. Ocean. Phys.*, **4**, 870-878.

Kitaigorodskii, S. A. (1969): Small-scale atmosphere-ocean interactions. Izv. Akad. Sci. USSR, *Atmos. Ocean. Phys.*, **5**, 1114-1131.

Kitaigorodskii, S. A. & M. M. Zaslavsky (1974): A dynamical analysis of the drag conditions at the sea surface. *Boundary-Layer Met.*, **6**, 53-61.

Koga, M. (1981): Detailed structure of breaking wind waves with droplet and bubble formation. Ph.D. dissertation at Division of Science, Tohoku University, 163 pp.

Koga, M. (1981a): Direct production of droplets from breaking wind waves - its observation by a multicolored overlapping exposure photographing technique. *Tellus*, **33**, 552-563.

Koga, M. & Y. Toba (1981): Droplet distribution and dispersion process on breaking wind waves. *Tohoku Geophys. Journ.* (Sci. Rep. Tokohu Univ. Ser. 5), **28**, 1-25.

Kondo, J., Y. Fujinawa & G. Naito (1973): High-frequency components of ocean waves and their relation to the aerodynamic roughness. *J. Phys. Oceanogr.*, **3**, 197-202.

Kunishi, H. (1963): An experimental study on the generation and growth of wind waves. *Bull. Disas. Prev. Res. Inst. Kyoto Univ.*, No. 61, 1-41.

Kunishi, H. & N. Imasato (1966): On the growth of wind waves by high-speed wind flume. *Disas. Prev. res. Inst. Kyoto Univ., Annals,* **9**, 1-10 (in Japanese with English abstract).

Longuet-Higgins, M. S. (1978a): The instabilities of gravity waves of finite amplitude in deep water. I. Superharmonics. *Proc. R. Soc. Lond.* **A360**, 471-488.

Longuet-Higgins, M. S. (1978b): The instabilities of gravity waves of finite amplitude in deep water. II. Subharmonics. *Proc. Roy. Soc. Lond.* **A360**, 489-505.

McLean, J. W. (1982): Instabilities of finite-amplitude water waves. *J. Fluid Mech.*, **114**, 315-330.

Melville, W. K. (1977): Wind stress and surface roughness over breaking waves. *J. Phys. Oceanogr.*, **7**, 702-710.

Melville, W. K. (1982): The instability of deep-water waves. *J. Fluid Mech.*, **115**, 165-185.

Mitsuyasu, H., R. Nakayama & T. Komori (1971): Observations of the winds and waves in Hakata Bay. *Rep. Res. Inst. Appl. Mech., Kyushu Univ.*, **19**, 37-74.

Monahan, E. C. (1971): Oceanic whitecaps. *J. Phys. Oceanogr.*, **1**, 139-144.

Monahan, E. C., C. W. Fairall, K. L. Davidson & P. J. Boyle (1983): Observed inter-relations between 10 m winds, ocean whitecaps and marine aerosols. *Quart. J. R. Met. Soc.*, **109**, 379-392.

Monahan, E. C. & I. Ó Muircheartaigh (1980): Optimal power-law description of oceanic whitecap coverage dependence on wind speed. *J. Phys. Oceanogr.*, **10**, 2094-2099.

Munk, W. H. (1955): Wind stress on water, a hypothesis. *Quart. J. Roy. Met. Soc.*, **81**, 320-322.

Nordberg, W., J. Conway, D. B. Ross & T. Wilheit (1971): Measurements of microwave emission from a foam-covered, wind-driven sea. *J. Atmos. Sci.*, **28**, 429-435.

Okuda, K. (1982): Internal flow structure of short wind waves. Part 1. On the internal vorticity structure. *J. Oceanogr. Soc. Japan*, **38**, 28-42.

Okuda, K., S. Kawai & Y. Toba (1977): Measurement of skin friction distribution along the surface of wind waves. *J. Oceanogr. Soc. Japan*, **33**, 190-198.

Phillips, O. M. & M. L. Banner (1974): Wave breaking in

the presence of wind drift and swell. *J. Fluid Mech.*, **66**, 625-640.

Ross, D. B. & V. Cardone (1974): Observations of oceanic whitecaps and their relation to remote measurements of surface wind speed. *J. Geophys. Res.*, **79**, 444-452.

Su, M.-Y., M. Bergin, P. Marler & R. Myrick (1982): Experiments on nonlinear instabilities and evolution of steep gravity-wave trains. *J. Fluid Mech.*, **124**, 45-72.

The SWAMP group (1983): Sea wave modelling project (SWAMP) Part 1. Proc. IUCRM Symp. on Wave Dynamics and Radio Probing of the Ocean Surface, May 13-20, 1981, Miami Beach (in press).

Toba, Y. (1961): Drop production by bursting of air bubbles on the sea surface. III. Study by use of a wind flume. *Mem. College Sci., Univ. of Kyoto, Ser. A*, **29**, 313-344.

Toba, Y. (1972): Local balance in the air-sea boundary processes. I. On the growth process of wind waves. *J. Oceanogr. Soc. Japan*, **28**, 109-120.

Toba, Y. (1978): Stochastic form of the growth of wind waves in a single-parameter representation with physical implications. *J. Phys. Oceanogr.*, **8**, 494-507.

Toba, Y. (1979): Study on wind waves as a strongly nonlinear phenomenon. Twelfth Symp. on Naval Hydrodyn., Nat. Acad. of Sci., Wash., D.C., 529-540.

Toba, Y. & M. Chaen (1973): Quantitative expression of the breaking of wind waves on the sea surface. *Rec. Oceanogr. Works in Japan*, **12**, 1-11.

Toba, Y., H. Kawamura & K. Okuda (1984): Ordered motions in turbulent boundary layers above and below wind waves. *Turbulence and Chaotic Phenomena in Fluids*, ed. by T. Tatsumi, North Holland Pub. Co., 513-518.

Toba, Y. & H. Kunishi (1970): Breaking of wind waves and the sea surface wind stress. *J. Oceanogr. Soc. Japan*, **26**, 71-80.

Toba, Y., H. Kunishi, K. Nishi, S. Kawai, Y. Shimada & N. Shibata (1971): Study on the air-sea boundary processes at the Shirahama Oceanographic Tower Station. *Disas. Prev. Res. Inst. Kyoto Univ, Annals*, **14B**, 519-531 (in Japanese with English abstract).

Toba, Y., M. Tokuda, K. Okuda & S. Kawai (1975): Forced convection accompanying wind waves. *J. Oceanogr. Soc. Japan*, **31**, 192-198.

Tokuda, M. & Y. Toba (1981): Statistical characteristics of individual waves in laboratory wind waves. I. Individual wave spectra and similarity structure. *J. Oceanogr. Soc. Japan*, **37**, 243-258.

Woodcock, A. H. (1953): Salt nuclei in marine air as a function of altitude and wind force. *J. Meteor.*, **10**, 362-371.

Wu, J. (1979): Oceanic whitecaps and sea state. *J. Phys. Oceanogr.*, **9**, 1064-1068.

Wu, J. (1980): Wind-stress coefficients over sea surface near neutral conditions-A revisit. *J. Phys. Oceanogr.*, **10**, 727-740.

ON CHARNOCK'S RELATION FOR THE ROUGHNESS AT SEA

LUTZ HASSE
Institut für Meereskunde,
Kiel, Germany

If you have not understood that the choice of characteristic variables defines the model, you have not understood dimensional analysis.

W. C. Swinbank, 1970

1. Introduction

The wind stress at the sea surface is a key parameter in air sea interaction. Oceanographers need to know it to determine the atmospheric forcing of the ocean's circulation, climatologists are interested in the related friction velocity to calculate the CO_2 flux through the surface, and meteorologists use the stress as a lower boundary condition. The energy input into the wave field, which we see also as the cause of whitecapping, depends on the roughness of the sea surface. Hence there has been a long-standing desire to parametrize friction, say, in terms of mean wind speed u.

Two parametrizations are in common use. One is via the drag coefficient C_D and the other via the roughness length z_0. If we set $u_* = \sqrt{\tau/\rho}$, where τ is the absolute value of the surface stress, and ρ is the density, then u_* is called the friction velocity, and:

$$u_*^2 = C_D u^2 \tag{1}$$

Likewise:

$$u/u_* = (1/\kappa) \log (z/z_0) \tag{2}$$

where κ is von Karman's constant and the integration constant z_0 is called the roughness length. Equation (2) is applicable only under conditions of neutral density stratification (similar equations are available for non-neutral stratification, but are not of interest here). Both equations allow the sea surface stress to be determined from mean wind speed u if C_D or z_o are known. z_o and C_D are equivalent in this respect and one can be determined if the other is known. To a simple minded observer it must seem strange that parametrization of the same process (in terms of mean wind speed) can be achieved by aid of a C_D which is a dimensionless variable, or alternatively from a z_o, which has the dimension of a length.

$$C_D = \kappa^2 (\log z/z_0)^{-2} \tag{3}$$

Considering the turbulent character of flow in the atmospheric surface layer over the sea and the non-local, stochastic character of the wavefield it is evident that both C_D and z_0 are used to characterize a complex process and probably are functions of a set of parameters. Hence much effort has been spent trying to relate C_D or z_o to simple parameters like wind speed, wave height, friction velocity etc., with reasonable success at least with respect to usefulness, and also with some success in the sense of better physical understanding. Unfortunately, most of the experimental determinations - with the exception of Smith (1980) and Large and Pond (1981) - have been obtained at moderate wind speeds. An extrapolation to higher wind speeds demands a knowledge of the controlling physical process. Wave breaking - as evidenced through whitecapping - is probably an important controlling parameter.

E. C. Monahan and G. Mac Niocaill (eds.), Oceanic Whitecaps, 49–56.

After more than a hundred years of marine meteorology, and intensive experimental work at sea during the last 30 years, the present perceptions of the status of sea surface stress parametrization vary widely. Some pessimists say that simple parametrization is impossible due to wave-turbulence interactions and the non-local character of the wave field. Others are overly optimistic in the sense that they take approximations as established truth and use them, e.g. for the modelling of atmospheric circulations. It is obvious that both the overly pessimistic and the overly optimistic hinder further research and a better understanding of the physical process involved. Hence reliable inferences on those states of the atmosphere-ocean system which are not covered by present experiments will be precluded.

It is the opinion of the author that one item, which has deceived the overly optimistic, because it looks so scientific, apparently a theoretical expression borne out by experiment, is the Charnock equation (where *g* is acceleration of gravity and *a* an empirical dimensionless constant),

$$z_0 = a\,(u_*/g)^2 \tag{4}$$

2. *The physical background of Charnock's equation*

Charnock (1955) rather casually introduced eq (4) as the simplest non-dimensional relation between u_* and z_0. No reason was given why g enters. Presumably, g is a parameter related to the wave field, assuming the flow to be aerodynamically rough. Since the Charnock equation should be valid also under neutral density stratification in the atmospheric boundary layer, g cannot enter as an atmospheric buoyancy parameter.

In wind tunnel work aerodynamically smooth and rough flow is distinguished. The roughness of the wall is produced by gluing sand of a certain grain size to the wall. Near the wall the turbulent intensity decreases as one approaches the wall and very near to the wall only molecular transport of momentum remains. The layer of predominately molecular transport is called the viscous sublayer. Its thickness, δ, depends on viscosity, ν, and on the friction velocity u_*:

$$\delta \approx 5\,\nu/u_* \tag{5}$$

As long as the roughness elements are submerged in the quasi-laminar flow, i.e. as long as these elements are contained within the region $z < \delta$ (where z is the distance from the wall), the flow is aerodynamically smooth. If the roughness elements protrude out of the viscous sublayer, the form drag of the roughness elements adds to the molecular momentum transfer, and the flow is called aerodynamically rough. With equation (2), the details of the flow near the wall are not considered explicitly. Rather, the logarithmic profile of fully turbulent flow is formally extended right to the wall such that $u(z_0) = 0$. When Prandtl (1912) introduced this device he called it a trick, which allows one to determine the integration constant in such a way that at the wall ($z = z_0$) the right value of the mixing length is provided. From experiments on aerodynamically smooth flow, z_0 was found to equal $11\nu/u_*$. For aerodynamically rough flow there was discovered an approximate relationship between z_0 and the sand grain size (h),

$$z_0 \sim h/30 \tag{6}$$

If the roughness elements are not of equal size and are not closely packed, an equivalent grain size must be determined experimentally. Considerable effort has been spent to determine z_0 from the geometric properties of roughness elements, with limited success in wind tunnel work (see Schlichting, 1982 for review), and with reasonable success in meteorology (Lettau, 1969). I believe that the relatively better results of Lettau stem from the fact that his roughness elements span a broader range of sizes (from 0.1 to 500 cm) than the ones in wind tunnel work.

The notion of aerodynamically rough or smooth flow has also been applied to flows at the air sea interface. In this sense the Charnock equation applies for rough flow.

2.1. Influence of capillarity

If we ask for the underlying physical model, the appearance of the acceleration of gravity in Charnock's equation obviously indicates that gravity waves play a role in determining the roughness of the sea surface. Using the hypothesis of aerodynamically rough flow it is evident that the larger gravity waves would not act as roughness elements. A typical roughness length at sea is 1.5×10^{-4} m. If the factor of 30 is used, this would correspond to a wave height of order 0.5 cm. This is much smaller than the typical wave height of wind-driven seas under average wind speeds. The obvious escape from the discrepancy is to postulate that only the small waves act as roughness elements, while the larger ones modulate the flow but exert no appreciable form drag.

With a steepness ratio of, say, 1:10 this would give a wavelength of about 5 cm, just a little longer than the 1.7 cm wavelength of the slowest deep-water wave, which is taken as the border between gravity and capillary waves. Hence it seems strange that surface tension does not appear in Charnock's equation. In fact, modification of Charnock's equation to include surface tension (γ) have been proposed. Wu (1980), on dimensional grounds, suggests

$$z_0 = (au_*^2 /g)\,(\mu u_*/\gamma)^{\beta-2} \tag{7}$$

where β is an empirical coefficient, whose value has been found experimentally to fall in the range $2 < \beta < 2.5$. In the case of surface waves we need not rely on dimensional arguments, since theory is available (Lamb, 1932). In the case of gravity and capillarity working simultaneously at the surface, the solution for the surface displacement is obtained if the restoring acceleration is written as

$$g_* = g(1 + (\gamma k^2/\rho g)) \tag{8}$$

where k is wave number, γ is the surface tension, and ρ the density of sea water. Hence, a revised Charnock equation would read:

$$z_0 = a'(u_*^2/g)/(1 + (\gamma k^2/\rho g)) \tag{9}$$

Slinn (1982) has derived this form in a similar way. Equation (9) illustrates the problem: the Charnock constant, a, depends on wave number, k, of capillary waves. If a′ is a universal constant, a is a weighted average of a′, weighted in an unspecified way with the capillary wave spectrum.

Only if we knew how the equivalent sand grain size can be determined from a given spectrum of wavelets, and if there were a known universal form of high frequency wave spectra, would we be able to predict the effective roughness of the sea surface. Hence, if capillary waves play a role as roughness elements, the usefulness of Charnock's equation is rather doubtful.

2.2. The r.m.s. Wave Height Argument

The rough-wall flow picture repeatedly has led researchers working with wind-water tunnels (where waves are usually small) to take the r.m.s. wave height as the average height of roughness elements and thereby determine z_0. Since the equivalent sand grain size is of order one centimeter, this inference cannot be true for sea waves.

Kitaigorodskii (1970) derived the Charnock equation in a more complicated way which included integration over the wave spectrum. He used the Phillips equilibrium wave spectrum for integration and he acknowledged that in actual sea state conditions the Phillips equilibrium form may not be reached, therefore the dimensionless quantity, $gz_0/u_*^2 = a$, may be variable. Without going into the details of his derivations, it seems that Kitaigorodskii's theory is questionable indeed. His main assumption is the validity of the analogy of the sea surface with an immobile rough surface. Specifically, he assumes that in a cartesian coordinate system moving with phase velocity c, the turbulent flow past mobile roughnesses will be analogous to the flow past an immobile rough wall. This assumption is rather doubtful when extended to a wave spectrum with a broad range of phase velocities. Even for a monochromatic wave, the picture is questionable since the boundary conditions for an immobile ridge and a mobile wave are different. With different boundary conditions one cannot expect the Navier-Stokes equations to have the same solution over mobile and immobile surfaces. Specifically, over an immobile surface the flow around roughness elements protruding from the viscous sublayer would build up a local pressure field. The immobile surface cannot respond to this pressure field. At the mobile sea surface, the flow over an obstacle also would produce a pressure field, but in this case the surface may give and wave motions are induced. So the pressure field in turn will change. As a consequence of this there is another feature distinguishing flows over mobile and immobile surfaces: surface (gravity) wave components are obviously coherent for many wave lengths (say greater than 20), while solid surface roughnesses usually are not. The difficulties one runs into when relating the roughness length to the r.m.s. wave height become evident when instead of wave height the integral over the wave spectrum is written, especially so when the integral is taken between the limits of zero and infinity. Swell would contribute to the r.m.s. wave height, but the energy flux between swell and air is obviously low, so swell cannot contribute significantly to the roughness.

3. Inferences from Wind Wave Generation Studies

During the last 15 years, considerable progress has been made in the understanding of wave generation, while the strongly nonlinear problem of wave energy dissipation is still a matter of research. JONSWAP has shown that the growth of long-waves is due to third-order wave-wave interaction (Hasselmann et mult. al., 1973). Wind-wave tunnel work has yielded reasonable agreement between the experimentally determined growth rate of initial wavelets and the Miles theory (Plant, 1982). The nondimensional growth rate of initial wavelets essentially depends on the square of the ratio u_*/c (where u_* is the friction velocity, and c is wave phase speed).* Studies of wave generation by the measurement of wave coherent pressure fluctuations at open water surfaces have been successfully conducted by Snyder, Dobson, Elliot & Long (1981). Our work has extended these measurements – with a Snyder pressure probe – to the rougher waters of the North Sea with practically the same result (Hasselmann, Bösenberg, Dunckel, Richter, Grünewald & Carlson, 1983). More recent measurements during the KONTUR experiment in 1981 at the JONSWAP site – again using the Snyder probe – have not yet been evaluated, but extend the observations to even greater wave heights.

Parametrization of the nondimensionalized out - of - phase wave induced pressure fluctuation β in terms of the parameter $\mu = u_5 \cos\theta/c$ has been proposed by Snyder et al (1981) to take the form:

$$\begin{aligned} &\beta = \text{const}\,(\mu - 1) && \text{for } \mu \geqslant 1 \\ &\beta = 0 && \text{for } \mu < 1 \end{aligned} \qquad (10)$$

where u_5 is the windspeed at 5m height and θ is the angle between wind and waves. The constant, given as 0.2 to 0.3 by Snyder et al (1981), is in excellent agreement (fig. 1) with the North Sea measurements (Hasselmann et al, 1983). The effect of currents has been discussed by Snyder et al (1981) and can be included through a modification of the parameter μ. The effect of short waves running against the wind, rarely observed at sea, is not included in (10), though there is a slight indication that these waves are damped, see fig. 1.

β determines the energy input into the waves since

$$1/E\cdot(\partial E/\partial t) = (\rho_a/\rho_w)\cdot\omega\cdot\beta \qquad (11)$$

*Note that the growth rates determined in the wind water tunnels and at open sea have slightly different physical meanings. In the wind water facilities, the growth of the initial wavelets is measured. This is proportional to energy transfer to the waves minus dissipation of wave energy. In the field, by measuring the wave coherent static pressure variations, only the input into the waves is measured.

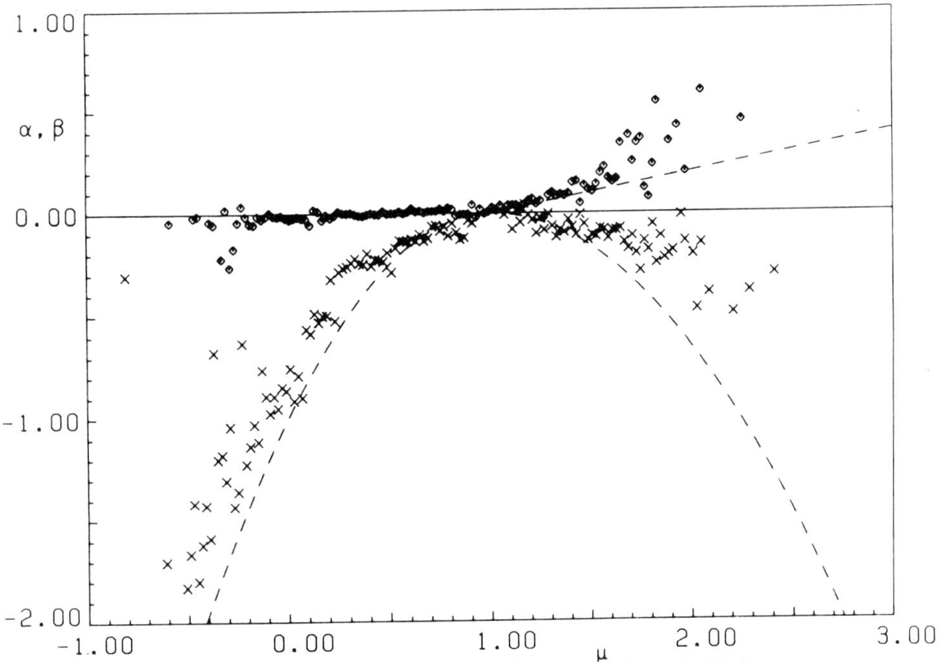

Figure 1. Dimensionless cross-spectrum $\alpha + i\beta$ between static pressure and wave height from measurements at the North Sea. The Quadspectrum is proportional to the relative energy transfer to waves of 'wave age' $\mu = (u_5 \cos\theta)/c$. (From Hasselmann et al, 1983).

where $E(\omega)$ is the spectral wave energy, ρ_a/ρ_w is the ratio of densities of air and water, and ω is the circular frequency. Since the energy and momentum (M) of waves are related by

$$E = \vec{C} \cdot \vec{M} \tag{12}$$

the wave induced momentum flux τ_w can also be obtained. Using 0.2 as the constant in (10), Snyder et al (1981) find that the wave supported stress at short fetches is typically a large fraction of the total wind stress.

From the wave generation studies two results are interesting in this context:

First, from the results reviewed by Plant (1982) there seems to be agreement that in wind wave tunnel studies the growth rate has a spectral maximum and that growth rates at frequencies above 20 Hz are negligible for all wind speeds up to $u_* = 1.20$ m/s. The fact that the maximum of growth rate falls below 20 Hz implies that capillary effects do not dominate, i.e. the growth rate is mainly sustained by gravity waves. Hence (as a tribute to Henry Charnock) the surface tension rightfully is neglected in the Charnock equation. By the same token, the effects of capillary waves cannot explain the apparent variability of experimental determinations of the Charnock constant.

Second, the parametrization (10) is a spectral one and thus tells us in what frequency range the main energy input into the waves takes place. By the dispersion relationship of gravity waves, $\omega^2 = kg$ and $c = \omega/k$, it therefore follows that $c = g/\omega$. Hence

$$\mu = (u_5 \cos\theta/g)\omega \tag{13}$$

and

(14) $\beta = \text{const}((u_5 \cos\theta/g)\omega - 1)$ for $\mu \geqslant 1$

This implies a linear increase with frequency of relative energy input into waves.

The absolute energy input can be calculated only if the actual wave spectrum is known. But typically wave spectra fall off at the high frequency side as ω^{-5} (see Phillips 1977, Hasselmann et mult al 1973). This rather sharp decline of the wave spectrum towards higher frequencies overides the relative increase of pressure-wave correlation (14). The combined effect is that the energy input into the waves decreases more or less as ω^{-4} at the high frequency side. In general, the energy input occurs mainly towards the larger gravity waves at the high frequency side of the spectral peak. If we use the Phillips spectrum

$$F(\omega) = \alpha g^2 \omega^{-5} \tag{15}$$

for an estimate, the maximum of the energy input is at

$$\mu_E = 4/3$$

The peak of the spectrum at large fetch is typically at $\mu = 0.88$ (based on the Pierson-Moskowitz spectrum). Hence, the peak of the energy input into waves is shifted by a factor of only 1.5 to higher frequencies from the peak of the wave spectrum for large fetch. Because $E = \vec{c}\vec{M}$ the spectral maximum of momentum flux is found at slightly higher frequencies. Using the Phillips spectrum (15) again, the maximum of momentum flux is at $\mu_M = 3/2$, also within the range of somewhat larger gravity waves. The spectral distribution of momentum flux as a function of fetch has been calculated by Snyder et al (1981) using the JONSWAP spectrum.

As a third result from the field studies, it should be mentioned that – in agreement with the wave generation studies detailed above – the air flow above the sea surface is modified by the presence of waves (Hasse et al, 1978 a). The logarithmic wind profile (or a diabatic variation thereof), which is essential in order to define and use z_0, is applicable only at heights above the crests of the waves.

4. Discussion

The results of wave generation studies at sea involving the measurement of wave coherent static pressure variations, as well as the different boundary conditions for mobile versus immobile waves, lead one to conclude that the roughness length at sea lacks one property which makes it a useful concept over land. For a rough surface over land, the roughness is a quasi geometric property of the surface; though some variation of roughness occurs (as a function of roughness Reynolds number), the roughness parameter z_0 can be viewed as a characteristic property of the surface. At sea, the roughness length is an abbreviation only for a complicated process and it is not possible to relate z_0 in a simple way to the geometric properties of the sea surface. The Charnock relation does not contain a parameter which has been found useful in describing the wave field or the momentum flux to the waves and probably is the wrong model. It is the opinion of the present author that the roughness length, and hence the Charnock equation, should not be used as a basis for parametrization at sea because it is deceiving. The use of a drag coefficient makes it much more evident that we are attempting to summarize a rather complicated process in a single coefficient.

Since gravity waves are important in energy and momentum transfer to the sea surface, variability of the wave field and its non-local character must enter in any parametrization of the momentum transfer to the sea. On the other hand, it has been possible to infer drag coefficients with a reasonable accuracy (25%) for the open ocean (Hasse et al., 1978 b). An explanation as to why that is feasible seems necessary. Hasselmann et al (1976) argued that it is possible to describe the wave field (of wind-driven waves) by a single parameter – say wave energy or spectral peak frequency – because of the nonlinear interaction between the waves which rapidly restores the spectrum to a quasi-equilibrium level.

At least for low to moderate wind speeds the time or space scale for the adjustment is smaller than the scale for variations of the mean wind field. Still, the scatter in their plots seems definitely larger than the one from experimental determinations of drag coefficients. As an explanation we point to the weighing implicit in equation (10): deviations from the equilibrium spectral form in the low frequency range affect the momentum transfer less than the total wave energy. Also the initial overshooting in the wave spectra noted during JONSWAP (Hasselmann et mult al, 1973) tends to act as compensation. The situation is probably less favourable with high wind speeds.

It is noted that the discussion of section 3 was essentially in terms of Fourier components. The more rare events like whitecapping are not treated as events, they are hidden in the Fourier components. The energy or momentum flux is determined mainly by the phase shift between pressure and wave field. This has been considered here as a kind of black box. The physical mechanism which produces such a phase shift is not identified. It is certainly possible that the more isolated events of wave breaking as studied by Longuet-Higgins and Smith (1985) are an important part of the process of wave growth and decay. Other effects, for example an asymmetric distribution of shorter waves on long waves as discussed by Longuet-Higgins (1969) or a variation of roughness along the larger waves as used in the model by Gent and Taylor (1976) are also possible candidates. Effects of surface films or bubble fields are conceivable if these, due to wave motion, become organized along the larger waves and modify the shorter waves. Research into these areas certainly will increase our understanding and our ability to estimate the momentum flux to the sea at higher wind speeds.

REFERENCES

Charnock, H., 1955: Wind stress on a water surface. *Quart. J. Roy. Meteorol. Soc.*, **81**, 639-640.

Gent, P. R. and P. A. Taylor, 1976: A numerical model of the air flow above water waves. *J. Fluid. Mech.* 77, 105-128.

Hasse, L., M. Grünewald and D. E. Hasselmann, 1978(a): Field observations of air flow above the waves. In: *Turbulent fluxes through the sea surface, wave dynamics and prediction.* Plenum Press, New York, 483-494.

Hasse, L., M. Grünewald, J. Wucknitz, M. Dunckel and D. Schriever, 1978(b): Profile derived turbulent fluxes in the surface layer under disturbed and undisturbed conditions during GATE. *Meteor-Forschungsergebnisse* **B** 13, 24-40.

Hasselmann, D., 1979: Energy and Momentum Flux to Nonresonant Forced Waves. In: Hasselmann and Favre, editors: *Turbulent fluxes through the sea surface, wave dynamics and prediction.* Plenum Press, New York, 457-468.

Hasselmann, D., J. Bösenberg, M. Dunckel, K. Richter, M. Grünewald and H. Carlson, 1983: Measurement of wave-induced pressure over surface gravity waves. In: *Wave dynamics and radio wave probing of the ocean surface.* Plenum Press, New York (In press).

Hasselmann, K. T. P. Barnett, E. Bouws, H. Carlson, D. E. Cartwright, K. Enke, J. A. Ewing, H. Gienapp, D. E. Hasselmann, P. Kruseman, A. Meerburg, P. Müller, D. J. Olbers, K. Richter, W. Sell and H. Walden, 1973: Measurements of wind wave growth and swell decay during the Joint North Sea Wave Project (JONSWAP). *Dt. Hydrogr. Z. Erg.-H.A.*, Nr. 12.

Hasselmann, K., D. B. Ross, P. Müller and W. Sell, 1976: A parametric wave prediction model. *J. Phys. Oceanogr.* **6**, 200-228.

Kitaigorodskii, S. A., 1970: *The physics of air-sea interaction* (in Russian). Engl. Transl. Isr. Progr. Sci. Transl. Jerusalem, 1973, 236pp.

Lamb, H., 1932: *Hydrodynamics.* 6th ed., Cambridge Univ. Press. 738pp.

Large, W. G. and S. Pond, 1981: Open Ocean Momentum Flux Measurements in Moderate to Strong Winds. *J. of Phys. Oceanogr.* **11**, No. 3, 324-336.

Lettau, H., 1969: Note on aerodynamic roughness-parameter estimation on the basis of roughness-element description. *J. Appl. Meteorol.* **8**, 828-832.

Longuet-Higgins, M. S., 1969: A nonlinear mechanism for the generation of sea waves. *Proc. Roy. Soc.* **A** 311, 371-389.

Longuet-Higgins, M. S. and N. D. Smith, 1986: New observations of whitecaps with a surface jump meter. This volume.

Phillips, O. M., 1977: *The dynamics of the upper ocean.* 2nd ed., 336pp. Cambridge Univ. Press.

Plant, W. J., 1982: A relationship between wind stress and wave slope. *J. Geophys. Res.* **87**, 1961-1967.

Prandtl, L., 1932: Meteorologische Anwendung der Strömungslehre. *Beitr. Phys. freien Atmosph.* **19**, 188-202.

Schlichting, H., 1982: *Grenzschicht-Theorie.* 8. Aufl., 843 pp. Verlag Braun, Karlsruhe. Engl. ed.: *Boundary-Layer Theory.* McGraw-Hill, New York.

Slinn, W. G. N., 1983: A potpourri of deposition and resuspension questions. 1-54. In: *Precipitation Scav-*

enging, Dry Deposition, and Resuspension, H. R. Pruppacher, R. G. Semonin and W. G. N. Slinn (eds), Elsevier, New York, 1983.

Smith, S. D., 1980: Wind stress and heat flux over the ocean in gale force winds. *J. Phys. Oceanogr.* **10**, 709-726.

Snyder, R. L., F. W. Dobson, J. A. Elliott and R. B. Long, 1981: Array measurements of atmospheric pressure fluctuations above surface gravity waves. *J. Fluid Mech.* **102**, 1-59.

Wu, J., 1980: Wind-stress coefficients over sea surface near neutral conditions - a revisit. *J. Phys. Oceanogr.* **10**, 727-740.

ACKNOWLEDGEMENTS

The generosity of Prof. R. Snyder who kindly provided the pressure probe heads used in our field studies is gratefully acknowledged. I wish to express my sincere gratitude to my colleagues who participated in the conduct and evaluation of the experiment, especially D. E. Hasselmann, M. Dunckel and J. Bösenberg.

BUBBLE CLOUDS: A REVIEW OF THEIR DETECTION BY SONAR, OF RELATED MODELS, AND OF HOW K_V MAY BE DETERMINED

S. A. THORPE
Institute of Oceanographic Sciences
Wormley, Godalming, Surrey GU8 5UB
United Kingdom

Clouds of subsurface bubbles caused by wind-waves breaking in deep water as whitecaps, or white horses, have been detected by upward-pointing sonars. The main results of experiments using a vertically-pointing narrow-beam sonar, a telesounder, and a dual-beam side-scan sonar, are described. Brief reference is made to the variety of physical phenomena, including internal waves, fronts, and Langmuir circulations, which can be detected through their effect on the clouds.

Several classes of models have been developed to describe the vertical distribution of bubbles below the sea surface. The analytical models, although useful as a guide to the potential value of models, are inadequate and numerical models have therefore been used to provide more realistic descriptions of the distributions and, in particular, of the vertical variation of the acoustic scattering cross-section per unit volume of the bubbles, M_V, a quantity measurable by the sonar system. Some discussion is given to the assumptions made in the models. Two principal objectives of the study are to calculate the gas flux into the water which occurs via the bubbles and to estimate the vertical turbulent diffusion coefficient, K_V.

1. Introduction

Subsurface bubbles produced by whitecaps, wind waves which break in deep water, have been detected by sonars mounted on the sea bed. Our purpose here is to review the principal results of studies using this detection technique. In addition to the information they give about the bubbles themselves, the sonar observations provide surprising and interesting insight into many processes which operate at, or just below, the sea surface, a region in which it is difficult to make prolonged or reliable observations with 'in situ' instruments because of the violence of the wave motions. The technique not only offers the exciting prospect of inferring information about the near-surface ocean in much the same way as meteorologists can deduce the motion and stability of the atmosphere by noting the movement and shape of clouds, but also of making measurements from which we may infer estimates of turbulent fluxes.

2. Observations

2.1. *Vertical beam sonars*

Although Kanwisher (1963) mentions that 'an echo sounder head floating at a depth of 30m and looking up showed foam being swept down to 20m' below the sea surface during a winter storm in the North Atlantic, Aleksandrov and Vaindruk (1974) appear to be the first to have described an experiment using a bottom mounted, upward-pointing, narrow-beam, sonar to detect the clouds of bubbles produced by breaking waves. They found that the depth to which bubbles were seen, typically a few metres, increases with wind speed, and concluded that the bubbles might be used to infer information about the turbulence in the water. This was a most important idea, and it signalled the prospect of significant advances in the detection and understanding of features of near-surface ocean dynamics.

Our own work began under rather fortuitous circumstances (see Thorpe, 1983a) in the fresh-

E. C. Monahan and G. Mac Niocaill (eds.), Oceanic Whitecaps, 57–68.

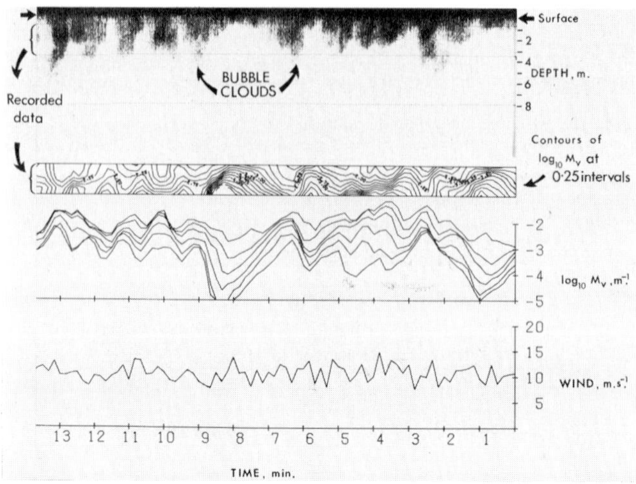

Figure 1. Bubbles observed using a vertically pointing sonar. The sonograph (top) shows clouds of bubbles below the surface. Below this are contours of log M_V and plots of M_V measured at six levels in the depth range bracketed at the left of the sonograph. The wind speed is shown at the bottom. The wind direction was southwesterly, the fetch exceeding 10 km, and the air temperature was 1.75K below the water temperature.

water Loch Ness, using a narrow beam device like that used by Aleksandrov and Vaindruk, although our sonar was moored in mid-water (Thorpe and Stubbs, 1979). The sonar was later moved to a site near Oban on the NW coast of Scotland and it was at this coastal sea-water site, some 700m off-shore, that most of the data were obtained. Figure 1 (top) shows a sonograph record obtained using the vertically pointing 248 kHz sonar mounted on a supporting framework which rests on the sea bed at a depth of about 34m. (More details are given in Thorpe, 1982, hereafter referred to as I). In this, and other figures containing sonograph records, time increases to the left. Clouds of subsurface bubbles are visible. The wind speed measured on shore is shown at the foot of the figure and averages about $11 \mathrm{ms}^{-1}$. At these wind speeds the bubbles form an almost continuous 'stratus' layer below the surface. Occasional clouds extend down to more than 4m.

The acoustic signal returning to the sonar from six different range intervals each about 45 cm in vertical extent (specified by time delays after the sound pulses were transmitted) and lying between the levels bracketed at the left of the sonograph and marked by fine horizontal lines, were recorded. They were averaged over 10s intervals and a calibration applied to convert them into estimates of the acoustic scattering cross section per unit volume of the bubbles, M_V. This is the principal measurement obtained by the sonar, and it is important to explain its physical significance. M_V is an integral of the scattering cross section of bubbles found at depth z and at time t;

$$M_V(z, t) = \int_0^\infty N(a, z, t)\sigma_1(a, z)\, da \qquad (1)$$

where $\sigma_1(a, z) = A/[(1-\omega_0^2/\omega^2) + \delta^2)]$ is the scattering cross section of a bubble of radius a at depth z, $A = 4\pi a^2$ is the bubble area, N is the number of bubbles per unit volume per unit radius, ω is the sonar frequency, δ is the damping coefficient and ω_0 is the resonant frequency of the bubble of radius a:

$$\omega_0 = \frac{1}{2\pi a}\left(\frac{3\gamma' p}{\rho}\right)^{1/2}$$

where γ' is the ratio of specific heats, p is the hydrostatic pressure at depth z and ρ is the density of water. When the bubble radius is much greater than the resonant radius, a_r, given by

$$a_r = \frac{1}{2\pi\omega}\left(\frac{3\gamma' p}{\rho}\right)^{1/2}$$

(about 14 μm at 1m depth for the 248 kHz sonar), σ_1 is approximately equal to A. Provided that the number of bubbles of radius near a_r is sufficiently small compared with those at larger radii, then we have

(2) $$M_v \doteqdot \int NA\,da$$

= surface area of the bubbles per unit volume at depth z.

Eq. (2) appears to be a good approximation if the bubble distributions, N, are similar in shape to those found by the camera techniques of Kolovayev (1976) and Johnson and Cooke (1979), and we may then identify M_v with the bubble area per unit volume (units, m^{-1}). The camera observations have been analysed by Wu (1981) who finds that $N \propto a^{-b}$ with $3.5<b<5$ for $a>80$ μm so that the integral (2) converges quite rapidly at large a.

Observational values of M_v at the six range, or depth, intervals are shown in figure 1 (third panel down) plotted on a log scale. The corresponding contours of $\log M_v$ are also shown (the second panel down in Figure 1) and may be compared with the sonograph display. Generally the contours are evenly spaced, suggesting that M_v decreases exponentially with depth, as is indeed the case (see below). At a fixed level M_v varies by about two orders of magnitude due to the variation in cloud density.

We have analysed several hundred hours of data (see I), have excluded periods in which the wind was variable or biological effects were present, and have drawn the following conclusions:

(a) Bubbles are not detected at wind speeds below 2.5 ms^{-1}.

(b) The mean depth, d_m (m), to which bubble clouds can be detected increases with wind speed, W_{10} (ms^{-1}), and air-water temperature difference, $\Delta\theta$ (K) ($0 \leqslant |\Delta\theta| < 4$), approximately as

$$d_m = 0.31(1-0.1\,\Delta\theta)(W_{10}-2.5) \qquad (3)$$

with a typical scatter of ±0.5m when estimates are based on hourly means. There is some evidence that, at wind speeds exceeding 10ms^{-1}, a non-linear, higher power, dependence of d_m on W_{10} may be appropriate and that at large fetch the coefficient may increase from 0.31 to about 0.4.

(c) In unstable conditions ($\Delta\theta<0$) the bubble clouds tend to have a vertical 'finger-like' structure, reminiscent of thermal plumes, whereas in stable conditions ($\Delta\theta>0$) a more 'billow-like' structure is dominant.

(d) Bubbles form an almost continuous 'stratus' layer below the surface when $W_{10} > 7$ ms^{-1}. At lower wind speeds the individual clouds more usually remain distinct.

(e) M_v decreases approximately exponentially with depth, z(m), at a rate, α, which decreases as W_{10} increases;

$$M_v \propto e^{-\alpha z} \qquad (4)$$

At $W_{10} = 4$ ms^{-1}, $\alpha^{-1} \doteqdot 0.4$m, whilst at $W_{10} = 10 ms^{-1}$, $\alpha^{-1} \doteqdot 0.7$m.

(f) At a fixed level below the surface, M_v increases rapidly with wind speed. The equation

$$M_v(1\,m) = 5.25 \times 10^{-6} \exp(0.63 W_{10}), (m^{-1}) \qquad (5)$$

provides a fair fit to the oceanic data at 1m depth in winds up to about 12 ms^{-1}. Somewhat higher values are found at large fetch for $W_{10} > 6$ ms^{-1}. The data from Loch Ness give significantly lower values.

Values of M_v calculated from Johnson and Cooke's (1979) distributions of bubble sizes observed using an underwater camera are in good agreement with (4) and (5).

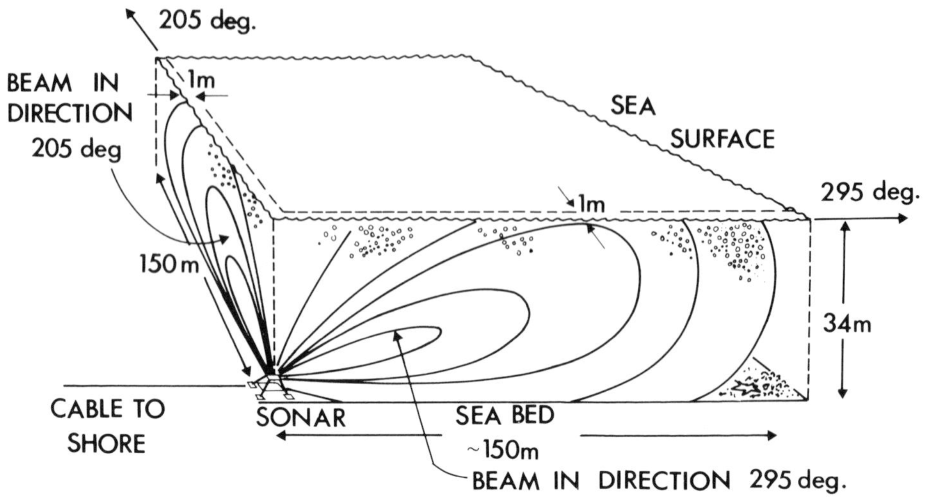

Figure 2. The two beam side-scan sonar on the sea bed some 700m off-shore. The sonar frequency is 248 KHz, pulse length 0.08 ms (about 12 cm) and the pulses were produced alternately by the two linear transducer arrays, each operating at 2 Hz. Two sonographs are produced (see Figures 3-5), in the 205 and 295 degree directions.

2.2. Side-scan sonar

It is not possible from records like that displayed in figure 1 to distinguish between bubble clouds which are advected through the vertical sonar beam and those which might be locally developing; the record includes contributions from both temporal and spatial effects and provides little information about the horizontal extent of the clouds. To investigate the plan-form of the bubble clouds we adopted the use of side-scan sonar, again mounted on the sea bed, first employing a telesounder multiple-beam device (Thorpe et al, 1982). This demonstrated that the moving targets displayed in figures 3-5 are indeed close to, but below, the surface, hence supporting the conclusion that they are bubble clouds. Figure 3 is an example of the sonographs obtained using two orthogonal fan-like beams, each in a vertical plane as shown in figure 2. (Thorpe and Hall, 1983, give a more complete description). These sonographs effectively provide a distance-versus-time plot of the bubble clouds in the two beam directions out to horizontal ranges of 150m from the sonar (although the horizontal range axis is hyperbolic). The lower lines at horizontal range 0m are a return from a side lobe directed towards the sea surface; surface waves can just be seen. This lobe has insufficient power for bubbles to be detected. The remainder of the record is dominated by streaks, many beginning with short, heavy, near-vertical lines which mark the advancing crests of breaking waves. A group of waves approaching the sonar downwind along one of the beam directions is marked in the figure. Waves reaching the centre of the group (where they have their largest amplitude) break for a short period of time whilst approaching the sonar at their phase speed, c; later the following and succeeding waves break. Each wave in breaking produces a short line (slope c indicating the wave phase speed) and a patch of bubbles, a bubble cloud, which is advected by the mean current, thus producing the streaks on the record. The slope of the line formed by the sequence of short, breaking wave, lines gives the speed of advance of the group of waves (approximately c/2; see Donelan, Longuet-Higgins and Turner, 1974; Thorpe and Humphries, 1980). The slope of

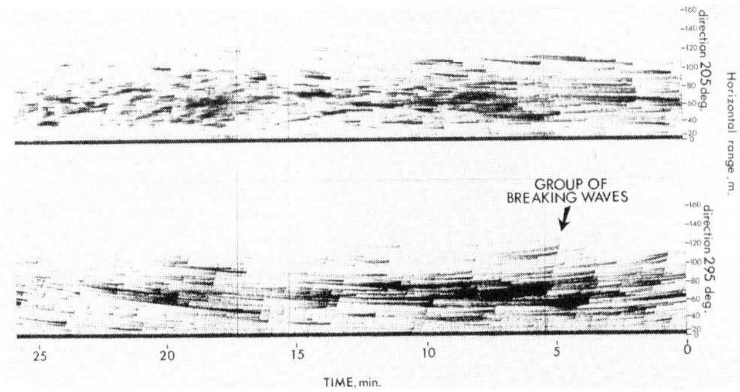

Figure 3. Sonograph from side-scan sonar. The range is measured along the surface from a position immediately above the sonar. The near-horizontal streaks are due to sound reflected from bubble clouds. The wind was 6.5 ms^{-1}, westerly. Groups of breaking waves can be seen approaching the sonar down the beam in the 295 degree direction.

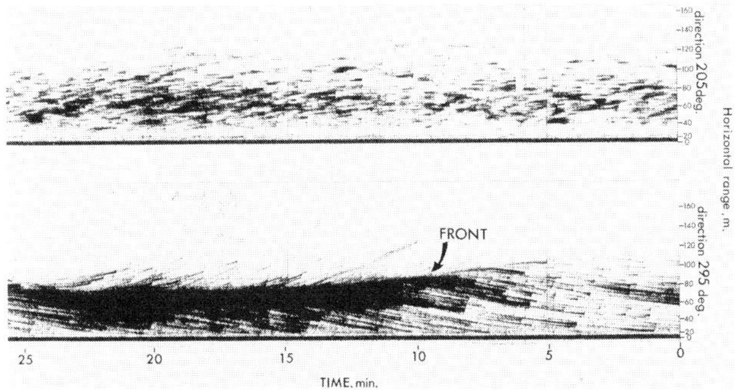

Figure 4. This is a continuation of Figure 3 (the Figures overlap for about 6 mins) and shows the arrival of a region of strong convergence, a front, in the 295 degree beam.

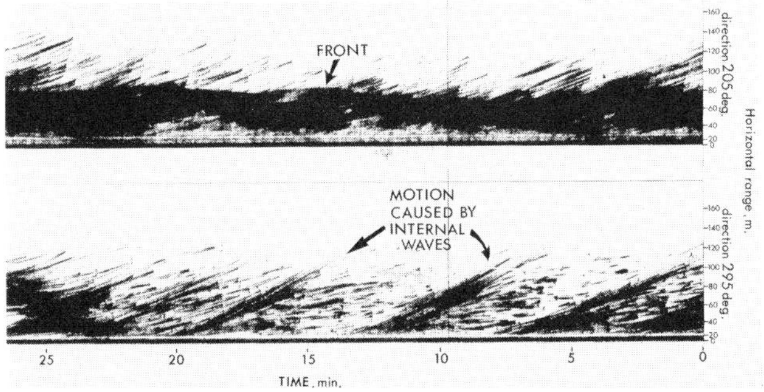

Figure 5. Side-scan sonograph showing (in the 295 degree beam) bubble cloud movements in surface currents caused by a train of internal waves approaching the sonar down the 295 degree beam. In this direction their wavelength is about 80 m and their speed 110 m in 8 mins, about 0.23 ms^{-1}. They appear to be impinging on the front seen in the 205 degree beam. Wind was 7.0 ms^{-1}, south-westerly.

the streaks gives the component of current in the beam direction. In the 205 deg beam direction the slope is near zero; there is no component of current in this direction. In the 295 deg beam direction the current is first towards the sonar and later away from it. The sonar can thus be used to measure surface currents.

Figures 4 and 5 show how the sonar can provide information about the horizontal variation of currents. In figure 4 a region of strong convergence, a front, appears in the 295 deg beam. Figure 5 shows a front, visible in the 205 deg beam, and a train of internal waves principally from the 295 deg direction. The dynamics of fronts and internal waves at the sonar site are discussed by Thorpe et al (1983).

The main conclusions of this study using side-scan sonar which bear on the properties of bubbles are described below (see Thorpe and Hall, 1983).

(a) Separate bubble clouds are created by individual waves which break, often several in succession, in groups of waves which propagate downwind.

(b) The breaking crests are seldom more than a few metres long (e.g. in the 205 deg direction in figure 3). Waves generally continue to break as they advance over a greater distance so that, at formation, the ratio of the downwind length to the crosswind width of the clouds is about 1.4. It is likely that this ratio will increase during the lifetime of a bubble cloud if, as is likely, there is significant vertical shear in the near-surface currents; it may also change if the turbulence acting to spread the cloud is non-isotropic in the horizontal. There is however no evidence in the sonographs to show these processes. The streaks fade with time.

(c) The horizontal size of bubble clouds increases with wind speed, the largest being about 4m in width in winds of 8 ms^{-1}.

(d) The bubble clouds produced by breaking waves appear initially to have some motion in the wave direction, but their motion becomes indistinguishable from the general drift of the surface within about 10s of their formation.

(e) The life-times of bubble clouds detectable by sonar are typically 1-5 mins (see the length of streaks in figure 3).

(f) In winds exceeding about 7 ms^{-1} (when a stratus layer of bubbles exists) or when heavy rain reduces the number of breaking waves but produces bubbles in a uniform manner over the whole water surface, bands of bubbles are formed. These have separations of about 5m, and usually continue for at least 5-10 mins. They are parallel to the wind direction, although they may be carried across wind by tidal currents. They appear to be caused by the convergence induced by Langmuir circulation.

(g) Individual bubble clouds become difficult to distinguish from the background bubbles in winds of 14 ms^{-1} (when you can't tell the clouds from the storm!).

(h) The horizontal advection of bubble clouds, and hence the near-surface currents, are influenced by internal waves and fronts. Bubble clouds appear most dense in regions of horizontal convergence (see the front in figure 4). The distribution of clouds of course reflects the distribution of breaking waves, and this may be affected by the currents induced by fronts or internal waves.

3. Models

3.1. Introduction; analytical models

We have developed several models to describe the vertical distribution of bubbles, and hence of M_V. These have the particular value that by comparing the results with the observed distributions of M_V we may infer the effective turbulent diffusion coefficients of the turbulence which carried the bubbles to depth. This inverse method offers the possibility of obtaining estimates of near-surface vertical diffusivity by remote sensing. We may, in addition, make estimates of the flux of gas between the bubbles and the water.

The models all assume that the subsurface distribution of bubbles results from a balance between

(a) turbulent transport from a source (or sources - the whitecaps) at the surface. The turbulence includes turbulence generated by the breaking waves, as well as that induced by shear, breaking internal waves, Langmuir circulation and so on;

(b) the buoyant rise of bubbles; and

(c) the change in radius of the bubbles due to
(i) diffusion of gases from the bubbles into the water and
(ii) compression of bubbles under hydrostatic pressure.

In a steady state the simplest equation representing these processes is

(6) $$-w_b \frac{dN}{dz} = K_V \frac{d^2N}{dz^2} - \sigma N ,$$

where w_b is the rise speed of bubbles in still water, K_V (here assumed to be independent of depth) is the turbulent diffusion coefficient, σ is a loss rate reflecting the processes in (c) above, and N is the number of bubbles per unit volume. We assume that the bubbles are all of approximately the same size and that w_b and σ are independent of z. We might, for example, suppose that (6) represents the diffusion of bubbles with radii near the peak of the distribution histograms observed by Johnson and Cooke (1979) and estimate w_b and σ accordingly. These are of course gross simplifications, by no means realistic, sufficient only to provide illustrations of the principle by allowing (6) to be solved analytically:

(7) $$N = N_0 e^{-\beta z} ,$$

where β is given by $K_V\beta^2 - w_b\beta - \sigma = 0$, or

(8) $$K_V = w_b\beta^{-1} + \sigma\beta^{-2}$$

If from observations we can measure β, and if also we can estimate w_b and σ, then K_V may be determined. In this simple model with bubbles of uniform radius, $M_V \propto N$ (by 2) and since (by 4) M_V is observed to be proportional to $\exp(-\alpha z)$, where α is determined by observations, we have $\alpha = \beta$. Since α decreases as W_{10} increases, the estimated value of K_V increases with wind speed (by 8) in accordance with expectation. In principle, this suggests that we might estimate K_V from the slope $S = -M_V^{-1}(dM_V/dz)$ of ln M_V at a given depth, or set of depths. The examination of this prospect, in particular the assessment of the accuracy of estimates which might be made, is an objective of the modelling.

Other analytical solutions, including the more plausible $K_V = ku_*z$, where u_* is a friction velocity and k is von Karman's constant, about 0.4, are discussed in I and Thorpe (1984a).

It is not possible to solve analytical models which include all the important physics, and to proceed further we must turn to numerical models.

3.2. Classes of numerical models

All these models include a finite distribution of bubble sizes. They differ principally in the way in which turbulence is represented. Three classes of models have been used;

(a) those in which turbulence is represented by an array of cells of differing vertical velocity, their space-averaged sum having a Gaussian distribution of velocities. These are rather unrealistic since they fail to represent the variation in vertical velocity during the probable lifetime of a bubble. They may however be of some value in providing estimates of the maximum depth to which bubbles might be carried in, for example, Langmuir circulation, and of the maximum lifetimes of bubbles (see I and Thorpe 1984c);

(b) those in which turbulence is represented by a diffusion coefficient, K_V, and in which a diffusion equation, similar to (6) but taking into account the presence of bubbles of different sizes, is solved for a given distribution of bubbles at z = 0. Solutions with $K_V = ku_*z$ and K_V constant with depth have been examined (see I and Thorpe 1984b); and

(c) those in which turbulence is simulated by a Monte Carlo or random walk technique. This is of particular value when the evolution of each particular bubble is to be followed from its inception and avoids the difficulty encountered in model (b) of specifying the bubble distribution at z = 0 when the water is supersaturated and bubbles may increase their radius. The model may be used with bubbles containing more than one gas and to examine the effect of Langmuir circulation on the distribution of bubbles (Thorpe, 1984c). It is of potential value in examining the development of the distribution of bubbles in a single cloud. Simulations have been made with the effective turbulent diffusion coefficient constant with depth (Thorpe, 1984a).

3.3. Assumptions

There are several assumptions inherent in the models. These are justified by estimates based on existing observations or by consideration of the bubble physics, and impose bounds on the conditions in which the models may be appropriate. Discussion of the assumptions is given in I, but it is pertinent to review some of them here as they lend insight into the nature of bubble clouds. In some cases (e.g. (b) below) we have extended the discussion of their range of validity beyond that given in I.

The principle assumptions are

(a) that the bubbles become covered by a surface-active film very soon after their generation, that this restricts tangential motion at the surface of the bubbles and hence affects the bubbles' rise speeds and the diffusion of gas; but that it does not totally inhibit gas transfer except perhaps for very small bubbles (see Detwiler and Blanchard, 1978; Detwiler, 1979; Johnson and Cooke, 1981);

(b) that fragmentation or coalescence of bubbles is negligible. Except near the water surface in the breaking waves the bubbles observed are so small that they should be stable and will not fragment.

The number of collisions per unit time and unit volume of neutral particles of radii a_j and a_k, of numbers per unit volume, n_j and n_k, in turbulent flow dissipating kinetic energy at a rate ϵ per unit volume and of viscosity ν, is

(9) $$N_{jk} = 1.3(\epsilon/\nu)^{1/2}(a_j + a_k)^3 n_j n_k$$

(Saffman and Turner, 1956). For bubbles of the same radii near the peak in the histograms of Johnson and Cooke,

$$a = a_j = a_k = 50\mu m, \quad n_j = n_k = n_0 .$$

Taking $M_v = 4\pi a^2 n$ to be given by (5) at 1m depth and

(10) $$\epsilon = u_*^3/kz$$

near the surface (Dillon et al., 1981), where $\rho u_*^2 = \rho_a C_D W_{10}^2$, with ρ_a the density of air and C_D a drag coefficient, about 1.3×10^{-3}, we find

$$F \equiv \frac{N_{jk}\tau}{n} = 3.05\times10^{-9} W_{10}^{3/2} \exp(0.63W_{10}), \quad (11)$$

where τ is a typical bubble lifetime, here taken as 1 min, and ν is taken to be $10^{-6}\,m^2 s^{-1}$. F represents the proportion of bubbles which might coalesce during the lifetime of a single bubble. It increases with W_{10}, but is less than 0.5% if $W_{10} = 16\ ms^{-1}$ (assuming that (5) is valid to these wind speeds).

It thus appears possible to neglect bubble coalescence due to turbulence except in very high winds or perhaps very close to the surface in the breaking waves, where in any case (10) will not be valid.

Coalescence between bubbles rising at different speeds is also assumed to be negligible. The rate of increase in radius of a bubble of size R rising through a cloud of smaller bubbles is given by

$$\frac{1}{R}\frac{dR}{dt} = \frac{\pi}{3R}\int_0^R \left(1+\frac{a}{R}\right)^2 (w_b(R)-w_b(a)) \times n(a)a^3 E\,da, \quad (12)$$

where w_b (a) is the rise speed of bubbles of radius a, n(a) da is the number of bubbles with radii between a and a+da per unit volume, and E is the capture efficiency (Rogers, 1979). If we assume that $n(a) = N_0 a^2 \exp(-ba)$ (a model distribution with a peak at $a = a_m = 2/b$ and total number $N = 2N_0/b^3$ bubbles, $w_b = 2ga^2/9\nu$ (appropriate for bubbles of sizes less than $80\mu m$) and $E = 1$ (a more realistic value is not known, but this will give an upper bound), then

$$\frac{1}{R}\frac{dR}{dt} = \frac{8\pi N g R^7 I(\lambda)}{27\nu a^3{}_m}, \quad (13)$$

where $I = \int_0^1 (1+y)^2(1-y^2)\,y^5 e^{-\lambda y}\,dy$, which is easily evaluated, and $\lambda = 2R/a$.

For $R = 70\mu m$, $a_m = 50\mu m$, $N = 5\times10^5$ bubbles m^{-3} (the value at 0.7m, $W_{10} = 11\text{–}13 ms^{-1}$, observed by Johnson and Cooke), we find that

$$\frac{1}{R}\frac{dR}{dt} = 1.3 \times 10^{-5}\,s^{-1} .$$

This compares with the estimates

$$\frac{1}{R}\frac{dR}{dt} = 6.7 \times 10^{-3}\,s^{-1}$$

due to gas flux from the bubbles, or

$$\frac{1}{R}\frac{dR}{dt} = 3.1 \times 10^{-4}\,s^{-1}$$

due to pressure effects (see I, equation 23 at $R = 70\,\mu m$, $z = 0.7\,m$, $x = 0$, 100% saturation), and is relatively negligible).

We shall in fact make a slightly more restrictive assumption, that the bubbles act individually and do not interact with each other or, dynamically, on the water. This appears valid provided (i) the mean separation between bubbles greatly exceeds their mean radius and (ii) that they do not, en masse, significantly change the density of the seawater.

Taking, as above, $M_V = 4\pi a^2 n$ to be given by (5), and $a = 50\mu m$, the mean separation of bubbles, l, is approximately $n^{-1/3}$. Hence,

$$l/a = (4\pi)^{1/3} M_V^{-1/3} a^{-1/3} ,$$

which decreases with wind speed, but is about 126 at $W_{10} = 16\ ms^{-1}$, and at 1m depth, sufficiently large for (i) to be reasonable. The net density of the fluid is $\rho\,(1-r)$ where

$$r = \frac{4}{3}\pi a^3 n(\rho-\rho_a)/\rho \doteqdot \frac{M_V}{3} a(\rho-\rho_a)/\rho .$$

Since r is approximately 2×10^{-6} at $W_{10} = 16 ms^{-1}$, the effect of bubbles on fluid density is negligible. Since, by similar reasoning, the volume fraction of bubbles in the water is small (approximately $aM_V/3$), the rise speed of individual bubbles should not be affected by the presence of surrounding bubbles (Batchelor, 1972);

(c) that turbulence does not distort the bubbles. This will be valid if the bubble radius is much less than the Kolmogorov scale $(\nu^3/\epsilon)^{1/4}$, as is generally the case. A bubble's vertical speed then is the sum of the rise speed in quiescent fluid and the vertical component of the turbulent velocity;

(d) that the bubbles are spherical, a good assumption provided their radii is less than about 400 μm. This is as observed by Johnson and Cooke (1979) at 0.7m; close to the surface in high winds the assumption may be invalid;

(e) that turbulence does not affect the gas transfer from the bubbles. The latter condition is valid provided the r.m.s. vorticity of the turbulent flow $(\epsilon/\nu)^{1/2}$ is less than order ϕ, where

$$\phi = \left(\frac{4}{81}\frac{g^2 D}{\nu^2}\right)^{1/3}$$

and D is the diffusivity of the bubbles' gas in the water and g is the acceleration due to gravity (Batchelor, 1980). Using (10) this condition becomes

$$W_{10} < O\left[\left(\frac{4}{81}\frac{g^2 D}{\nu^2}\right)^{2/9} \frac{(kz\nu)^{1/3}}{(\frac{\rho_a}{\rho} C_D)^{1/2}}\right]$$

or, taking $D = 2 \times 10^{-9}\ m^2 s^{-1}$, $g = 9.81\ ms^{-2}$, and other values as before,

$$W_{10} < O(34.4 ms^{-1})$$

when $z = 1m$. It hence seems reasonable to ignore the effect of turbulence on the gas flux from the bubbles except in extreme winds;

(f) that a steady state and a uniform distribution of sources are maintained. These are assumptions which need careful consideration, for in practice the input is intermittent, when and where waves break. Although generally the time interval between successive inputs of bubbles into water at a fixed position is much less than their time of decay, this is not true in low winds when wave breaking is very infrequent and when, as we have seen, there is no stratus layer of bubbles underlying the surface. Work is at present in hand to examine the variation in time of the population of bubbles in a cloud after its creation. More information about the bubbles produced by breaking waves, and about the frequency of breaking, is needed before a reliable model of the mean distribution of bubbles produced at intermittent sources can be constructed. It is possible that some progress might be made by examining the relation between the bubble clouds and their surface counterpart, the floating foam. In the models we have developed so far, the wave breaking is supposed to be effectively continuous at the surface.

A related assumption is that the observed bubble distributions do not result primarily from turbulence which is very different from a spatially averaged mean: that is, that the bubbles, being produced in a patch of intense turbulence in the breaking waves, are not subjected to the effect of this anomalously high turbulent level for a large part of their lifetime. This appears reasonable on two counts; first that the bubble lifetime is much longer than the typical periods of the wind waves, and second that the motion of bubble clouds is observed to be indistinguishable from that of the surroundings in a period less than 10s after their formation. The assumption, and (a) above, is most likely to fail for large bubbles which, returning rapidly to the surface, have lifetimes shorter than 10s. Fortunately it appears that the majority of bubbles penetrating to more than 0.5m or so are small (Johnson and Cooke, 1979) and have longer lifetimes.

It is further implicit in the assumption of a steady state that the mean saturation level of gases in the water surrounding the bubbles does not change. This is a reasonable approximation provided that the time over which significant changes in saturation levels occur in the near-surface mixing layer by diffusion across the sea surface, by the bubbles themselves, by turbulent mixing at the base of the mixing layer, or by chemical or biological processes, is much greater than the bubble lifetimes. This appears to be generally a valid assumption for oxygen and nitrogen which together compose about 99% of the bubbles' gas and hence determine the changes in the bubbles' volume;

(g) that the turbulent diffusivity of the bubbles, K_v, is equal to that of momentum, K_m. Observational confirmation of this assumption is lacking, although it appears valid in sediment flow (Newberger and Caldwell, 1981), a closely related problem, and is a common assumption in other similar situations (e.g. in radiation fog; see Brown, 1980). This assumption is inherent in the selection of the formulation, $K_v = ku_*z$, and is necessary if we are to infer properties of turbulent momentum transfer from the distribution of bubbles.

3.4. Conclusions

The principal conclusions of the modelling concern the possibility of estimating K_v from the slope s, or the length scale $d = 1/s$ (m). This is particularly attractive as a reference value since it is independent of the absolute magnitude of M_v itself, a quantity which it is difficult to determine precisely. (It would require very accurate calibration of the sonar).

The solid line in figure 6 shows the variation of d, calculated using a model of type B, with $K_v = ku_*z$. The points are from observations in which d is estimated from the vertical distribution of M_v, while K_v is determined at z = 2m by assuming that u_* can be related to wind speed by use of a drag coefficient. The vertical line shows the effect of changing the saturation level assumed in the model by 3%. The observations are reasonably consistent with the model predictions, particularly if we assume (as was likely) that the water was supersaturated with gas, up to values of K_v of $100 cm^2 s^{-1}$ or wind speeds of about $10 m s^{-1}$. In stronger winds a larger value of K_v than that produced by the linear variation with depth is needed to explain the observations at 2m depth (Thorpe, 198 b).

Further conclusions are that

(a) the predicted maximum depths of bubbles based on vertical velocities observed in Langmuir circulations by Filatov et al (1981) are in general agreement with observations (I, figure 15b);

(b) the models provide fair predictions of bubble lifetimes;

(c) the flux of oxygen entering the water from the bubbles is insufficient to explain the gene-

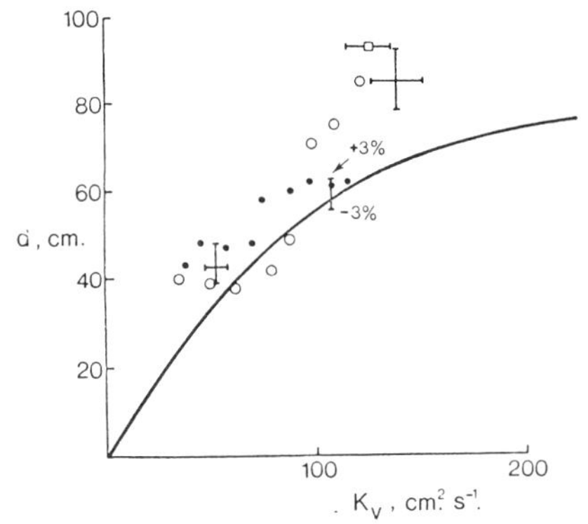

Figure 6. The variation of d with $K_V = ku_*z$. The solid line is from a model of type B, while the points are from observations: o from Oban, • from Loch Ness and □ Johnson and Cooke (1979). The crosses represent error estimates in the data, and the vertical bar shows the effect of changing the saturation level by ±3%.

ral 3% supersaturated levels found in the near-surface ocean water and described by Broecker and Peng (1982); see Thorpe (1984d).

(d) There is a general agreement between the predicted variation in the total number of bubbles with depth and that observed by Johnson and Cooke (1979). Whilst the observed distribution maintains a peak near 50μm radius, the models however find that the bubble distributions have a peak which tends to a lower radius as depth increases (see for example I, figure 26). The most obvious process we have neglected which might reduce this trend is bubble coalescence. We have argued that this is negligible. However (13) depends on N. If, in the bubble clouds, the bubble density is very much higher than occurs on average (i.e. the distribution is very skewed, and this is plausible since M_V increases by two orders of magnitude in the intermittent clouds) it is possible that the neglect of coalescence might account for the difference between theory and observation. There is an urgent need for further careful observations of the distribution of bubble sizes, especially at radii less than 50μm, and of the temporal variation of the distribution. It is particularly important that care is taken to obtain data which are not affected by mechanical generation of bubbles by the sensing device, or biassed because the instrument tends to remain in areas of convergence (e.g. in wind rows where more bubbles are likely), and that the sampling of the intermittent bubble clouds and breaking waves is adequate. Observations in fresh water lakes when bubble size distributions may be very different (e.g. see Scott, 1975) would be valuable.

In summary then, much has been learnt about the distribution of bubbles using sonar. It is potentially a powerful tool to probe the near-surface ocean. Recent trials of a ship-towed inverted side-scan sonar have produced promising results. More observations are needed, especially at a site remote from shore, and preparations are now being made to deploy an internally-recording, upward-pointing, narrow-beam, sonar at the edge of the Continental Shelf.

ACKNOWLEDGEMENTS

This review was prepared during a visit to the School of Oceanography, Oregon State University, Corvallis, in summer, 1983, and I wish to express my gratitude to my hosts, particularly Doug Caldwell and Tom Dillon, for inviting me and for providing such pleasant surroundings and generous facilities. Partial support was received from the Office of Naval Research, contract N00014-79-C-004.

REFERENCES

Aleksandrov, A. P. & Vaindruk, E. S. 1974. In 'The Investigation of the variability of hydrophysical fields in the ocean' (ed R. Ozmidov), pp 122-128. Moscow: Nauka Publishing Office.

Batchelor, G. K. 1972. Sedimentation in a dilute dispersion of spheres. *J. Fluid Mech.* 52, 245-268.

Batchelor, G. K. 1980. Mass transfer from small particles suspended in turbulent flow. *J. Fluid Mech.* 98, 609-623.

Broecker, W. S. & Peng, T.-H. 1982. Tracers in the Sea. Lamont-Doherty Geological Observatory, New York. 690 pp.

Brown, R. 1980. A numerical study of radiation fog with an explicit formulation of the microphysics. *Quart. J. Roy. Met. Soc.* **106**, 781-802.

Detwiler, A. 1979. Surface-active contamination on air bubbles in water. In Surface contamination: genesis detection and control, Vol. 2 (ed. K. L. Mittal) pp 993-1007. New York: Plenum Press.

Detwiler, A. & Blanchard, D. C. 1978. Ageing and bursting bubbles in trace-contaminated water. *Chem. Engrg. Sci.* **33**, 9-13.

Dillon, T. M., Richman, J. G., Hansen, C. G. & Pearson, M. D. 1981. Near-surface turbulence measurements in a lake. *Nature* **290**, 390-392.

Donelan, M., Longuet-Higgins, M. S. & Turner, J. S. 1974. Periodicity in whitecaps. *Nature* **239**, 449-451.

Filatov, N. N., Ryanzhin, S. V. & Zaycev, L. V. 1981. Investigation of turbulence and Langmuir circulation in Lake Ladoga. *J. Great Lakes Res.* **1**, 1-6.

Johnson, B. D. & Cooke, R. C. 1979. Bubble populations and spectra in coastal waters: a photographic approach. *J. Geophys. Res.* **84**, C7, 3761-3766.

Johnson, B. D. & Cooke, R. C. 1981. Generation of stabilised micro-bubbles in seawater. *Science* **213**, 209-211.

Kanwisher, J. 1963. On the exchange of gases between the atmosphere and the sea. *Deep Sea Res.* **10**, 195-207.

Kolovayev, P. A. 1976. Investigation of the concentration and statistical size distribution of wind produced bubbles in the near-surface ocean layer. *Oceanology* **15**, 659-666.

Newberger, P. A. & Caldwell, D. R. 1981. Mixing and the bottom nepheloid layer. *Marine Geology* **41**, 321-336.

Rogers, R. R. 1979. A short course in cloud physics. Pergamon Press, Oxford, 235 pp.

Saffman, P. G. & Turner, J. S. 1956. On the collision of drops in turbulent clouds. *J. Fluid Mech.* **1**, 16-30.

Scott, J. C. 1975. The role of salt in whitecap persistence. *Deep Sea Res.* **22**, 653-657.

Thorpe, S. A. 1982. On the clouds of bubbles formed by breaking wind-waves in deep water, and their role in air-sea gas transfer. *Phil. Trans. Roy. Soc. Lond.* **A 304**, 155-210.

Thorpe, S. A. 1983a. Bubble Clouds. *Weather* **38**, 66-70.

Thorpe, S. A. 1984a. A model of the turbulent diffusion of bubbles below the sea surface. *J. Phys. Oceanog.* **14**, 841-854.

Thorpe, S. A. 1984b. On the determination of K_V in the near-surface ocean from acoustic measurements of bubbles. *J. Phys. Oceanog.* **14**, 855-863.

Thorpe, S. A. 1984c. The effect of Langmuir circulation on the distribution of submerged bubbles caused by breaking wind waves. *J. Fluid Mech.* **142**, 151-170.

Thorpe, S. A. 1984d. The effect of bubbles produced by breaking wind-waves on gas flux across the sea surface: submitted to *Annales Geophysicae.*

Thorpe, S. A. & Hall, A. J. 1983. The characteristics of breaking waves, bubble clouds, and near-surface currents observed using side-scan sonar. *Continental Shelf Research* **1**, 353-384.

Thorpe, S. A., Hall, A. J. & Hunt, S. 1983. The bouncing internal bores of Ardmucknish Bay. *Nature* **306**, 167-169.

Thorpe, S. A. & Humphries, P. N. 1980. Bubbles and breaking waves. *Nature* **283**, 463-465.

Thorpe, S. A. & Stubbs, A. R. 1979. Bubbles in a freshwater lake. *Nature* **279**, 403-405.

Thorpe, S. A., Stubbs, A. R., Hall, A. J. & Turner, R. J. 1982. Wave-produced bubbles observed by side-scan sonar. *Nature* **296**, 636-638.

Wu, J. 1981. Bubble populations and spectra in near-surface ocean: summary and review of field measurements. *J. Geophys. Res.* **86**, C1, 457-464.

BUBBLE POPULATIONS: BACKGROUND AND BREAKING WAVES

BRUCE D. JOHNSON
Oceanography Department,
Dalhousie University, Halifax,
Nova Scotia, Canada

Marine bubble populations can be described as being composed of wave-produced bubble patches superimposed on a background population. This background population comes from the decay of the bubble patches, from secondary sources such as biological activity, droplet impact, and sediment outgassing, and from stable microbubbles.

A radio-controlled bubble camera is described, as is its use in following the decay of a laboratory bubble injection event. The existence of bubbles in the background population at higher sea states is suggested to be consistent with a source associated with the formation and decay of whitecaps.

Many processes of meteorological, geochemical and biological interest are associated with the presence of bubbles in the near surface of the ocean. While the role of bubbles has been described in processes such as aerosol formation (Blanchard and Woodcock 1957) surface coagulation (Sutcliffe et al 1963, Johnson and Cooke 1980), fractionation of organic and inorganic materials (Wallace and Duce 1975, Gershey 1983) and gas exchange (Thorpe 1982, Merlivat and Memery 1983), a quantitative evaluation of these processes awaits the determination of bubble populations and their dynamics.

One outgrowth of the studies on oceanic bubble populations is the identification of a background population - a population for which the source must be determined as well as the significance. A description of the sources of the background population will be addressed here. However, since it is an important tool in these studies, the bubble camera will first be described along with the rationale for its development.

1. The Photographic Method

Early attempts to measure bubble populations involved both simple and complex methods based on bubble traps, acoustics and photography. While some of these methods have been used effectively, most face apparent limitations. For example, the 60 cm length of a trap employed by Kolovayev (1976) provided the means of concentrating bubbles, however the delay time necessary for the smallest bubbles to rise before measurement raises questions about the effects of bubble dissolution, growth and coalescence. According to Stokes Law, a bubble 25μm in radius requires 8 minutes to traverse the 60 cm tube length of Kolovayev's trap, but the same bubble requires only 3 minutes to dissolve at 10 cm depth in air-saturated water.

While some success accompanied early acoustic studies such as those of Medwin (1977), the data that emerged have presented certain difficulties for interpretation, i.e., in particular, the large bubble concentrations observed at low sea states where wave injection must be ruled out as the primary source. If microbubbles provide the answer, as has recently been suggested (Johnson and Cooke 1981), then the effect of organic films on their acoustic response must be addressed (see figure 1).

Because of the apparent limitations and uncertainties associated with the then existing methods for determining bubble populations,

E. C. Monahan and G. Mac Niocaill (eds.), Oceanic Whitecaps, 69–73.

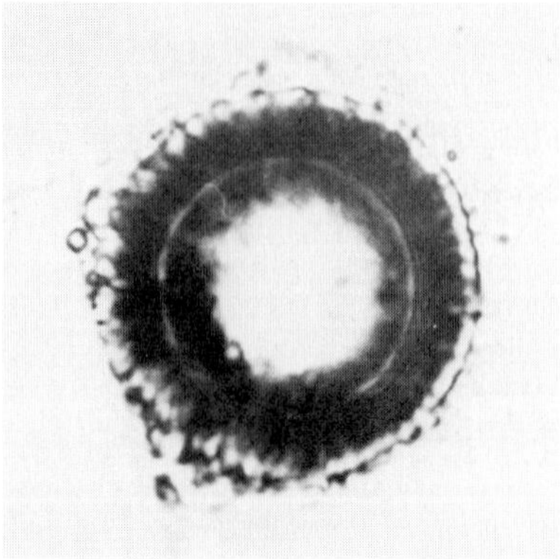

Fig. 1. A bubble of 25 μm radius that is in process of dissolving. It was initially about 60 μm in radius. The material on its surface was collected during 20 seconds of rise in unfiltered seawater.

a photographic method was developed (Johnson and Cooke 1979). With this method three images corresponding to the specular reflections of the light provided by three strobes served to identify bubbles, and provided the basis for measurement. The depth of field was established on the basis of intersection of the three dots - a strategy that, in conjunction with the use of a high resolution high acuity film that permitted low magnification (1/3 X), served to maximize the sample volume. With this technique bubbles as small as 17 μm in radius were identifiable.

Since the initial studies, the fixed interval timer that triggered the camera has been replaced with a radio control unit. This radio control device can be coupled to an above surface camera, allowing the simultaneous recording of events above and below the water surface. This system provides the means for photographing at short intervals the breaking of a wave and the attendant decay of the bubble patch while simultaneously obtaining photographs below the water surface for determination of bubble populations.

Further modification of the photographic technique was required when the film that had previously been used was no longer available from the manufacturer. While an alternative film has been found, the resolution is not as good, and photographs now are made at 1/2 X instead of 1/3 X magnification and bubbles of 32 μm in radius are now the smallest measurable.

Calibration of the new system was effected as previously described (Johnson and Cooke 1979), i.e., from photographs of the controlled movement of a bubble laden fiber, 15 μm in cross section, through the zone of focus. As before, the depth of field, and hence sampling volume, is a function of bubble size.

2. The Background Bubble Population

Under conditions of higher sea states, wave injection must be a major source of bubbles for the background population. While whitecap coverage is certainly an indicator of the formation and decay of an injected population, significant concentrations of smaller bubbles are still present in the water parcel long after the whitecap ceases to be visible.

To demonstrate the long residence times of these smaller wave generated bubbles, a beaker of seawater was poured onto the surface of a pool of air-saturated seawater. The character of the water surface at the point of bubble injection was followed with photographs at 1 second intervals, and simultaneously photographs were made 15 cm below the surface with the bubble camera. Figure 2 shows the bubble distributions that were determined for 1, 2, 8 and 11 seconds after the bubble injection. While above surface evidence of the episode disappeared within 4 seconds, significant numbers of bubble images appeared in the last photograph of the series, 30 seconds after the event.

An indication of the persistence of small bubbles in the near-surface ocean can be obtained through an examination of what might be called the critical depth curve. This critical depth curve is the locus of points describing the size and position of a bubble which after it is released at depth in quiescent water, rises, and through convective diffusion dissolves completely as it reaches the interface. While describing the delineation between bubbles that reach the surface and break and those that dissolve, this critical depth curve also describes the maximum residence times of bubbles under

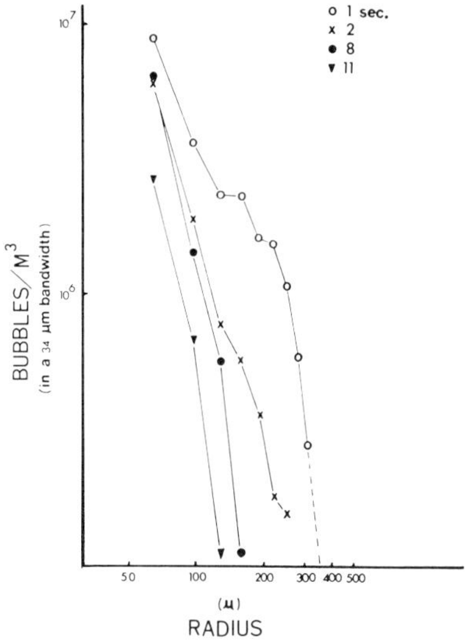

Fig. 2. Curves showing the decay of a bubble population that was injected by pouring a beaker of water on to the surface of a tank. (The numbers of bubbles greater than 320 μm are not included, but in the distribution at 1 second, bubbles as large as 800 μm in radius were present).

quiescent conditions.

3. *The Critical Depth Curve*

The equation describing the rate of change of the bubble volume, V, is readily derived from the ideal gas law and appears as

$$\frac{dV}{dt} = \frac{RT}{P_B}\frac{dn}{dt} - \frac{V}{P_B}\frac{dP_B}{dt} \tag{1}$$

where R is the gas law constant, n is the number of moles of gas in the bubble, and P_B is the bubble internal pressure. Substituting for P_B

$$P_B = P_0 + \frac{2\gamma}{r} - \rho gz \tag{2}$$

and converting dV/dt to dr/dt gives

$$\frac{dr}{dt} = \frac{3RT\frac{dn}{dt} + 4\pi r^3 \rho g \frac{dz}{dt}}{4\pi r\,[4\gamma + 3rP_0 - 3\rho grz]} \tag{3}$$

where g is the gravitational constant, ρ is the water density, γ is the surface tension and P_0 is atmospheric pressure plus the hydrostatic pressure at the point of bubble release.

Various versions of (3) and the equations that follow have appeared in the literature (e.g. Le Blond, 1969; Thorpe, 1982; Merlivat and Memery, 1983). Most vary in choices of expression for dn/dt (see, e.g., Clift, Grace and Weber 1978; Thorpe, 1982). An expression for dn/dt presented by Levich (1962) for Re<<1, for high Peclet Numbers, and for a bubble with an immobile interface is

$$\frac{dn}{dt} = -7.98D^{2/3}u^{1/3}r^{4/3}c \tag{4}$$

where u is bubble velocity and D is gas diffusivity. A first order correction to (4) was presented by Acrivos and Goddard (1965)

$$\frac{dn}{dt} = -(7.85\,D^{2/3}u^{1/3}r^{4/3} + 5.78Dr)c \tag{5}$$

and is the expression that will be used here.

The expression for dn/dt in which nitrogen and oxygen are considered is

$$\begin{aligned}\frac{dn}{dt} &= -(7.85D_{N_2}^{2/3}U^{1/3}r^{4/3} + 5.78D_{N_2}r)C_{N_2} \\ &\quad - (7.85D_{O_2}^{2/3}U^{1/3}r^{4/3} + 5.78D_{O_2}r)C_{O_2} \\ C_{N_2} &= K_{N_2}[Y_{N_2}(1 + P_o - \rho gz + 2\gamma/r - P_{H_2O}) - P_{N_2})] \\ C_{O_2} &= K_{O_2}(1 - Y_{N_2})[(1 + P_o - \rho gz + 2\gamma/r - P_{H_2O}) - P_{O_2}]\end{aligned} \tag{6}$$

where K is the solubility, and y is the mole fraction. Subscripts N_2 and O_2 refer to parameters for nitrogen and oxygen respectively.

When the expression for dr/dt, Stokes law for dz/dt, and an expression for following y_{N_2} are stepped through iteratively by computer a solution for a rising bubble is obtained.

The critical depth curves for seawater of 35‰

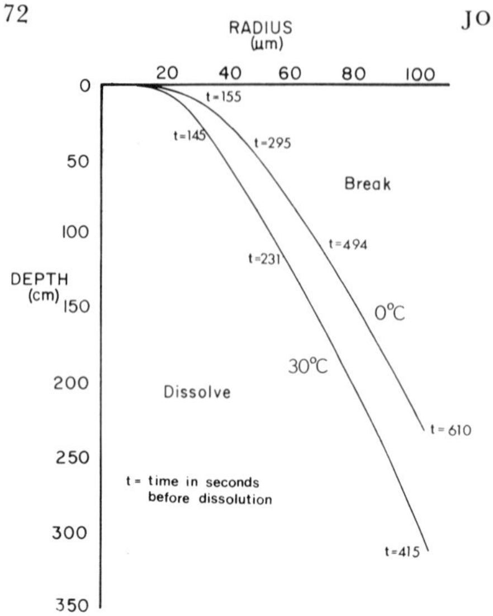

Fig. 3. Critical depth curves for air bubbles rising in air-saturated seawater, 35‰ salinity, at 0°C and 30°C. These curves describe the boundary between those bubbles that reach the interface and break, and those that dissolve before reaching the interface. They also represent the longest residence times for bubbles rising in quiescent conditions.

salinity, saturated with nitrogen and oxygen, and at 0°C and 30°C, appear in figure 3. Again, these curves describe the trajectory of a bubble through space, size and time, and represent the maximum periods of bubble residence in quiescent seawater. Thus, from figure 3, a bubble of 100 μm radius released from a depth of 225 cm in seawater at 0°C requires almost 610 seconds to dissolve, and a bubble with an initial radius of 35 μm takes 205 seconds to dissolve from 27 cm depth. At 30°C, residence times are shorter, but a bubble of 88 μm, when released from a depth of 250 cm, still requires 415 seconds to rise to and dissolve at, the interface.

A characteristic time for whitecap bubble injection might be taken from Monahan et al. (1981).

(7) $$t_i = \frac{\tau}{W(U)}$$

where τ is the time constant for decay of a single whitecap and W(U) is the fraction of sea surface covered by whitecaps as a function of the 10 meter elevation wind speed. This characteristic time is 175 seconds for a τ of 3.5 seconds and W(U) of 2% (corresponding to U of about 10 m/second, Monahan and O'Muircheartaigh 1981). For a U of 15m/second, t_i is about 70 seconds.

These times are certainly within the order of magnitude of the residence times of bubbles in the near surface (figure 3) - a result that suggests that whitecapping can support significant background bubble populations.

4. *Alternative Background Bubble Sources*

Bubbles in the background population may also exist because of production through biological processes, droplet impact, outgassing of sediments, or growth from stable bubble nuclei in regions of gas supersaturation. Also, stabilization, or retardation of bubble dissolution, may be important in enhancing the background population.

Blanchard and Woodcock (1957) observed bubble formation by droplet impact, and described the resulting torroid-shaped cloud composed of bubbles primarily of less than 25μm in diameter. To test this mechanism of bubble formation described by Blanchard and Woodcock, a photograph was made of such a torroid-shaped cloud and was found to contain images of over 600 bubbles. The greatest portion were found to be less than 25μm, with none larger than 48μm.

In order to have a complete understanding of bubbles in the ocean, we must also understand stable microbubbles - their production, behavior and distribution. Despite their probable importance in acoustics, cavitation and bubble nucleation, we know little of the stabilizing mechanism, and less of the surface active materials that are responsible for stability. In one study, Johnson and Cooke (1981) found that as many as 93% of dissolving bubbles became stabilized at diameters of up to 13.5μm. In that study microbubbles remained stable for longer than 22 hours at 1 atmosphere of pressure, or until the hydrostatic pressure exceeded about 0.83 meters of water.

As this microbubble study of Johnson and Cooke was conducted on only a few different water samples and at a single season, it is probable that the maximum size, pressure stability, and persistance can exceed the observed limits. Indeed, the anomalous acoustic data of Medwin (1977) may require such an explanation.

In conclusion, while a number of sources for the bubbles in the background population have been identified, at least at higher seastates, the bubbles from the decay of earlier whitecapping events must be an important contributor. However, further effort must be expended to determine the relative significance of the various sources at all seastates and to assess the importance of the background population in various processes of oceanographic and atmospheric interest.

REFERENCES

Acrivos, A., and J. D. Goddard, 1965: Asymptotic expansions for laminar forced-convection heat and mass transfer. Part 1. Low speed flows. *J. Fluid Mech.* 23, 273-291.

Blanchard, D. C. and A. H. Woodcock, 1957: Bubble formation and modification in the sea and its meteorological significance. *Tellus,* **9**, 145-158.

Clift, R., J. R. Grace, and M. E. Weber, 1978: *Bubbles, drops, and particles.* Academic Press, 380 pp.

Gershey, R. M., 1983: Characterization of seawater organic matter carried by bubble-generated aerosols. *Limnol. Oceanogr.* **28**, 309-320.

Johnson, B. D., and R. C. Cooke, 1979: Bubble populations in coastal waters: a photographic approach. *J. Geophys. Res.,* **84**, 3761-3766.

Johnson, B. D., and R. C. Cooke, 1980: Organic particle and aggregate formation resulting from the dissolution of bubbles in seawater. *Limnol. Oceanogr.* 25, 653-661.

Johnson, B. D., and R. C. Cooke, 1981: Generation of stabilized microbubbles in seawater. *Science,* **213**, 209-211.

Kolovayev, D. A., 1976: Investigation of the concentration and statistical size distribution of wind-produced bubbles in the near-surface ocean, *Oceanology,* Engl. Transl., **15**, 659-661.

LeBlond, P. H., 1969: Gas diffusion from ascending gas bubbles. *J. Fluid Mech.,* 35, 711-719.

Levich, V. G., 1962: *Physicochemical hydrodynamics.* Prentice-Hall, 700 pp.

Medwin, H., 1977: In situ acoustic measurements of microbubbles at sea. *J. Geophys. Res.,* **82**, 971-975.

Merlivat, L. and L. Memery, 1983: Gas exchange across an air-water interface: experimental results and modeling of bubble contribution to transfer. *J. Geophys. Res.,* **88**, 707-724.

Monahan, E. C., and I. G. Ó Muircheartaigh, 1980: Optimal power-law description of oceanic whitecap coverage dependence on wind speed. *J. Phys. Oceanogr.* **10**, 2094-2099.

Monahan, E. C., P. A. Bowyer, D. M. Doyle, M. P. Fitzgerald, I. G. Ó Muircheartaigh, M. C. Spillane and J. J. Taper, 1981: Whitecaps and the marine atmosphere – Report No. 3, University College, Galway, Ireland.

Sutcliffe, W. H., Jr., E. R. Baylor, D. W. Menzel, 1963: Sea surface chemistry and Langmuir circulations. *Deep Sea Res.,* **10**, 233-243.

Thorpe, S. A., 1982: On the clouds of bubbles formed by breaking wind-waves in deep water, and their role in air-sea gas transfer. *Phil. Trans. R. Soc. Lond.,* **A 304**, 155-210.

Wallace, G. T., Jr., and R. A. Duce, 1975: Concentration of particulate trace metals and particulate organic carbon in marine surface waters by a bubble flotation mechanism. *Mar. Chem.* **3**, 157-181.

ON RECONCILING OPTICAL AND ACOUSTICAL BUBBLE SPECTRA IN THE MIXED LAYER

FERREN MACINTYRE
Centre for Atmospheric-Chemistry Studies,
Graduate School of Oceanography,
University of Rhode Island,
Narragansett RI 02882-1197

Acoustically estimated size spectra of bubbles in the upper 15 m of the oceanic mixed layer increase logarithmically as diameter decreases, at least to 30 μm (8μm in tap water); numbers fall off slowly with depth. Optically determined estimates, in contrast, show a strong peak above 80μm; numbers fall off rapidly with depth. Existing theory is insufficiently developed to exclude either of these interpretations. Re-examination of the acoustical and optical responses of small bubbles suggests that classical acoustical theory overestimates small bubble numbers by as much as an order of magnitude, while the optical devices employed are 1000 times less sensitive to small bubbles than they might be.

But these corrections do not bring agreement. The tentative conclusion is that optical and acoustical techniques count fundamentally different objects. Optical methods detect buoyant, specular objects and probably miss the smaller bubbles; acoustical methods detect anything that contains gas, including neutrally buoyant non-specular bubble 'ghosts'. There is great need for intercomparison studies employing several detection techniques simultaneously.

A new optical method for sizing small bubbles provides 1-μm resolution between 2 and 24μm diameter. (These limits are not absolute, but simply represent those explored in this paper). The technique requires only descriptive estimates of the colour of light scattered at several angles, say, 25, 35, and 45 degrees.

Introduction

There is reasonable agreement between the several optical estimates of bubble-size spectra in the mixed layer. Glotov et al. (1962) allowed bubbles to rise to the top of a 60-cm pipe and photographed them against a glass surface, separating bubbles from particles by illuminating them with 3 symmetrically placed lights so that bubbles showed as sets of three spots. Medwin (1970) used a similar technique but incorporated 90-degree light scattering without a bubble trap, and with a very small sampling volume that resulted in very poor counting statistics. Kolovayev (1976) (who seems to be the (P./D.) A. Kolo(b/v)a(y)ev of the bibliographies, transmogrified by transliteration) extended Glotov et al.'s work using the same trap. Johnson and Cooke (1982) repeated Medwin's optical approach with a larger sampling volume and better statistics, measuring bubbles down to 17-μm diameter. Merrill and Mattocks (1981) counted buoyant oceanic objects rising through a light beam, (but for logistical reasons obtained very little oceanic data, which has remained unpublished). They report an 'indication' of a peak above 50 μm (Merrill, pers. comm. 1983). Cipriano & Blanchard (1981) generated bubbles in the laboratory in a simulated breaker, and raised a slab of bubbly water into the view of camera for photography, obtaining spectra very similar to Kolovayev's. Blanchard and Woodcock (1957), in contrast, counted bubbles in surf by catching them in a petri dish and examining them through a microscope, and, curiously, found a spectrum which is similar to the acoustically determined counts. Many of these spectra are shown in Fig. 1.

Wu (1981) has drawn attention to the basic differences between optically and acoustically determined spectra and also to the fact that optical spectra decrease with depth far faster than

E. C. Monahan and G. Mac Niocaill (eds.), Oceanic Whitecaps, 75–94.

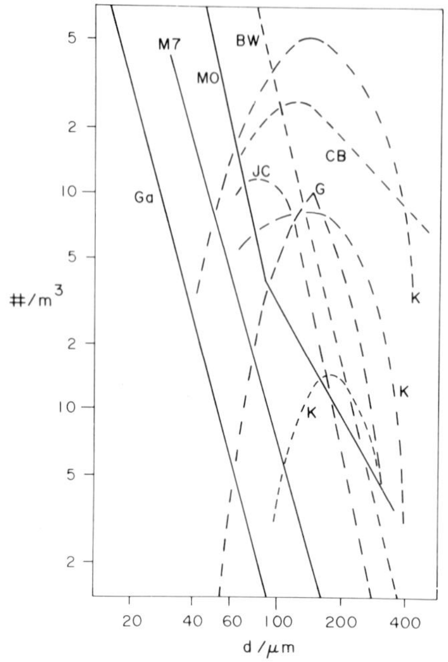

Fig. 1. Bubble size spectra obtained by acoustical (solid lines) and optical (dotted lines) techniques. The two methods probably measure different objects at the small end of the range. The spectra are for oceanic conditions except as noted. Ga=Gavrilov, 1969 (tap water); M0=Medwin, 1970; M7=Medwin, 1977a; BW=Blanchard and Woodcock, 1957 (surf); G =Glotov et al., 1962; K=Kolobayev, 1976; JC=Johnson and Cooke, 1982. Only spectral shape is important; vertical comparisons are not meaningful because of differences in depth, wind speed, etc.

acoustical spectra. Although Glotov et al. (1962) claim that their optical and acoustical measurements show 'adequate agreement', this strikes me as a generous interpretation, since they do not support their claim by resolving their acoustical data into a spectrum. In any case, their data do not speak to the question of the numbers of small bubbles present in the environment, since they show that their apparatus does not collect them. The photographically determined spectrum they offer peaks with extraordinary precision under two curves labelled 'rise rate' and 'dissolution', and not surprisingly, the resulting peak-shape differs from that found by Kolovayev (1976), and particularly by Johnson and Cooke (1982) in the absence of a 60-cm rise tube.

To account for the large numbers of small bubbles Medwin found in the absence of apparent production mechanisms, Mulhearn (1981) proposed that the objects Medwin detected acoustically were persistent, neutrally buoyant objects containing gas, adsorbed organic material, and particles of various sorts, of the sort described by Fox and Herzfeld (1954), Lieberman (1957), Turner (1961) and Johnson and Cooke (1980). My own very preliminary results using a 'Medwin machine' are not strikingly different from Medwin's (1970), but it is premature to draw conclusions from my data at this time (like Medwin I find differences between attentuation- and backscatter-derived bubble counts attributable to an inadequate evaluation of the damping coefficients, as discussed below).

The problem at hand is to attempt to reconcile the acoustic and optical approaches, or, failing that, to explain the reasons for the gross differences between the bubble spectra obtained using the two techniques.

Acoustical Theory

One might suspect that the acoustic results are discrepant, since viscoelastic phenomena were not addressed by Medwin's approach except parenthetically (Medwin 1970), and it is now well known that ocean bubbles are not the clean-surfaced bubbles of the laboratory. But based on the work described below, this deficiency appears insufficient to account for the whole difference. The general sense of the acoustic results appears to be valid.

The acoustical response of oceanic bubbles is that of oscillation between the kinetic energy of radially moving water, and the potential energy of the compressed gas. This produces the 'giant monopole' resonance which gives a gas bubble an acoustical cross section some thousand times its geometrical cross section. The first-order theory of this process has been ably reviewed by Medwin (1977a) and only some relevant results will be repeated here. The two parameters of concern are the resonant frequency, and attentuation.

The zero-order expression for the resonant frequency f_0 (Hz) is

$$2\pi f_0 = (3\gamma P/\rho)^{1/2}/d$$

where γ is the ratio of specific heats of the gas (1.4 for air), P the ambient pressure, ρ the water density, and d the bubble diameter.

To take into account first-order corrections, the ratio of specific heats is multiplied by a complicated factor b to account for the 'polytropic equation of state', in which the gas expands adiabatically near the bubble center but isothermally next to the water. A second factor

$$\beta = 1 + \Delta P/P$$

multiplies the ambient pressure to take into account the effect of surface tension, with ΔP the pressure jump across the interface. (We ignore here a small interaction term between β as defined here and b. See Medwin 1977a for details.) The resulting first-order expression is

$$f = f_0\,[\beta b]^{1/2}$$

Fortuitously, b and β have opposite trends so that their product remains near unity, being 1.2 at 2 μm diameter, 1.0 at 5.5 μm, 0.96 at 20 μm, and essentially 1.0 at and above 200 μm.

There are three mechanisms for attenuating sound. The only one important for bubbles larger than 6.5-mm diameter (and below 1 kHz) is re-radiation: Energy striking the bubble as a plane wave excites a radial oscillation which scatters sound in all directions. Below this size (and above this frequency) the dominant mechanism is irreversible heat loss associated with isothermal compression near the bubble wall. This mechanism retains its pre-eminence for three orders of magnitude, down to a 6.5-μm bubble, resonant at 1 MHz. Only below this diameter, and above this frequency, does viscous dissipation in the bulk liquid dominate. The damping coefficients associated with these classical mechanisms are shown in Fig. 2.

The effects of viscoelasticity are two. The elasticity-surface tension and related properties-increases the restoring force and hence the resonant frequency, while the viscosity increases the attenuation. However, the term 'viscoelastic' encompasses a wide range of non-ideal fluid behavior, the simplest example of which is a liquid with a complex surface modulus containing both a real restoring component (surface tension) and an imaginary dissipative component (surface viscosity), but no elastic behavior in the bulk. This case has recently been analyzed in detail (Glazman 1983), revealing that a new component of the elasticity introduces another first-order frequency effect, while the viscous component introduces only second-order damping effects.

The easiest approach to the problem is to consider the radial component of the hydrodynamic force, which we will write as a pressure jump across the interface. This directly influences the resonant frequency; if in addition $\dot{r} = dr/dt$ introduces a phase difference between the pressure and the radius terms (i.e., between pressure and volume), the integral of PdV has a finite area and energy is dissipated as heat. The pressure-jump equation for an oscillating bubble has accumulated terms over time, and for didactic purposes may now be written:

ΔP	=	$(2\sigma/r)$	+	$2E(r_0/r)^2/r$	+
pressure jump		surface tension, Laplace		surface elasticity, Glazman 1983	

$4\mu\dot{r}/r$	+	$4\kappa\dot{r}/r^2$
bulk shear viscosity, Navier-Stokes		surface dilational viscosity, Scriven 1960

where r_0 is the equilibrium radius, the elasticity $E = d\sigma/d\ln\Gamma$ with Γ the surface concentration (Levich, 1962; Davies and Rideal, 1961; Lucassen-Reynders, 1967, 1973; Joly, 1972; Rusanov and Krotov, 1979), and μ and κ the coefficients of bulk shear and surface dilational viscosity. Implicit in this formulation, to keep the elasticity term simple, is an insoluble surface film. (The case of a soluble surfactant is complicated by the fact that we must ultimate-

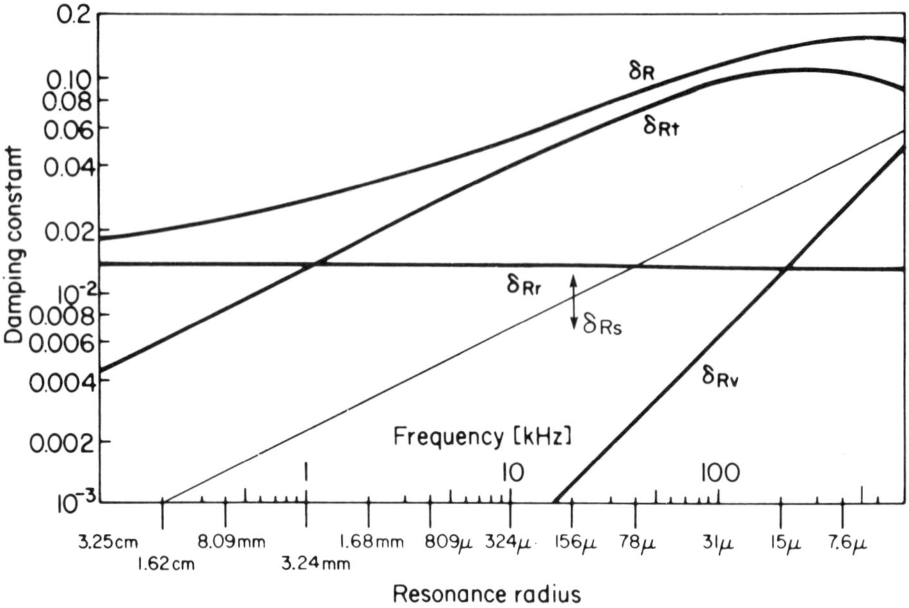

Fig. 2. The damping coefficients at resonance for sound re-radiation δr, isothermal expansion δt, and bulk-liquid viscous dissipation δv are well established. The damping associated with surface viscosity has the slope of δs, but its vertical position is unknown. (After Medwin 1977b).

ly deal with the bulk-liquid-to-surface 'diffusion' of long polymer chains with occasional surface-active sites, a problem which appears not to have been studied.) In the static case, the $\dot{r}$ terms vanish, the elasticity term is zero by virtue of constraints on its derivation, and we are left with the familiar Laplace form.

We can immediately dismiss the Scriven term if the surface viscous dissipation is by a monomolecular surface layer, for the energy argument of Levich and Krylov (1969) shows that a monolayer has so little heat capacity that if it absorbed enough viscous energy to be experimentally detectable, it would vaporize. The presence of the adjacent water (an infinite heat sink) is no help: Assuming a no-slip boundary condition between the monolayer and the water, the energy dissipated by traction into the subjacent water so far exceeds the dissipation by the monolayer that only the bulk viscosity of the water is important. We retain the Navier-Stokes term.

There is a second contribution to the 'surface viscosity', which is the dissipation of energy as molecules adsorb and desorb from the surface, exchanging with the bulk solution. For the relatively soluble materials in seawater, this term may become appreciable. Our theoretical work suggests that this mechanism contributes in an important way to the below-resonance absorption of sound by small bubbles, but that at resonance (where the acoustical bubble counters operate) the effects are small.

The Glazman term is new (1983), and takes into account the change of surface tension with radius $\delta\sigma/\delta r$ and hence with surface concentration whenever the area changes without change in surfactant amount. The way in which this comes about is sketched in Fig. 3. For a classical surfactant lipid like the 14-carbon myristic

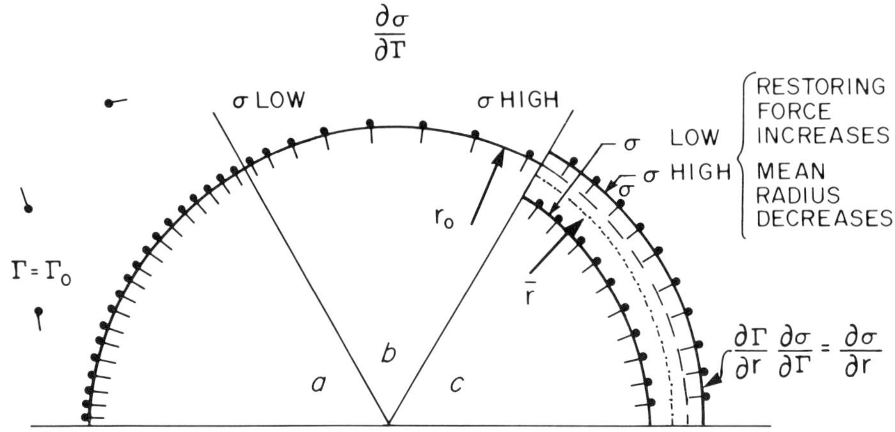

Fig. 3. The dynamic effect of insoluble surfactants on bubble oscillation. In sec. (a) the bubble is covered with an insoluble monolayer, in (b) there is a concentration gradient and a resulting surface-tension gradient. In (c) the bubble is oscillating: at large diameter the surface concentration is low and surface tension high; at small diameter, the concentration is large and the tension low. Accordingly the restoring force is increased at both extrema, and the mean oscillatory radius $\bar{r}$ is smaller than the equilibrium resting radius $\dot{r}$.

acid, the surface elasticity can become several times larger than the surface tension itself as the film compresses toward a solid surface phase. The net effect is to increase the bubble stiffness and the resonant frequency. However, two things appear to militate against the importance of the Glazman term for oceanic bubbles. First, the entire ΔP term is small compared to the ambient pressure (including the atmosphere), so this new correction is at most a 10% effect (rather that the 40% suggested by Medwin on the basis of Fox and Herzfeld's 1954 hueristic analysis); second, the elasticity of glycoprotein films, for instance, is small compared to that of lipids, and it may be that oceanic coatings, thick though they be, will cause but small frequency changes.

Still to be considered is the effect of a thick surface phase, such as that observed around oceanic bubbles by Lieberman (1957) and Johnson & Cooke (1979, 1981). This question divides into two parts: the behavior of the thick surface, and the statistical distribution of shell thicknesses of oceanic bubbles (for newly formed bubbles will be relatively clean and old bubbles will be heavily coated). The first part can be approached through the theory of thin shells (Fox and Herzfeld (1954) tried a solid shell, now understood to be inappropriate; Avetisyan (1976), and Yount (1979) have used fluid shells), or, for sufficiently small bubbles, through the behavior of a fluid with bulk viscoelasticity. But to answer either question we need better data than are currently available. The question of what is the distribution of shell thickness is one that has not previously been asked, let alone answered. For the behavior of liquid shells, we must choose a phenomenological model, and this in turn needs rather more data than are presently available on the rheological coefficients of the skins of ocean bubbles.

Viscoelastic fluids show a much wider range of behavior than is encompassed by the two-parameter model described above. Frequently, the effects of sluggish molecular orientation become important, introducing phase lags between stimulus and response, with attendant dissipation of energy. Mechanical analogs of rheological models of fluids and surfaces contain springs (restoring forces) and dashpots (dissipative elements) in series, parallel, and various combinations (Joly (1972) gives models covering a wide variety of dynamic behavior). Although Dragcevic et al. (1979, 1981) have made a brave start, the appropriate model for oceanic surfactants is not known because measurements of their viscoelastic coefficients lag far behind determination of their composition, which also remains largely unknown. One might anticipate rheological behavior comparable in variability to that of the chemical melange at the surface of seawater. (Van Vleet and Williams (1983) have recently examined the elastic properties of the ocean surface and compared them with the behavior of model surfactants on seawater. Their data support the now conventional wisdom that lipids are largely absent from the natural sea surface, which appears to be dominated by proteins, polysaccharides, marine humates (presumably polymerized, cross-linked and photo-oxidized triglycerides), and waxes.

The light line in Fig. 2 labelled δRs shows the trend of the damping coefficient associated with the Scriven surface-viscosity term for a thick layer, but until film thickness and dilational viscosity are known the vertical location of this line remains unknown. It should be emphasized that the four theoretically pure surface viscosity coefficients of Goodrich (1981) - normal and tangential components of surface shear and dilational viscosities - are abstractions whose effects cannot be separated by any experiment yet devised. In fact, most experimental measurements of 'surface' viscosity have trouble separating surface effects from those of the underlying bulk fluid (Joly, 1972 II; Goodrich and Allen, 1972). It may well be that spheres are the most appropriate geometry for experiments aimed at elucidating the effects of surface phenomena upon bubble acoustics.

Acoustics offers two ways to count bubbles: by attenuation of sound, and by backscatter. The two mechanisms depend in somewhat different ways upon the components of the pressure-jump equation above, and it may prove possible to exploit this distinction to deduce additional features of oceanic bubbles.

If energy is viscously dissipated in thick shells, present acoustical theory overestimates the number of bubbles necessary to attenuate sound. The frequency dependence of this overestimate has not been explored in detail, but it certainly increases as diameter decreases, and at 30 μm diameter, present theory may overestimate numbers by an order of magnitude. However, the discrepancy between optical and acoustical measures is much larger than this.

To conclude this section, it appears that the omission of viscoelasticity in the classical acoustical theory may introduce important overestimates of the numbers of bubbles at small diameters. Although all the corrections considered above shift the acoustical results in the direction of the optical results, they appear inadequate to produce agreement, suggesting that corrections are also required in optical theory. In terms of natural bubble populations, this analysis supports the idea of large numbers of small gas-containing resonant objects, possibly bubbles but more probably their semi-collapsed, film-covered, non-buoyant 'ghosts'.

Optical Theory

We begin by observing that optical devices are hindered in three ways from detecting the small bubbles found acoustically, those bubbles which need be neither buoyant (whence they will not be detected by optical devices which require them to rise across a beam or into a trap), nor specular (whence they will not be detected by optical devices which depend upon clean reflection). Thirdly, we will see below that the intensity of light scattered at 90 degrees is very weak for small bubbles, making their optical detection improbable by some devices.

The optical properties of water drops suspended in air have long attracted attention. A suite of papers in *J. Opt. Soc. Am. 69*, 1068-1132 (1979) describes recent efforts, and there are several accessible reviews (Minnaert, 1959;

Boyer, 1959; Greenler, 1980).

But the optical properties of air bubbles in water have been comparatively neglected since an early paper on bubbles much larger than the wavelength of light by Davis (1955). Marston and his co-workers (Marston and Kingsbury, 1981; Kingsbury and Marston, 1981a, b; Marston et al., 1982) have recently taken up this challenge, but have devoted most of their attention to angles near the critical angle (82.82 degrees) or to comparing the results of applying the (approximate but fast) geometrical-optics with the results from the (exact but very time-consuming) Mie-equation approach. MacIntyre & Blanchard (in preparation) exhaustively treat bubbles under 25-μm diameter for scattering between 10 and 90 degrees, using the exact Mie equations, and we will draw heavily upon these results.

It may seem strange that 1-μm bubbles are visible to the naked eye, which can resolve only to 100 μm. However, we do not see an actual image of the bubble, but as for a distant star, a bright spot broadened by diffraction and scattering in the eye. The intensity of the color scattered from small bubbles is also startling, appearing much brighter, for instance, than a rainbow. Considering objects much larger than the wavelength of light, so that geometrical optics applies, an idea of the expected intensity difference can be obtained by examining the reflection from an infinitesimal spot on a drop and bubble of the same diameter. The fraction of unpolarized light reflected is given by

$$r = 0.5\,[\tan^2(\phi-\phi')/\tan^2(\phi+\phi') + \sin^2(\phi-\phi')/\sin^2(\phi+\phi')]$$

where ϕ and ϕ' are the angles of incidence and refraction. This function is nearly flat at about 0.025 up to $\phi = 40$ deg, and then climbs to unity at 48.6 degrees (the water-to-air critical angle), or, for air-to-water, more deliberately to unity at 90 degrees.

In Fig. 4(a) a beam of intensity 1.000 hits a large sphere at 66 degrees. If the sphere is a bubble, the light is totally reflected at a forward angle of 48 degrees. If the sphere is a drop, most of the light is refracted and some follows the indicated path to emerge as a rainbow (backward angle, 40-42 degrees) with only 8% of the initial intensity. The 88% emitted at 45.5 degrees provides the sparkle of water fountains.

Figure 4(b) - a small bubble - shows the more important pathways that contribute - by virtue of their different lengths - to an interference pattern at a given angle. (The Mie Equations automatically account for these paths and all minor paths also). The complicated details of the refracted spectra result from interference effects between the various rays of Fig. 4(b), plus additional diffraction.

We will now summarize the Mie approach (without repeating the details, which are available in van de Hulst (1957) and Wiscombe (1980), whose notation we follow). Basically, Mie gave an infinite-series solution to Maxwell's equations at a boundary between two media with complex refractive indices. Since we are dealing with spheres, we expect the solution to contain Bessel (actually, Riccati-Bessel) functions and their derivatives, which provide the dependence upon bubble diameter and light wavelength. In addition, the solution contains the derivatives of Legendre polynomials, which provide the dependence of the solution upon scattering angle. Because of the low absorption of light by water and air in the optical range, it is possible to simplify the full Mie equations somewhat and consider only a real refractive-index ratio m = m(air)/m(water), which is approximately 3/4. We are further interested only in the farfield solution at distances large compared to the bubble diameter.

The Mie equations depend upon two parameters only: m, and the size parameter ka, (where $k = 2\pi/\lambda$, and a is the bubble radius, so that ka is the ratio of circumference to wavelength λ). The length of the 'infinite' series involved grows with ka, and in addition each term in the series also grows, so that the computation time devoted to the central equations increases as $(ka)^2$. The number of necessary terms N in the infinite series is of course not infinite. But rather more attention must be paid than usual to N, since the answers are inexact if too few terms are taken, and numerical instability sets in rapidly if too many are taken. Fortunately, Wiscombe (1979) has found a simple expression

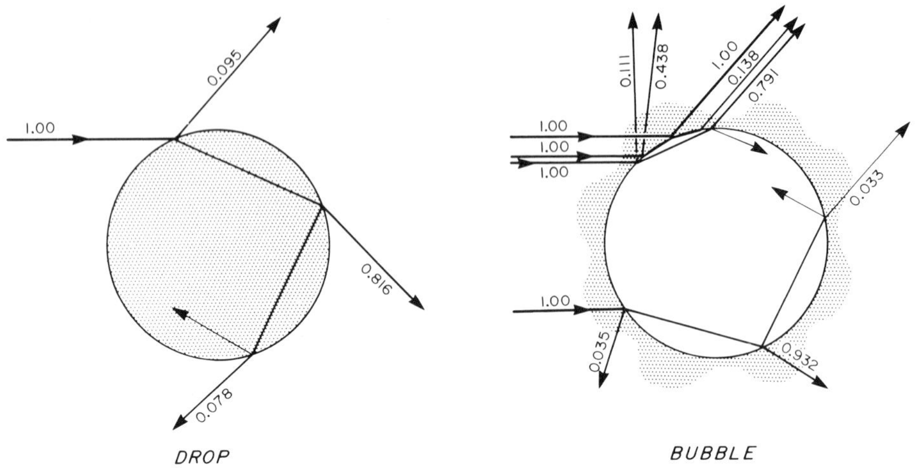

Fig. 4. (a) Beam intensities from geometrical optics, for ray paths through drops (m=4/3) and bubbles (m=3/4) at 66 degrees, showing the high intensity of light scattered from a bubble (1.00) and the low intensity of the rainbow (0.091). (b) Some of the more important ray paths through a bubble. An equal amount of scattered energy does not go through the bubble, but is simply diffracted. In addition, there are surface waves, and tunnelling effects, which are not taken into account in ray optics, but which do influence the color of the scattered light.

for N for the values of ka which concern us, namely:

$$N = ka + 4.05(ka)^{1/3} + 2$$

As a result of the N^2 time dependence, and the necessity for repeating the entire calculations across the spectrum, and frequently across a range of diameters, most serious light-scattering calculations are done on large computers such as the Cyber 205 and CRAY-1.

Nevertheless, the IBM Personal Computer is bigger, faster, and smarter than the IBM 704 on which the Mie equations were run in the 1960s (Matiejevic et al., 1960). It does not support vectorization, yet many of the techniques developed by Dave (1969) and Wiscombe (1979) for use on supercomputers proved helpful when scattering calculations were done on the IBM PC. The 8087 Numerical Coprocessor of the IBM PC is 80 bits wide, and carries 18-digit precision, more than sufficient for the task, and enables the IBM PC to do in minutes what the CRAY-1 can do in milliseconds. In this application the 8087 accelerated calculations by a factor of 115. (A program which ran in 9.5 hrs in non-8087 MMSFORTH required 5:07 minutes with the 8087. The same program in BASIC would have taken about 8 days). In addition, the number of digits which agreed with Wiscombe's test values increased from 2 to 6 (all that Wiscombe provides). Agreement was found for the scattering functions S_1 and S_2 at ka=10, m=1.5 and ka=100, m=1.5 (but, as discussed below, not at the next available test ka=1000, m=1.5).

The major, time-consuming, portion of the program calculates the strictly physical properties of the scattered light, the most fundamental of which are the two complex components of the scattering matrix which survive spherical symmetry, S_1 and S_2. Without going too deeply into material covered better elsewhere, we have

$$S^{\pm} = \sum_{n=1}^{N} [(2n+1)/n(n+1)]\ [a_n \pm b_n]\ [\pi_n \pm \tau_n]$$

$$S_1 = (S^+ + S^-)/2$$

$$S_2 = (S^+ - S^-)/2$$

from which the real-valued polarized intensities i_1 and i_2 are obtained by

$$i_1 = |S_1|^2$$

$$i_2 = |S_2|^2$$

π_n and τ_n are the aforementioned derivatives of Legendre polynomials of order n which give the angular distribution, while a_n and b_n are functions of Ricatti-Bessel functions of order n and depend upon ka, that is, upon wavelength and diameter.

The calculations divide naturally into two parts. For a given wavelength, the scattered polarized intensities i_1 (electric vector parallel to the surface) and i_2 (perpendicular vector) are calculated at a number of scattering angles α or diameters d, and stored on disc, so that at the end of the calculation one has two large arrays $i_1(\lambda, \alpha)$ and $i_2(\lambda, \alpha)$ or $i_1(\lambda, d)$ and $i_2(\lambda, d)$. Figure 5 shows a 3-dimensional view of the unpolarized intensity $i_0 = (i_1 + i_2)/2$ as a function of wavelength across the visible spectrum and over diameter from 4 to 24 μm. The irregular nature of the intensity along any given diametral slice of this surface indicates that the resulting light will be colored. This immediately takes us out of the realm of pure physics and involves us with the response of the human eye, a field known as the 'psychophysics' of color.

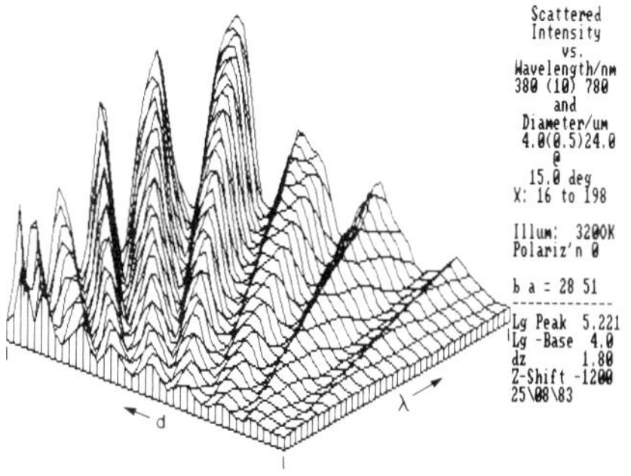

Fig. 5. Light scattered at 15 degrees as a function of wavelength and diameter. The circaxial edges of these graphs (those seen when viewed with a, the angle of rotation of the plot, near 45 degrees) is marked by a line along the base, which the back edges do not have. The data at the right label first the x, or 'front right' axis (wavelength), then the y or 'front left' (diameter). The axis parameters show the limits, and, in parentheses, the increment. As the marks along the edges indicate, every other curve is omitted for clarity, so that the apparent increment is twice the printed value. The parameter b is the vertical (downward) viewing angle. 'X' shows the range of the size parameter ka. Also shown are the nature of the illuminating light, the logarithms of the highest point of the surface, and of the baseline, and some additional plotting parameters.

Psychophysical Color Theory

The second part of the problem is to synthesise the intensity distribution across the spectrum into the single visible color which the eye sees. These calculations are rapid, but it is something of a challenge to present the results in easily interpretable form. The best approach is one which apparently has not been used before, consisting of plotting the locus of the resultant color on the CIE chromaticity diagram shown in Fig. 6. The necessary calculations are described in texts on color science (Committee on Colorimetry, 1953; Wyszecki and Stiles, 1967; Bouma, 1971).

Psychophysical perception of color is not simple. The three-primary-color theory has been with us since Aristotle, and has some foundation in retinal pigments, but Land (1954a, b, 1983) has shown that if the visual field is sufficiently complicated, nearly all colors can be perceived with only two illuminating wavelengths, even when these are very close together. But for easy mathematics, even the

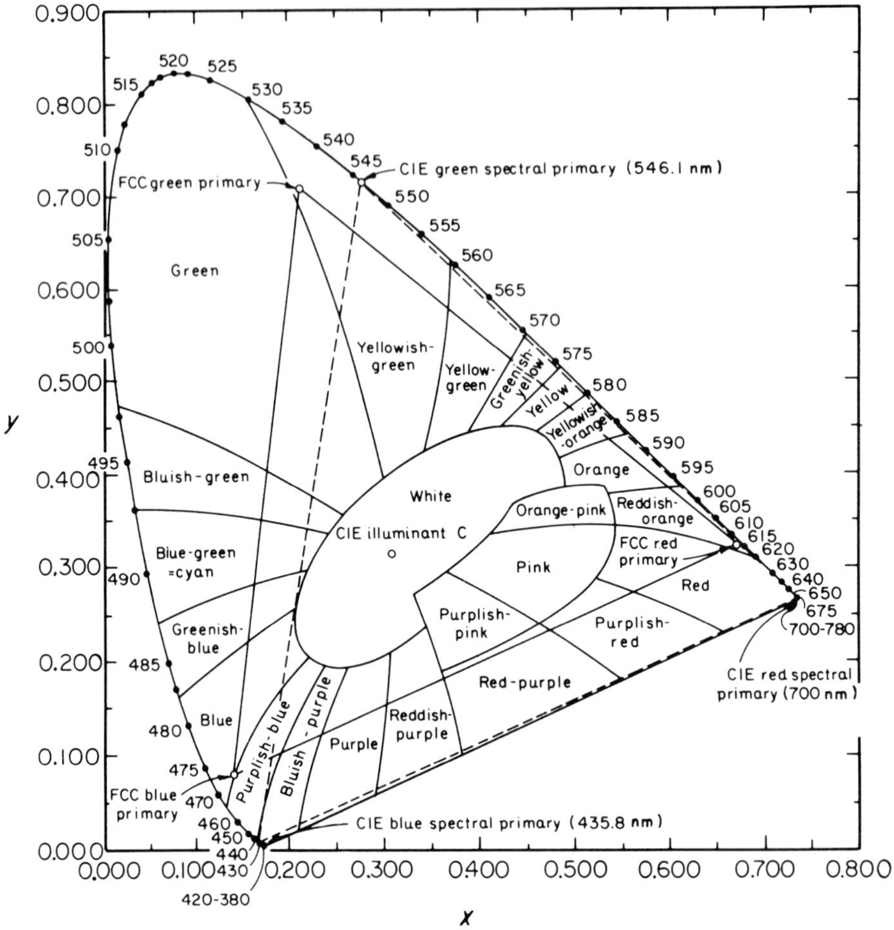

Fig. 6. CIE Chromaticity diagram for illuminant C, approximating daylight. Colors between C and the spectral locus represent mixtures of white light and a pure color from the locus; colors inside the non-spectral triangle 380–780–C represent white light minus the wavelength on the spectral locus on the other side of the white point. (From Fink, 1982.)

Aristotelian theory must be modified. We follow the 1931 CIE convention: To avoid negative values of the primary colors, it is necessary to postulate three 'hyper-primaries': a hyper-red X, a hyper-green Y, and a hyper-blue Z, all of which lie beyond the ability of the eye to perceive. Using these, all perceptible colors lie somewhere in X-Y-Z space. The actual values of X, Y, and Z corresponding to a given color can be calculated as described below. But to allow plotting on the chromaticity diagram, the intensities of the hyper-primaries are given up, and only the color retained, by normalization into 'chromaticity coordinates' x and y, through

$$x = X/(X+Y+Z)$$
$$y = Y/(X+Y+Z)$$
$$z = Z/(X+Y+Z)$$

Y, by design, has the useful property of being the total intensity. Since $x+y+z=1$, z itself becomes redundant, and the color can be described completely by x and y, which designate a single point of Fig. 6.

All colors detectable to the eye are enclosed by the experimentally determined 'spectral locus' of Fig. 6, which is marked with the spectral wavelengths. Color versions of this diagram are both hard to find and misleading. Plate 24 of 'The Science of Color' (Committee on Colorimetry 1953) reproduces a rather segmented oil painting; Bouma (1971) has a more continuous representation facing page 72; while Weiskopf (1968) depicts more accurately the limited area accessible with printers' ink.

The bounding spectral locus represents the purest colors visible to a mythical, if average, 'standard observer'. The interior represents mixtures, tending toward white at the center. But the effective center, or 'achromatic point', is mobile, and shifts with the color of the illuminating light, which may take on a wide range of actual colors while still being interpreted as achromatic. Such a shifted achromatic point is shown in Fig. 7, where it appears at $x = 0.423$, $y = 0.399$, the color of a 3200-degree-Kelvin black-body, approximating a slide-projector bulb. Figure 7 also shows the domains of the

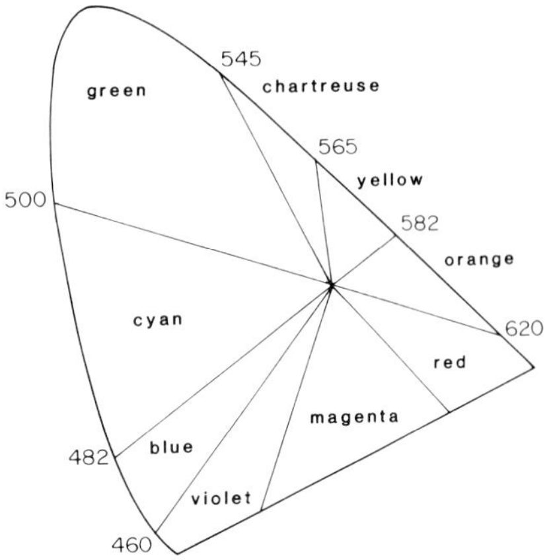

Fig. 7. Shifting the achromatic point by changing the illuminant to the 3200-degree projector bulb shifts all perceptions of equivalent wavelength and color purity. The boundaries define the color names used in this paper. They should probably curve in the manner of Fig. 6, but I have found no published data, and arbitrarily accept these as valid for present purposes.

color names used in this paper. Our chromaticity loci showing the color of light scattered by bubbles will be drawn on a portion of this diagram.

In addition to the spectral locus, the CIE convention describes three 'color-matching functions' of the standard observer ($\bar{x}_\lambda$, $\bar{y}_\lambda$, $\bar{z}_\lambda$), corresponding to the experimentally determined response of the human eye. These color-matching functions – which may seem rather arbitrary – are shown in Fig. 8. We employ these only after multiplication by the energy of the illuminating source to provide the 'weighted color-matching functions' ($\bar{x}_\lambda P_\lambda$, $\bar{y}_\lambda P_\lambda$, $\bar{z}_\lambda P_\lambda$), which are available in tables (Wyszecki and Stiles, 1967; Table 3.19).

Converting a spectrum into a single color involves multiplying the scattered spectral intensity found above by the weighted color-matching functions. This is done separately for each of the three color-matching functions, which are then separately integrated over the visible spectrum (380–780 nm), to produce three 'tri-

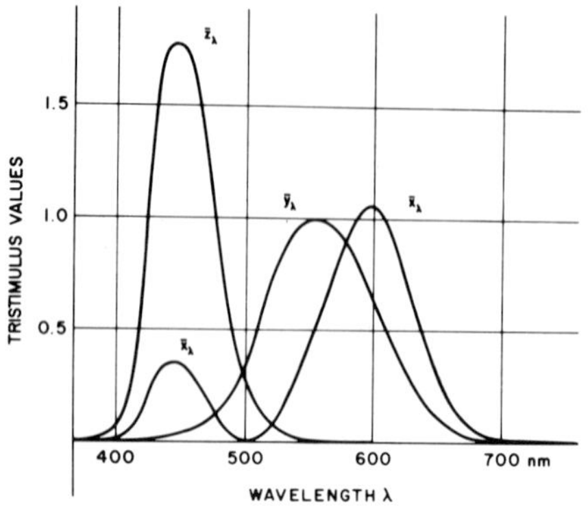

Fig. 8. CIE color-matching functions. These are experimentally determined properties of the normal human eye, transformed to a coordinate system of maximum convenience. (From Wyszecki and Stiles, 1967).

stimulus values' of the scattered light:

$$X = \int_{380}^{780} i_1(\lambda)(\bar{x}_\lambda P_\lambda)d\lambda = \sum_{380}^{780} i_1(\lambda)(\bar{x}_\lambda P_\lambda),$$

with similar formulas for Y and Z.

i_1 and $x_\lambda P_\lambda$ are measured at the center of intervals $\Delta\lambda$ wide. Trials with 80-, 40-, 20-, and 10-nm wavelength intervals suggested that 10-nm spacing yielded resolution on the chromaticity diagram which was compatible with the ability of the human eye to resolve colors under laboratory conditions. Because polarization effects are small, there is little reason to keep i_1 and i_2 separate, so the data are converted to unpolarized light by replacing i_1 above with $i_0 = (i_1 + i_2)/2$.

Color names are naturally spaced more densely with wavelength where perception is most sensitive. This leads to asymmetry in the assignment of names to wavelengths around the spectral locus. With the basic three-fold symmetry of the red-green-blue primaries and the complementary cyan-magenta-yellow set, one would expect 3n additional common names between the six named colors, for 12 (or 24) evenly spaced named colors. However, Fig. 6 has 19 named divisions, while the well known Munsell system has 20 (numbered) divisions at this level of precision. Furthermore, the boundaries between colors are matters of idiosyncratic opinion, often strongly held. The simplest resolution of this problem is to look down the barrel of a spectrophotometer and choose your own color boundaries: the author's are given in Fig. 7, and are reproducible to 5 nm (2 nm at the strongly opinionated boundaries). The nonspectral colors of course cannot be examined in this manner.

The 'higher-order Tyndall spectra' of small spheres with $m>1$ attracted attention long ago (Johnson and LaMer, 1946), and LaMer (1950) looked at small liquid drops, giving the angular locations of the red bands. (In contrast to small bubbles, where the bands occur only above 25 μm, collections of droplets show 8 red bands below 2-μm diameter.) Kerker et al. (1966) made red and green calculations for very small cylinders and spheres (0.8-2.8 μm diam.) of glass and ice, reporting results in terms of equivalent wavelength and purity rather than plotting a chromaticity locus. If we think of x and y as the Cartesian coordinates of the chromaticity diagram, purity and equivalent wavelength are the polar coordinates. They have the virtue of relating rather directly to everyday experience. These coordinates are available from x and y through the transformations:

$$\text{purity} = (x - x_w)/(x_s - x_w)$$
$$= (y - y_w)/(y_s - y_w)$$
$$\text{equivalent wavelength} = f[(x - x_w)/(y - y_w)]$$

where w denotes the coordinates of the achromatic point and s is the intersection of the radius vector and the spectral locus. Unfortunately, because of the arbitrary shape of the spectral locus, the function f and the points x_s, y_s must be specified piecemeal, making these transformations clumsy to automate (although trivial by eye), so purity and equivalent wavelength have gone out of fashion with the advent of computerized algorithms. Exact values of r and θ can be found in MacIntyre and Blanchard

(1984), for those who wish to make these transformations.

Chromaticity-diagram graph paper is available from the MIT Press as 'diagram 23 of Hardy's *Handbook of Colorimetry* (1936)' for $1/sheet, and is, conveniently, almost exactly the same size as the natural display on the IBM color monitor.

Light Scattering from Small Bubbles

To provide a feeling for how light interacts with bubbles, representative graphs are given for scattering at 45, 35, and 25 degrees in Fig. 9. The number of ridges increases as the scattering angle decreases, but the fine structure so apparent at 45 degrees is smoothed out in the forward direction. The ridge-and-valley structure appears to radiate from the origin at 0 wavelength and 0 diameter. When intermediate angles are examined the ridges can be seen to swing counter-clockwise as they increase in number. Figure 4 belongs in this same set.

The vertical scales on these plots are arbitrary to provide the maximum possible resolution, rather than to allow easy comparison of absolute intensities between plots. In compensation, the logarithm of the maximum value is given for each plot as 'Lg Peak', while a sense of the linear scaling can be obtained from 'dz', a factor which multiples the raw data. The vertical baseline is similarly scaled automatically, and is always negative: its logarithm is given as 'Lg -Base'. The vertical zero is for all practical purposes at the bottoms of the lowest valleys. 'Z-shift' merely moves the graph vertically to stay on the screen.

Although these intensity patterns are interesting, they are not easy to relate to visual observations. Far more useful for this purpose are the 'chromaticity loci' of Figs. 10 A, B and C, which integrate the spectral data of Fig. 9 and reduce each spectrum to a single observable color. It is left as an exercise for the reader to convince himself that a given spectrum might reasonably be expected to appear as the color given.

Slightly more complex than Fig. 9 are the spectra at constant diameter, of which we show samples at 6, 12, and 18μm in Fig. 11 and 24 μm in Fig. 13. In this case the ridge pattern is

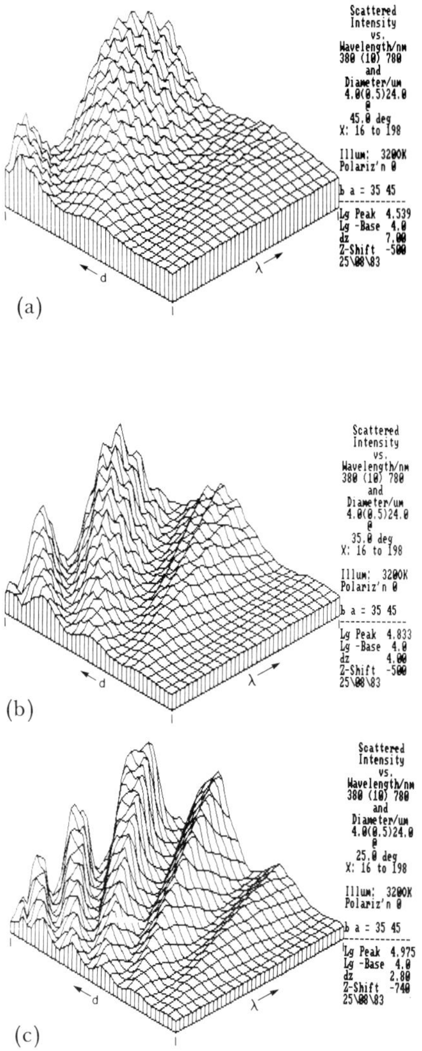

Fig. 9. Scattered intensity at (a) 45 degrees, (b) 35 degrees, and (c) 25 degrees from the light-beam direction, as a function of diameter. The pattern of ridges and valleys appears to radiate from the origin. The same information, displayed differently, is contained in Fig 10. See also Fig. 5.

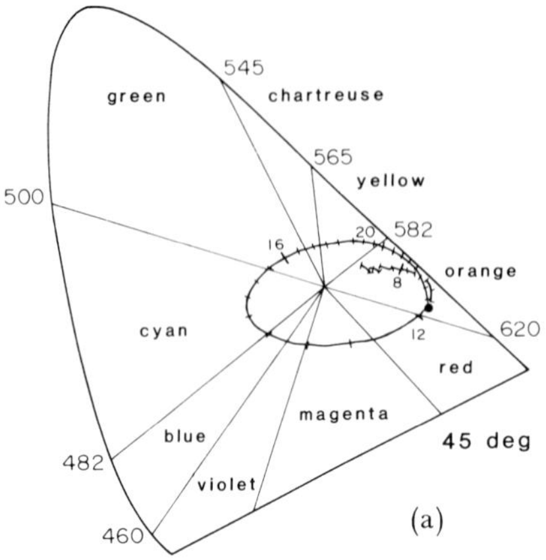

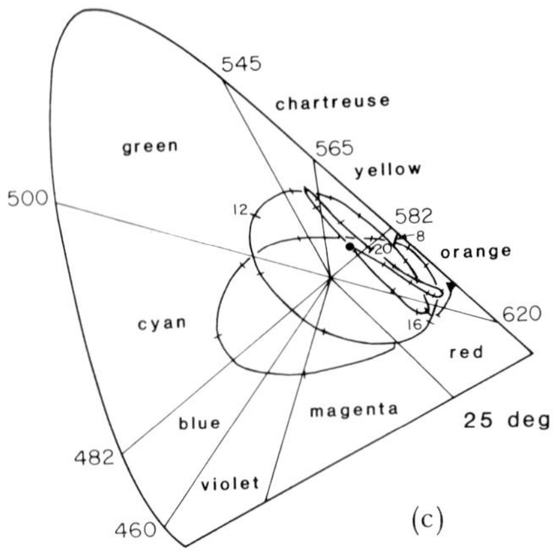

Fig. 10. Chromaticity loci of Fig. 9, showing the observed color as a function of diameter at given scattering angles. Intensity information has been lost, but covers about a 64-fold range, with the lowest values for small bubbles and in the magenta, and highest values for large bubbles in the yellow-green. The curves start at 4μm at the triangle and end at the circle at 24μm, with ticks at 0.5-μm intervals.

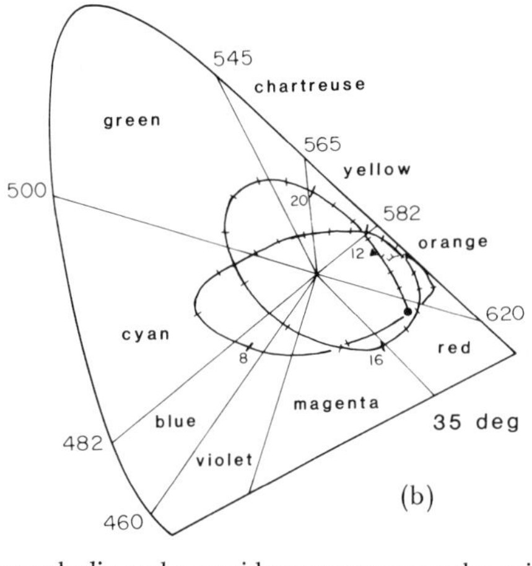

hyperbolic and new ridges appear near the origin. Note that to maintain a view down the valleys, the graph has been rotated 110 degrees, and we are looking at the back of the wavelength axis along the left of these plots. 20-degree scattering is at the right amongst the peaks. Secondary structure, best seen at the left of the 12-μm plot, increases with primary structure here.

Chromaticity loci corresponding to Fig. 11 for constant diameter are given in Fig. 12.

Although patterns are visible in the chromaticity loci, they are seldom regular enough to allow confident extrapolation, and give no insight into underlying regularities in the spectral data. Nevertheless, to help make sense out of them, we summarize the regularities of the chromaticity loci (deduced from rather more data than are shown in the figures).

Patterns: a) Constant Angle

1) The forward-scattering pattern below 20 degrees rotates around the white point too rapidly for convenient color resolution.

2) Between 25 and 50 degrees the pattern is one of CW rotation about the white point with increasing diameter (CCW, of course, if you are watching a bubble dissolve).

3) The number of revolutions about the white point decreases with increasing scattering angle, from 3-1/2 (4–24 μm) at 25 degrees to 3/4 at 55 degrees. This sometimes allows unequivocal size determination from color alone with about three observations at different

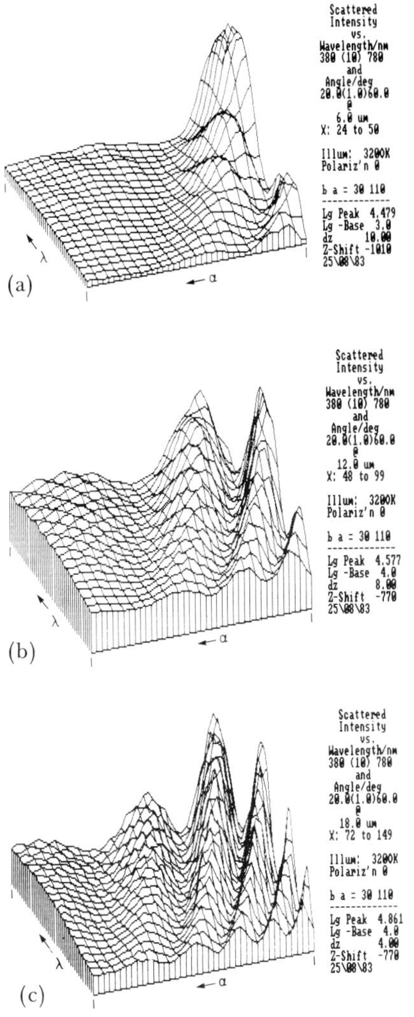

Fig. 11. Scattered intensity at (a) 6 μm, (b) 12 μm, and (c) 18 μm, as a function of scattering angle. The ridge pattern is hyperbolic. See also Fig. 14.

angles as shown in Table 1.

If sunlight, or CIE standard illuminant C, is used, the white point shifts toward the center of the diagram, and the frequency of orange decreases in Table 1, making it possible to distinguish 3-, 4-, 22-, and 23-μm bubbles also. It should also be possible to use the red-green oscillation as a sizing tool, perhaps by counting the number of oscillations over a certain angular range, or by noting the locations of the yellow transitions. This method has the advantage of not requiring subjective estimates of color.

Patterns: b) Constant Diameter

1) Small bubbles are almost white nearly everywhere. Under 3200-deg illumination, the 'white' area is similar to that of Fig. 5., but shifted to follow the achromatic point (which, however, is no longer centrally located within the 'white' area). A 2-μm bubble emerges from this area briefly between 12-16 deg, where it is a golden orange.

2) Forward scattering is white. The locus begins at the color point of the illuminant and moves toward the yellow, so that the first visible color is yellowish. For daylight illumination this occurs at:

μm	2	4	6	8	10	12	14	16
deg	6	4	3.5	3.5	2.5	2	1.8	1.5

3) The shape of the locus with increasing angle can be described in broad terms as having three stages. The first stage is variable: an irregular hypotrochoid for small bubbles, the last feeble

Table 1. Sizing of small bubbles by color at selected angles.

D=	2	3	4	5	6	7	8	9	10	11	12	13	14	15	16	17	18	19	20	21	22	23	24
25°	y	o	o	m	o	y	o	o	o	m	g	ch	y	r	r	r	ch	ch	o	r	o	o	r
35°	o	o	o	o	c	r	b	c	g	y	y	o	o	r	m	b	c	g	y	y	o	o	y
45°	o	o	o	o	o	o	o	o	o	o	r	m	c	c	g	g	y	y	0	o	o	o	o

The abbreviations stand for red, orange, yellow, chartreuse, green, cyan, blue, violet, magenta, and white, with the color boundaries as shown in Fig. 7.

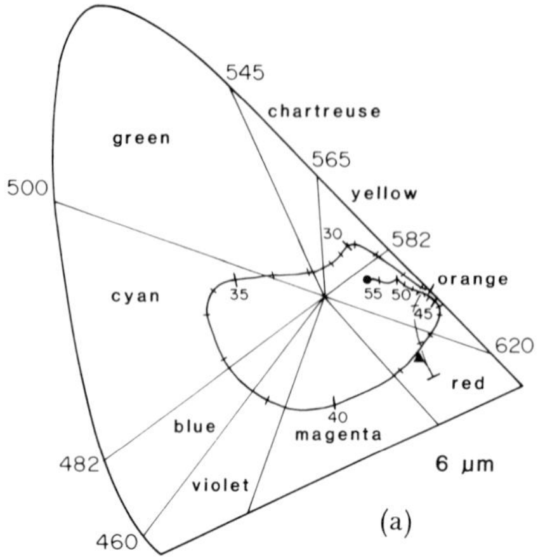

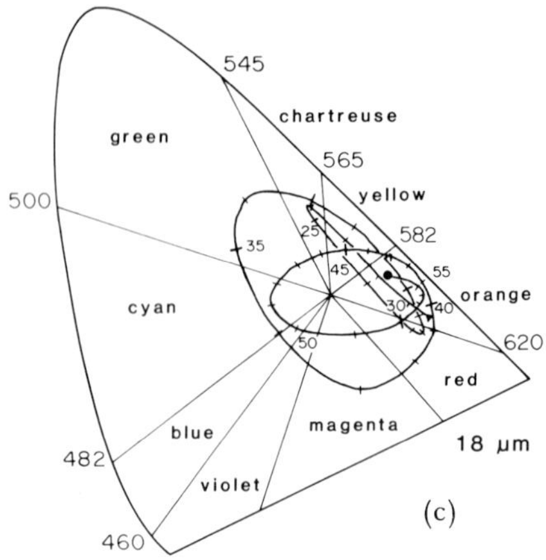

Fig. 12. Chromaticity loci of Fig. 11. The curves start at 20 degrees at the triangle and end at the circle at 60 degrees, with ticks at 1-degree intervals.

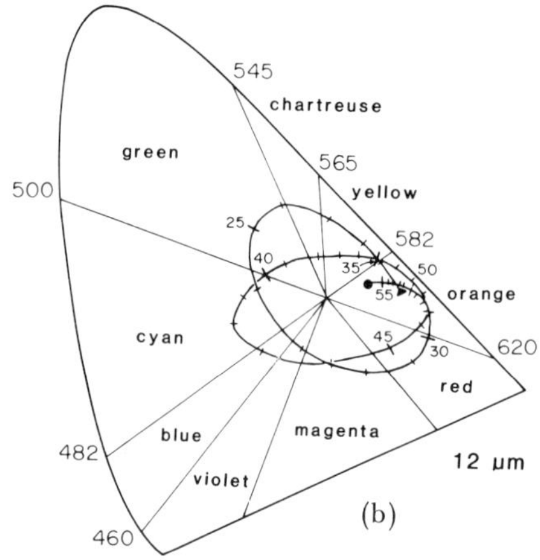

cusps of which can be seen between 20 and 33 degrees on the 6-μm plot (Fig. 12a); or red-green oscillation as seen between 20 and 33 degrees for the 18-μm bubble (Fig. 12c). The second stage is counter-clockwise rotation around the white point in the band of maximum brilliance. The 12-μm plot (Fig. 12b) shows this stage. Finally there is convergence to white through orange and yellow, the beginning of which appears in all parts of Fig. 12.

4) The CCW rotation is the most suitable for size determination.

The scattered colors are generally near the position of maximum brilliance on the chromaticity diagram. At higher saturation (closer to the spectral locus) the bandwidth and intensity are lower; closer to the achromatic point the illumination is high but the color is washed out.

The pattern moves toward larger angles as bubbles grow, so that smaller angles are more sensitive to the color changes of small bubbles and larger angles to those of larger bubbles. Yellow and orange are the commonest colors, because the locus leaves the white point at small angles, and returns to it at large angles, via orange and yellow. Yellow is also passed during the phases of red-green oscillation, but it is a subtle color and will be seen as white unless attention is paid.

The rarest colors are magenta through cyan. A workable scheme might want to pay particular attention to these colors, and to the shift of the pattern with diameter. Each angle is most sensitive to certain ranges of diameters which loop out through the rare colors. Thus, from Fig. 9, 45 degrees is diagnostic for 12.25-16μm, 35 degrees for 4.75-6.25 and 10.5-11.75 μm, and 25 degrees for 7.25-9.5 and 16-18 μm. By careful choice of angles it might be possible to obtain 0.5-μm resolution throughout the 2-24-μm range with three angles and only off-hand estimates of color.

Even in the absence of reference guides, the color boundaries are reproducible, for a given observer, to within 5 nm, although it seems to be harder to estimate wavelength in the center of a color. But matching to within 2 nm is possible almost everywhere in the spectral colors if a reference (e.g., the beam from a monochromator) is visible near the bubble.

Caveats and Limitations

I stop at 24-μm diameter because a numerical instability of presently undetermined origin becomes evident in the calculations at this point. This can be seen along the ka=190 line of Fig. 13, where the pattern of ridges and valleys suddenly breaks as though sheared by an earthquake fault line. (Half of the lines have been omitted to show the 'fault line' to better advantage). For larger bubbles, this break moves upward across the spectrum (i.e., the instability always occurs at ka=190), while the high-ka end errupts into an irregular oscillation orders of magnitude larger than the desired wavelike patterns.

Such instabilities are well known in the recursive calculation of the Bessel function j, which approaches zero for n approximately equal to ka (Abramowitz and Stegun, Introduction, Sec. 7, 1972). Wiscombe (1979) shows how to avoid this problem, and I have made full use of his techniques as I converted his CRAY-1 FORTRAN program into MMSFORTH.

However, the instability at ka=190 is not related to the Bessel-function problem, since comparison with Lentz's (1976) tabulation shows complete agreement of j and y (with the 6 digits Lentz offers) for ka=10, 100, and 1000. Further comparisons show complete agreement with Wiscombe's (1979) test values for S_1, S_2, total intensity, and degree of polarization (again, to 6 digits) for ka=10 and 100, but not for the next available value, at ka = 1000. (In seeking the source of this instability I copied Wiscombe's FORTRAN program directly into the campus PRIME, only to find that the results are much worse: here using 'quadruple precision' I have so far gotten no more than 2-digit agreement with test values near the axis. To quote Edsger Dijkstra, 'The sooner we can forget that FORTRAN ever existed, the better, for as a vehicle of thought, it is no longer adequate'. (Olsen, 1983)).

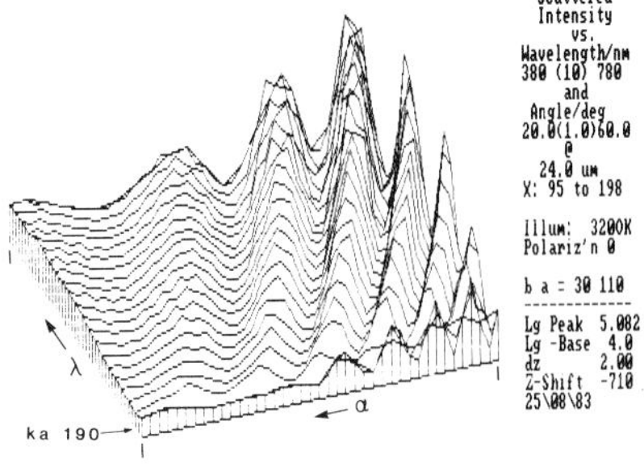

Fig. 13. Scattered intensity of 24-μm bubble, showing the numerical instability at ka=190 appearing like a left-lateral earthquake fault at 400 nm. This puts an effective upper limit on bubble diameter at this time, but has no effect on smaller bubbles.

Although it offends one's sense of tidiness to use a program with a known (if momentarily unlocatable) instability, the results for ka<190 agree not only with the test values of Lentz and Wiscombe, but also with theoretical predictions of Marston & Kingsbury (1981). More to the point, the color patterns agree with experimental data of Blanchard (MacIntyre, Struthwulf and Blanchard, 1984). I have complete confidence in the present data up to a diameter of 24μm.

Conclusions

Pending the development of a better rheological model of an organically coated bubble in seawater, it appears that the classical acoustical approach of Medwin (1977) will change no more than 10% after correction for surface elasticity. This error is comparable to the intrinsic uncertainty of the method at present. Taking account of thick skins, and of off-resonance absorption, may introduce rather larger corrections.

However, acoustical methods count anything containing resonant gas cavities (which are not necessarily spherical in shape) and in the upper ocean this includes many objects which are not immediately recognizable as bubbles, although they probably represent the collapsed remains of true bubbles. Such objects, being neturally buoyant, are of little interest to whitecap or air-sea interaction studies, since they will not rise to the surface and will not eject particles into the atmosphere. They should be of great interest to marine biologists, since they represent the direct conversion from dissolved to particulate organic material, they are the right size to be retained by filter-feeding zooplankton, and similar bubble-created particles have been shown to be a satisfactory maintenance food for brine shrimp (Baylor and Sutcliffe, 1963). They are also of interest to marine bacteriologists, since they provide surfaces for bacterial growth and represent an abundant bacterial food source. Like fecal pellets, bubble 'ghosts' may provide microscopic regions in which chemical potentials (e.g., of oxygen, methane, formaldehyde) may differ greatly from those of the immediately adjacent seawater.

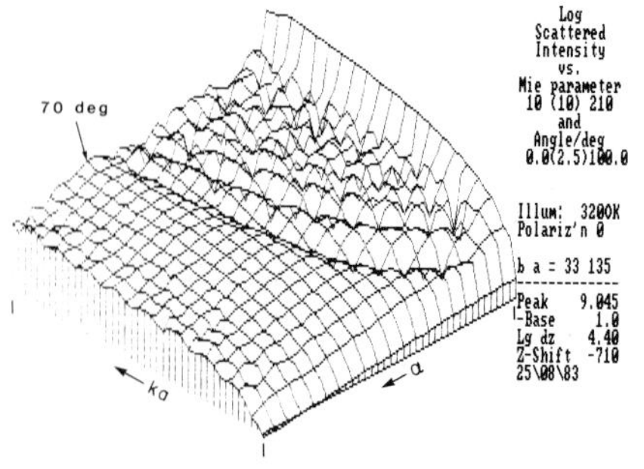

Fig. 14. Logarithmic plot of scattered intensity as a function of size parameter ka and scattering angle. Forward scattering (0 degrees) is along the back right; 90-degree scattering is the second line in from the left. Note that the total intensity range here exceeds 10 orders of magnitude, and that along any ka line there is at least a 100,000-fold change, suggesting why 90-degree scattering may not pick up the smaller bubbles very Note also the irregular nature of the surface: white light spans a two-fold range in ka (10–20, or 35–70, for instance), which will smooth out the response to some degree, but laser light is not to be used incautiously, although the curve at 70 degrees seems smoothly monotonic.

In the color-vs.-scattering-angle approach, we appear to have found a novel optical way to measure the diameters of small bubbles. Because we have not addressed the problem of bubbles covered by a layer of organic material whose refractive index differs from water, and cannot yet estimate the effect such a layer would have upon the scattered colors, this new approach is not yet ready for use in the ocean. However, similar calculations have long since been made for two-phase particles (Matijevič et al., 1962; Kerker et al., 1966), and present no difficulties in principle other than requiring twice the calculation time. Furthermore, since the material coating bubble 'ghosts' is neither smooth nor homogeneous, even a two-layer model may not be applicable to oceanic bubbles. However, for clean water in the laboratory the method would appear to be quite powerful.

Perhaps the most important conclusion reached is that the incautious use of light scattering to measure the diameter of small bubbles is not to be recommended. The irregular nature of the scattered intensity shown in the logarithmic plot of Fig. 14 underscores a need for careful evaluation of interference effects. In particular, monochromatic sources such as lasers, used with a non-axial detector, will show a response which is highly non-linear with bubble diameter. Conversely, in a sufficiently well defined system, these very irregularities can be used to measure subtle changes in bubble size.

Finally, the remarkable gain in scattered intensity available from forward scattering has not been taken advantage of in studies of oceanic bubble populations. (The circa 90-deg scattering angle of Medwin (1970) and Johnson and Cooke (1982) was chosen for overriding reasons of focal depth and three-spot resolution, which are incompatible with small-particle detection.) This raises some question–shared by P. Marston (pers. comm. 1983) who is apparently the only other theoretician working on the optics of small bubbles–about the interpretation of the small-diameter cut-off in oceanic bubble populations deduced from optical measurements, in a region where the acoustical methods show increasing numbers. Previous attempts to use optical and acoustical techniques simultaneously (Glotov et al., 1962; Medwin, 1970) have failed to resolve this question.

REFERENCES

Abramowitz, M. and I. A. Stegun, 1972. *Handbook of Mathematical Functions.* (Dover, NY).

Avetisyan, I. A., 1976: Effect of polymer additives on the acoustic properties of a liquid containing gas bubbles. *Sov. Phys. Acoust.* 23, 285-288.

Baylor, E. R. and W. H. Sutcliffe Jr., 1963: Dissolved organic matter in seawater as a source of particulate food. *Limnol. Oceanogr.* **8**, 369-371.

Blanchard, D. C. and A. H. Woodcock, 1957: Bubble formation and modification in the sea and its meteorological significance. *Tellus* **9**, 145-158.

Bouma, P. J., 1971: *Physical Aspects of Color.* (St Martin's Press, NY). 2nd ed.

Boyer, C. B., 1959: *The Rainbow: From Myth to Mathematics.* (Thos. Yoseloff).

Cipriano, R. J. and D. C. Blanchard, 1981: Bubble and aerosol spectra produced by a laboratory 'breaking wave'. *J. Geophys. Res.* **86**, 8085-8092.

Committee on Colorimetry of the Optical Society of America, 1963: *The Science of Color.* (Opt. Soc. Am., Washington DC), 385 pp.

Dave, J. V., 1969: Scattering of electromagnetic radiation by a large absorbing sphere. *IBM J. Res. Dev.* **13**, 302-313.

Davies, J. T. and E. K. Rideal, 1961: *Interfacial Phenomena.* (Academic Press, NY).

Davis, G. E., 1955: Scattering of light by an air bubble in water. *J. Opt. Soc. Am.* **45**, 572-581.

Dragcevic, D. and V. Pravdic: 1981. Properties of the seawater-air interface. 2. Rates of surface-film formation under steady-state conditions. *Limnol. Oceanogr.* **26**, 492-499.

Dragcevic, D., M. Vukovic, D. Cukman and V. Pravdik: 1979. Properties of the seawater-air interface. Dynamic-surface-tension studies. *Limnol. Oceanogr.* **24**, 1022-1030.

Fink, D. G., D. Christiansen (eds.), 1982: *Electronic Engineers' Handbook.* (McGraw-Hill, NY).

Fox, F. E. and K. F. Herzfeld, 1954: Gas bubbles with organic skin as cavitation nuclei. *J. Acoust. Soc. Am.* **26**, 984-989.

Gavrilov, L. R., 1969: On the size distribution of gas bubbles in water. *Sov. Phys. Acoustics* **15**, 22-24.

Glazman, R., 1983: Effects of adsorbed films on gas-bubble radial oscillations. *J. Acoust. Soc. Am.* **74**, 980-986.

Glotov, V. P., P. A. Kolobaev and G. G. Neuimin, 1962: Investigation of the scattering of sound by bubbles generated by an artifical wind in seawater and the statistical distribution of bubble sizes. *Sov. Phys. Acoustics* **7**, 341-345.

Goodrich, F. C., 1981: The theory of capillary excess viscosities. *Roy. Soc. London Proc.* **A 374**, 341-370.

Goodrich, F. C. and L. H. Allen, 1972: The theory of absolute surface-shear viscosity. V. The effect of finite ring thickness. *J. Colloid Interface Sci.* **40**, 329-336.

Greenler, R., 1980: *Rainbows, halos, and glories.* (Cambridge U. Press).

Hardy, A. C., 1936: *Handbook of Colorimetry.* (Technology Press, Cambridge, MA). 87 pp.

Johnson, B. D., and R. C. Cooke, 1976: Nonliving organic particle formation from bubble dissolution. *Limnol. Oceanogr.* **21**, 444-446.

Johnson, B. D. and R. C. Cooke, 1980: Organic particle and aggregate formation resulting from the dissolution of bubbles in seawater. *Limnol. Oceanogr.* **25**, 653-661.

Johnson, B. D. and R. C. Cooke, 1981: The generation of stabilized microbubbles in seawater. *Science* **213**, 209-211.

Johnson, B. D. and R. C. Cooke, 1982: Bubble populations and spectra in coastal waters: A photographic approach. *J. Geophys. Res.* **84**, 3761-3766.

Johnson, I. and V. K. LaMer, 1946: The determination of particle size of monodispersed systems by the scattering of light. *J. Am. Chem. Soc.* **69**, 1184-1192.

Joly, M., 1972: Rheological properties of monomolecular films. I. Basic concepts and experimental methods. In E. Matiejevic (ed.), *Surface and Colloid Science* (Wiley-Interscience, NY), vol. 5, 1-78.

Joly, M., 1972: Rheological properties of monomolecular films. II. Experimental results. Theoretical interpretation. Applications. In E. Matiejevic (ed.), *Surface and Colloid Science* (Wiley-Interscience, NY), vol. 5, 79-194.

Kerker, M., W. A. Farone and W. F. Espenscheid, 1966: Color effects in the scattering of white light by cylinders and spheres. *J. Colloid Interface Sci.* **21**, 459-478.

Kerker, M., L. H. Kauffman and W. A. Farone, 1966: Scattering of electromagnetic waves from two concentric spheres when the outer shell has a variable refractive index. *J. Opt. Soc. Amer.* **56**, 1053-1056.

Kingsbury, D. L. and P. L. Marston, 1981: Mie scattering near the critical angle of bubbles in water. *J. Opt. Soc. Am.* **71**, 358-361.

Kingsbury, D. L. and P. L. Marston, 1981: Scattering by bubbles in glass: Mie theory and physical-optics approximation. *Appl. Opt.* **29**, 2348-2350.

Kolovayev, D. A., 1976: Investigation of the concentration and statistical size distribution of wind-pro-

duced bubbles in the near-surface ocean. *Oceanologia USSR Engl. Transl.* **15**, 659-661.
LaMer, V. K., E. C. Y. Inn and I. B. Wilson, 1950: The methods of forming, determining, and measuring the size and concentration of liquid aerosols in the size range 0.01 to 0.25 microns diameter. *J. Coll. Sci.* **5**, 471-496.
Land, E. H., 1959: Color vision and the natural image I; II. *Proc. Nat. Acad. Sci.* **45**, 115-129; 630-645.
Land, E. H., 1959: Experiments in color vision. *Sci. Am.* **200**(5), 84-99.
Land. E. H., 1983: Recent advances in retinex theory and some implications for cortical computations: Color vision and the natural image. *Proc. Nat. Acad. Sci.* **80**, 5163-5169.
Lentz, W. J., 1973: A new method of computing spherical Bessel functions of complex argument with tables. ECOM 5509 AD 767223.
Levich, V. G., 1962: *Physicochemical Hydrodynamics.* (Prentice-Hall, Englewood Cliffs, NJ).
Levich, V. G. and V. S. Krylov, 1969: Surface-tension-driven phenomena. *Ann. Rev. Fluid Mech.* **1**, 293-316.
Lieberman, L., 1957: Air bubbles in water. *J. Appl. Phys.* **28**, 205-211.
Lucassen, J. and E. H. Lucassen-Reynders, 1967: Wave damping and Gibbs elasticity for non-ideal surface behavior. *J. Colloid Interface Sci.* **25**, 496-502.
Lucassen-Reynders, E. H., 1973: Interaction in mixed monolayers. III. Effect on dynamic surface properties. *J. Colloid Interface Sci.* **42**, 573-580.
MacIntyre, F., Struthwulf, M. and D. C. Blanchard, 1983: Color effects in light scattered from small air bubbles in water. *J. Opt. Soc. Amer.* (in preparation).
Marston, P. L. and D. L. Kingsbury, 1981: Scattering by a bubble in water near the critical angle: interference effects. *J. Opt. Soc. Am.* **71**, 192-196.
Marston, P. L., D. S. Langley and D. L. Kingsbury, 1982: Light scattering by bubbles in liquids: Mie theory, physical-optics approximations, and experiments. *Appl. Sci. Res.* **38**, 373-383.
Matijevic, E., M. Kerker and E. F. Schulz, 1960: Light scattering of coated aerosols. *Farad. Soc. Disc.* 1960: 178-184.
Medwin, H., 1970: In-situ acoustic measurements of bubble populations in coastal ocean waters. *J. Geophys. Res.* **75**, 599-611.
Medwin, H., 1977: Counting bubbles acoustically: a review. *Ultrasonics* **15**, 7-13.
Medwin, H., 1977: In-situ acoustic measurements of microbubbles at sea. *J. Geophys. Res.* **82**, 971-976.
Merrill, J. T. and C. A. Mattocks, 1981: Ambient bubble-size measurements using an electro-optical instrument. (Unpublished paper).
Minnaert, M., 1959. *Light and colour in the open air.* (G. Bell & Sons).
Mulhearn, P. J., 1981: Distribution of microbubbles in coastal waters. *J. Geophys. Res.* **86**, 6429-6434.
Olson, S., 1984: The sage of software. *Science 84*, 5, 74-80.
Rusanov, A. I. and V. V. Krotov, 1979: Gibbs elasticity of liquid films, threads and foams. *Prog. Surf. Membr. Sci.* **13**, 415-524.
Scriven, L. E., 1960: Dynamics of a fluid interface. *Chem. Eng. Sci.* **12**, 98-108.
Turner, W. R., 1961: Microbubble persistence in fresh water. *J. Acoust. Soc. Am.* **33**, 1223-1233.
Van de Hulst, H. C., 1957: *Light Scattering in Small Particles.* (Wiley-Interscience, NY).
Van Vleet, E. S. and P. M. Williams, 1983: Surface-potential and film-pressure measurements in seawater systems. *Limnol. Oceanogr.* **28**, 401-414.
Weiskopf, V., 1968: *Light and Color.* (Freeman, San Francisco).
Wiscombe, W. J., 1979: Mie scattering calculations: Advances in technique and fast, vector-speed computer codes. NCAR/TN-140+STR.
Wiscombe, W. J., 1980: Improved Mie scattering algorithms. *Appl. Optics* **19**, 1505-1509.
Wu, J., 1981: Bubble populations and spectra in near-surface ocean: Summary and review of field measurements. *J. Geophys. Res.* **86**, 457-464.
Wyszecki, G. and W. S. Stiles, 1967: *Color Science.* (Wiley-Interscience, NY).
Yount, D. E., 1979: Skins of varying permeability: A stabilization mechanism for gas-cavitation nuclei. *J. Acoust. Soc. Am.* **65**, 1429-1439.

ACKNOWLEDGEMENTS

Duncan Blanchard reported the observations which prompted the work on optical theory. Conversations with Philip Marston, Warren Wiscombe, and Don McSperron (NBS) were of great assistance.

Tom Dowling speeded up my computations by creating or advising on every issue I requested for the 8087 numerical-coprocessor routines in MMSFORTH. Questions about MMSFORTH should be referred to Miller Microcomputer Services, 61 Lakeshore Rd., Natick, MA 01760.

This work was supported by NSF grant OCE 81-17849.

THE CONTRIBUTION OF BUBBLES TO GAS TRANSFER ACROSS AN AIR-WATER INTERFACE

L. MEMERY & L. MERLIVAT
Laboratoire de Géochimie Isotopique
D.P.C. / C.E.N. - Saclay
91191 Gif-sur-Yvette Cedex France

A first theoretical approach to gas transfer by bubbles is undertaken. It is found that transfer velocity increases when solubility decreases. Bubble over-pressure leads to water supersaturation at equilibrium, this supersaturation being more significant for less soluble gases. Although the transfer velocity remains roughly constant for a variable concentration far from equilibrium, its range of variation becomes infinite near equilibrium. Attention is turned directly to the flux itself: it is shown that the flux is a linear function of the concentration gradient. At least for tracers, the coefficients of this function are entirely defined by the physico-chemical properties of the gas and by the bubble distribution.

INTRODUCTION

Attempts to model gas transfer across an air-water interface have until now involved the assumption of a smooth surface separating two homogeneous phases (Deacon (1977), Hasse and Liss (1980)). However, in ocean conditions waves often break and create bubbles. Moreover, a number of results from wind tunnel experiments (H. C. Broecker (1980), Merlivat and Memery (1983)) suggest a significant increase in transfer velocity under conditions of high wind speeds and breaking waves. In both invasion and evasion experiments this phenomenon is more pronounced for gases with low solubility. For these reasons, a first parameterization of gas exchange through bubbles has been undertaken.

All the equations are based on the following assumptions: the bubbles rise very nearly at their terminal velocity; the bubbles are created instantaneously at their initial depth Z_0 with their initial radius R_0 and with the same composition as the atmosphere from which they are formed.

BUBBLE GAS TRANSFER

Equations.

The transport velocity and the rate of mass transfer of a bubble depend on its radius.

Small bubbles behave like solid spheres. The drag coefficient C_D is equal to $24/Re\ (1 + 3/16\ Re)$ (Batchelor, 1967): $Re = 2ru/\nu$ is the Reynolds number, r the bubble radius, u its velocity and ν the viscosity of the water (for $C_D = 24/Re$, $u = 2gr^2/9\nu$). For bubbles that are not too large, $C_D = 48/Re\ (1\text{-}2.2/Re^{1/2})$ (Moore, 1963) (with $C_D = 48/Re$, $u = gr^2/9\nu$).

For large bubbles ($Re > 700$), u reaches a maximum value, independent of the radius: $u \cong 35$ cm/s (Levich, 1962).

By definition:

$$\frac{dn_i}{dt} = k_i r^2 (C_i - s_i P_i) \qquad (1)$$

where n_i is the number of moles of the gas component in the bubble, C_i its concentration in the water, P_i its partial pressure in the bubble, s_i

E. C. Monahan and G. Mac Niocaill (eds.), Oceanic Whitecaps, 95–100.

its solubility and k_i the mass transfer coefficient.

For a large Peclet number ($Pe = ru/D_i$, D_i is the molecular diffusivity of the gas) and a small Reynolds number, $k_i = 8(\pi D_i u/6r)^{1/2}$ and for large radii, $k_i = 8(\pi D_i u/2r)^{1/2}$ (Levich, 1962).

For solid bubbles, $k_i = 8(D_i^2 u/r^2)^{1/3}$.

As $P_i v = n_i RT$ ($v = 4/3\pi r^3$ = bubble volume), after due substitution, the following equation can be obtained:

$$\frac{dr}{dt} = \frac{\frac{3RT}{4\pi r^2} \sum_i \frac{dn_i}{dt} + r\rho g u}{3P - \frac{2\sigma}{r}} \tag{2}$$

where the total pressure in the bubble P is equal to $P_0 - \rho gz + 2\sigma/r$ (P_0 = atmospheric pressure, ρ = density of the water, σ = surface tension.

Discussion.

The solution of those equations for a single bubble shows:

a) in almost all the cases, if r_0 is not too small ($> 50\mu m$), the radius can be assumed to be constant. The variation in the radius is significant only if a large concentration gradient for the air gases (O_2 and N_2) prevails between the two phases: this variation is naturally due to gas diffusion.

b) the solubility s is an important physicochemical parameter in gas transfer through a bubble: the total amount of gas exchanged during the lifetime of a bubble increases with s. Moreover, the increase of this amount with r_0 is more pronounced as s becomes large (Merlivat and Memery, 1983).

DEFINITION OF TRANSFER VELOCITY

If ψ (r, x, t) is the bubble distribution considered, the global flux of gas is given by:

$$F_B = \int \psi (r, x, t)\, dn/dt\, (r, x, t)\, dr dx dt.$$

Assuming a steady state regime and horizontal homogeneity, it can be shown that

$$F_B = \iint S (r, z)\, N (r, z)\, dr dz$$

(Memery, 1982) where S is the bubble source and N(r,z) the amount of gas transferred during the lifetime of the bubble defined by its initial radius r and its initial depth z.

Assuming that the radius of the bubbles remains constant in the transport equation (Garrettson, 1973), $S = u\partial\psi/\partial z$. Thus,

$$F_B = \iint u(r) \frac{\partial \psi}{\partial z} (r,z)\, N (r,z)\, dr dz \tag{3}$$

Data reviewed by Wu (1981) show that the general shape of a bubble distribution can be represented by:

$$\psi = N_0 u^m r^{-a} e^{kz} \quad \text{with } m \cong 4.5 \text{ and } 3<a<5$$

By the definition of transfer velocity, it follows that (Merlivat and Memery, 1983):

$$k_B = - \frac{F_B}{S(C - sP)} \tag{4}$$

where S is the area of the air water interface (without bubbles), C the concentration of the gas in the water and P its partial pressure in the atmosphere.

APPROXIMATIONS

The following approximative equations have been set up in order to make possible an easier (qualitative) understanding of the results of the exact solutions of the equations (1.2.3.4).

The changes in radius are assumed to be small enough to be negligible when calculating the amount of gas transferred.

(1) can be written:

$$dn/dt = kr^2 s\, (\chi_w P_0 - 3nRT/4\pi r^3)$$

where χ_w is the molar fraction of the gas in the water: $C = s\chi_w P_0$.

If χ_a is the molar fraction of the gas in the air,

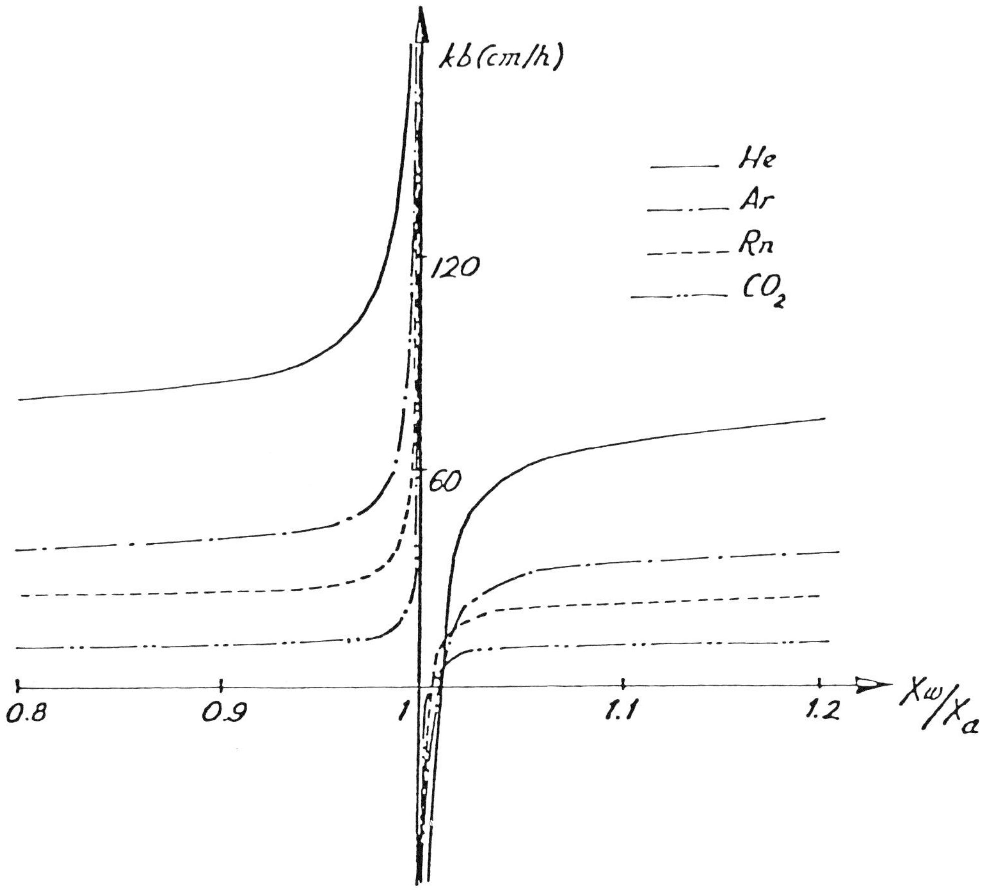

Fig. 1. Variation of the transfer velocity k_B with $x = \chi_w/\chi_a$. The bubble distribution considered is given by $\psi = \psi_0 r^{-3} e^z$ with $\psi \sim 5 \times 10^6$ bubbles/m^3 when $z = 0$.

$$n(t=0) = 4\pi r^3 \chi_a P/3RT$$
$$= 4\pi r^3/3RT\, \chi_a (P_0 - \rho gz + 2\sigma/r)$$

when $t \to \infty$, $dn/dt \to 0$ or $n \to 4\pi r^3 \chi_w P_0/3RT$,

thus, for a bubble reaching the surface at time t=-z/u,

$$N(r,z) = -Ar^3 [P_0(\chi_w - \chi_a) + (\rho gz - \frac{2\sigma}{r})\chi_a]$$
$$\times (1 - e^{\frac{z}{u\tau}})$$

where $A = 4\tau/3RT$ and $\tau = Ar/sk$.

In this case, (3) is equivalent to:

(5) $$F_B = -p(\chi_w - \chi_a)P_0 + q\chi_a P_0$$

where $p = A \int\int u\, \partial\psi/\partial z\, r^3 (1-e^{z/u\tau})\, drdz$
$q = A \int\int u\, \partial\psi/\partial z\, r^3\, p_1/P_0\, (1-e/^{z/u\tau})\, drdz$
with $p_1 = 2\sigma/r - \rho gz$

q represents the influence of the over pressure p_1 in the bubbles.

(6) $$k_B = \frac{1}{Ss}(p + \frac{\chi_a}{\chi_a - \chi_w} q)$$

DISCUSSION

Fig. 1 shows the variations in k_B with $x = \chi_w/\chi_a$ for four different gases: CO_2, Rn, Ar and He.

The main results which can be inferred from those curves are the following:

a) there is a vertical asymptote for $x = 1$. This is unquestionably due to the overpressure in the bubbles. The approximative relation (6) indicates clearly that as $q \neq 0$, $K_B \to \infty$ when

$$\chi_a \to \chi_w.$$

b) For the same reasons when $x > 1$, k_B may be negative. It is possible to have at the same time supersaturation of the liquid and transfer towards this liquid phase, in other words, the flux of gas and the gradient of concentrations may have opposite signs. From (6), $k_B < 0$ when $1 < x < 1 + q/p$. If x_0 is defined by

$$k_B(x_0) = 0$$

the calculations show that x_0 increases as the solubility decreases.

c) When x becomes different from 1, the transfer velocity remains constant for a variable concentration gradient. The transfer velocity for invasion ($x < 1$) is higher than for evasion ($x > 1$) because of the overpressure in the bubbles which facilitates the gas transfer towards the water.

d) Finally, the less soluble the gas, the higher the value of k_B. Note that from the simplified equation, $k_B \to p/sS$ when $x \to \infty$

$k_B \to (p+q)/sS$ when $x \to 0$.

GAS FLUX AND BUBBLES

Because of the vertical asymptote, the notion of transfer velocity does not appear to be well suited to the study of bubble gas exchange, especially near the equilibrium in solubility. The approximate formula (5) suggests that the relation between the flux and x is linear. As a matter of fact, the exact solutions show that at least for $0 < x < 2$ and for tracers ($\chi_a < 10^{-2}$), the function $F_B(x)$ is nearly linear: the coefficients (slope and origin) are entirely defined by the physico-chemical properties of the gas and by the bubble distribution.

For a flat air-water interface, the flux F_I satisfies the following equation:

$$F_I = k_I s S P_0 (\chi_a - \chi_w).$$

Thus gas flux across a plane interface and gas flux through bubbles can be described in similar terms, whereas the transfer velocities associated with these types of exchange are not comparable.

When there are bubbles, the actual equilibrium is obtained for $F_T = F_B + F_I = 0$ or $F_B = -F_I$: the flux through bubbles is balanced by the flux across the air-water interface. It can be shown that x_{eq}, the value of x for which $F_B(x_{eq}) = -F_I$, increases when the solubility decreases (fig. 2).

APPLICATIONS AND CONCLUSION

Wind Tunnel Experiments

Two sets of experiments were performed:

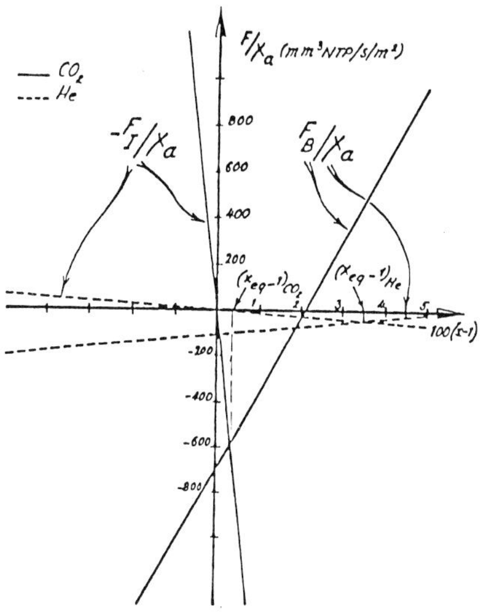

Fig. 2. Variation with x - 1 of the normalized flux through bubbles F_B/χ_a and the opposite of the normalized flux across the interface F_I/χ_a with $F_I = 50$ cm/h. Same bubble distribution as in Fig. 1.

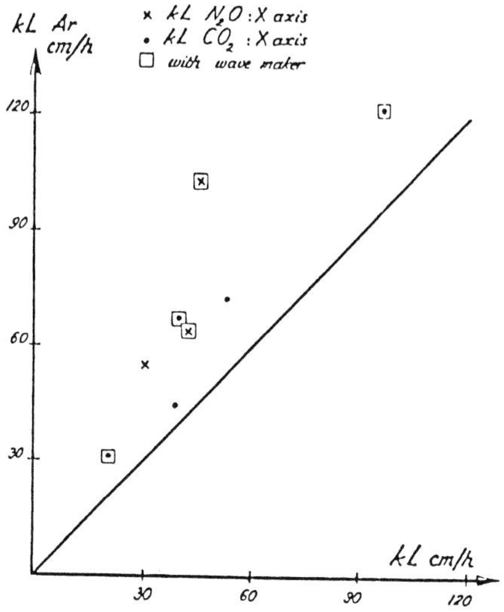

Fig. 3. Experimental results of transfer velocity.

1) In the large air sea interaction simulating facility (length: 40 m) of the IMST at Marseilles, two gases were used for evasion experiments, CO_2 and Ar.

2) In the small facility (Scale: 1/5 of the large one), invasion experiments with N_2O and Ar have been run.

Under the experimental conditions those gases were not chemically reactive, and their physico-chemical properties differ markedly (solubilities vary by a factor of as much as 100).

Moreover, in each facility, a wave generator at the upstream end of the water tank can be used to generate gravity waves and thus at high wind speeds, to increase the role of breaking waves.

Fig. 3 gives the values of k_L for Argon and N_2O or CO_2 measured under identical experimental conditions. Because of the very similar values of the gas diffusion coefficients in water, in the case of a flat interface, the results should be represented by points very close to the diagonal (which is the case for the experiments at lower wind or wave maker regimes). The deviations observed can be explained by the influence of the bubbles. During the experiments, the concentration gradients were very significant ($x > 1.3$ for invasion and $x < 0.6$ for evasion): thus the notion of transfer velocity can be used and the higher values of k_L found for Ar are the consequence of the role of a new parameter, the solubility. As a matter of fact, it has been shown that the lower the solubility, the higher the transfer velocity. Calculations show that in those experiments the bubbles can account for up to 60% of the total value of k_L.

Ocean Field Measurements.

This study shows the problems involved in computing the exchange rate of one gas using data obtained for another gas:

a) The assumption of a proportionality between transfer velocity and some power of the Schmidt number becomes meaningless under high wind regimes when there are bubbles: the solubility must be taken into account.

b) If for Radon the concentration gradient is large enough to allow the use of the notion of transfer velocity, gases in the ocean are often near equilibrium: for instance, the concentrations of CO_2 measured in surface waters in the Atlantic and Pacific oceans (Takahashi and Azevedo, 1982) show that $0.8<x<1.2$, that is to say the transfer velocity of gases via bubbles is highly variable. In order to determine the gas flux in those cases, it seems better to use a formula identical to (5).

The predictions of this modeling need to be compared quantitatively with experimental data: gas transfer experiments with simultaneous bubble distribution measurements (in progress) have been run at Marseilles and other experiments (with CO_2, Ar and He) are planned. In addition, the modeling will be developed in order to apply it to conditions prevailing at the atmosphere-ocean interface.

REFERENCES

Batchelor, G. K., 1967. *An introduction to fluid dynamics.* Cambridge Univ. Press.

Broecker, H. C., 1980. Effects of bubbles upon the gas exchange between atmosphere and ocean. Symposium on *Capillary waves and gas exchange,* Universität Hamburg, Heft 7, 127-129.

Deacon, E. L., 1977. Gas transfer to and across an air-water interface. *Tellus* 29, 363-374.

Hasse, L. and P. S. Liss, 1980. Gas exchange across the air-sea interface. *Tellus* 32, 470-481.

Levich, V. G., 1962. *Physical chemical hydrodynamics.* New York, Prentice-Hall.

Mémery, L., 1982. Influence de la formation de bulles créées par le déferlement sur le transfert de masse à l'interface air-eau. Rapport CEA No. 2, 306.

Merlivat, L. and L. Mémery, 1983. Gas exchange across an air-water interface: experimental results and modeling of bubble contribution to transfer. *J. Geophys. Res.* **88**, 707-724.

Moore, D. W., 1963. The boundary layer on a spherical gas bubble. *J. Fluid Mech.* **16**, 161-176.

Takahashi, T. and A. E. G. Azevedo, 1982. The ocean as CO_2 reservoir, in: *Interpretation of climate and photochemical models, ozone and temperature measurements,* Reck and Hummel eds., American Institute of Physics, N.Y., 83-109.

OCEANIC AIR BUBBLES AS GENERATORS OF MARINE AEROSOLS

FRANÇOIS RESCH
Laboratory of Physical Oceanography,
University of Toulon, 83130 La Garde,
France

The description of the physical pattern of air bubbles produced by whitecaps and wave breaking is of great importance for predicting the production of liquid marine aerosols at the sea-surface. Relevant parameters are: bubble size, concentration, injection depth and age. The transfer function giving the liquid droplet production rate in relation to the bubble parameters is of prime interest for any physical modelling.

In order to estimate the oceanic bubble parameters an optical method was selected. A prototype was especially designed and built incorporating a technique based on the principle of light-scattering at an angle of ninety degrees from the incident light-beam. The output voltage is a direct function of the bubble diameter. Results were obtained in a large sea-air interaction simulating facility. In view of these initial results, a second generation of probe is being presently tested. Improvements in the precision and technical performance as well as the use of the probe in field experiments have been of particular concern.

The bubble bursting phenomenon is studied by means of a photographic method. Holograms of bubble bursting are obtained with a high-power pulsed laser. This technique allows an illumination of the transient phenomenon with high light intensity during a very short period of time (10 ns). For the first time, photographs of holograms show the onset of the bubble bursting and the production of film droplets.

I. Introduction

It is well known that the physical mechanisms responsible for particulate exchange between the oceans and the atmosphere are rather complex in nature.

These transfers are mainly due to the formation of liquid aerosols produced by air bubbles bursting at the sea surface. These air bubbles are themselves produced by the wave-breaking which occurs when the wind is blowing over the sea (see figure 1). A more comprehensive picture of marine aerosol generation is given in figure 2, where the production of marine liquid aerosols is schematically represented as a 'cascade' or chain process (Resch, 1982). In the present study we will focus our attention on steps (2) and (3) of the production process.

Therefore the determination of the characteristics (size, concentration, injection depth) of the oceanic air bubbles produced by whitecaps and wave-breaking is of great importance for predicting the production of liquid aerosols at the sea surface. At the same time, this prediction is possible only if a transfer function giving the liquid droplet production rate as a function of the bubble characteristics is proposed.

These two goals were, for the past years, the main objective of the research program of the Laboratory of Physical Oceanography of Toulon. As no standard measurement techniques were available, a new probe was especially designed and built using an optical technique. In the meantime, the film-drop/jet-drop partition from bubble bursting was approached by a photographic technique.

The stages of development of these two studies are reported here.

II. Experimental determination of the characteristics of oceanic air bubbles

In order to estimate the oceanic bubble parameters – i.e. bubble size, concentration and in-

E. C. Monahan and G. Mac Niocaill (eds.), Oceanic Whitecaps, 101–112.

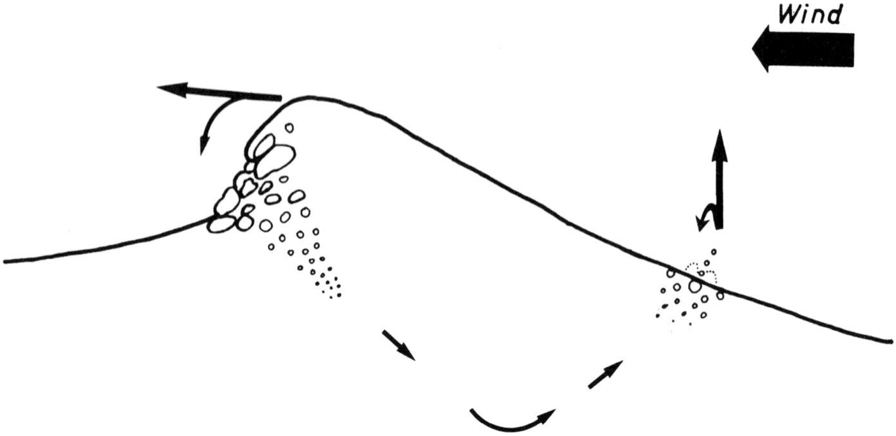

Figure 1. Air entrainment in sea water and aerosol injection into the atmosphere.

jection depth - it was decided to develop a new probe which utilised an optical technique. Although a few measurement techniques have already been used to determine the oceanic bubble characteristics (an acoustic technique by Medwin (1970), a trapping technique by Blanchard and Woodcock (1957) and by Kolovayev (1976) and a photographic technique by Johnson and Cooke (1979)) it turns out that more systematic measurements using a reliable and proven technique are needed. This is the conclusion of a detailed review of the subject by Wu (1981).

These considerations led to the development of a new kind of probe based on the well-known principle of light scattering at an angle of ninety degrees from the incident light beam. Under these conditions the output voltage is a direct function of the bubble diameter.

A prototype was designed, built, calibrated and tested by Avellan and Resch (1983). This prototype showed that the use of this technique was promising but early measurements of bubbles produced by wave-breaking in a large air-sea interaction simulating facility * led to statistically unstable results which are difficult to interpret. The technical performance of the prototype was markedly improved by Baldy and Resch (unpublished). The principle of the data acquisition and processing system was entirely re-shaped by Baldy leading to statistically stable and reliable results. A new series of experiments was carried out in the simulating facility and has yielded scientific results of great interest. The data have been processed but the results are not yet published.

Since much was learned from this second type of probe it was decided to produce a third type of probe which would be suitable for 'in situ' measurements where the environmental conditions are much more severe than those of the laboratory. This version has been developed by Richou, Vigliano and Resch.

II-1 Characteristics of the probe

When a bubble is passing through the 'sens-

*This part of the program was carried out at the Institut de Mécanique Statistique de la Turbulence, University of Aix-Marseille, 13003 Marseille, France.

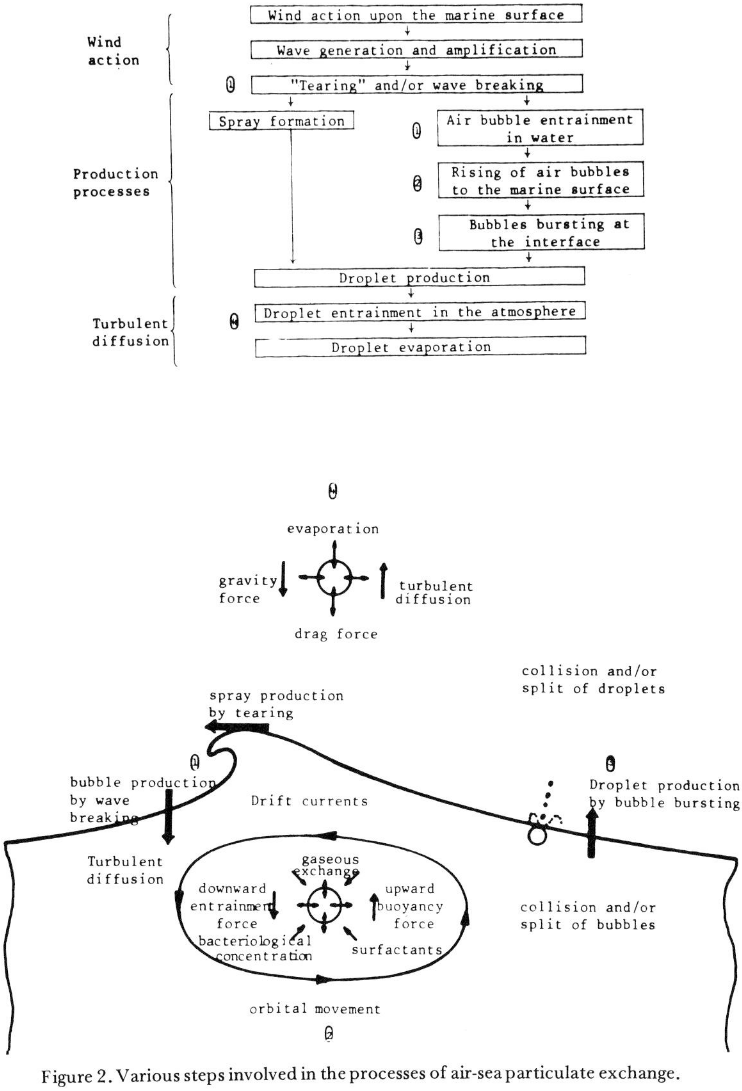

Figure 2. Various steps involved in the processes of air-sea particulate exchange.

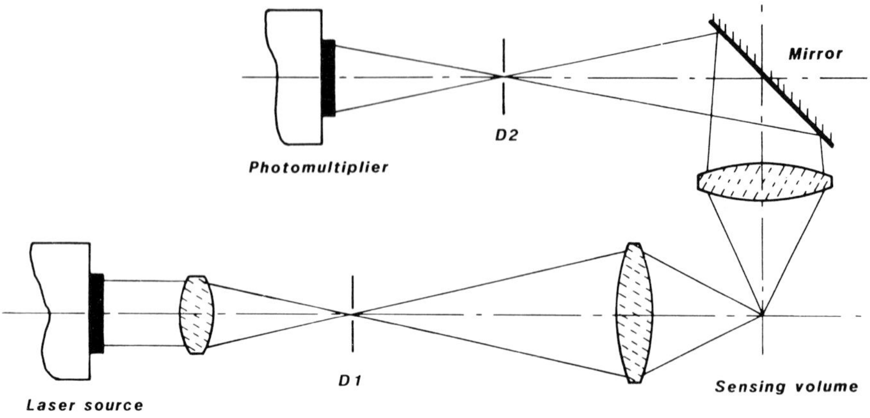

Figure 3. Schematic diagram of the optical part of the probe.

ing' volume, a part of the incident light is scattered in a given direction, chosen here at ninety degrees from the incident beam direction, as shown in figure 3. In the diameter range considered - 50 μm to 3000 μm - the output signal, which is a direct function of the scattered flux, is proportional to the square of the bubble diameter.

The slit D1 and the diaphragm D2 determine the sensing volume which is approximately represented by a cylinder of 500 μm diameter and 300 μm height (figure 4). The diaphragms D1 and D2 can be set to select only the central part of the gaussianly distributed laser light beam intensity in order to provide a uniform light intensity distribution in the sensing volume.

The light source, the optical components, the photoamplifier and its associated electronics are gathered in a rectangular stainless steel case of 5 X 10 X 30 centimeters (figure 4).

It should be mentioned that it is not the actual bubble diameter which is detected but only the image of the diaphragms through the lenses in the bubble. If d is the bubble diameter, then the apparent diameter of the bubble is k d, where $k \ll 1$, which allows bubbles with a diameter larger than the 'sensing' volume to be detected. For a more complete description, see Avellan and Resch (1983).

II-2 Modified second type of the probe

The development of the above described prototype took longer than expected, which in retrospect is not so abnormal for a new kind of sensor. Once we were persuaded that this technique was operational in the sense that an instrument so designed could be used to get reliable measurements of oceanic air bubbles produced by wave-breaking, it became necessary to improve its technical performance and look carefully at the processing of the data obtained with such a probe. In fact it is very easy to get an output signal from any probe, but much more difficult to interpret the signal in terms of a physical phenomenon.

i) Sensing volume

Without any estimate of the bubble concentration in the open ocean, an arbitrary small

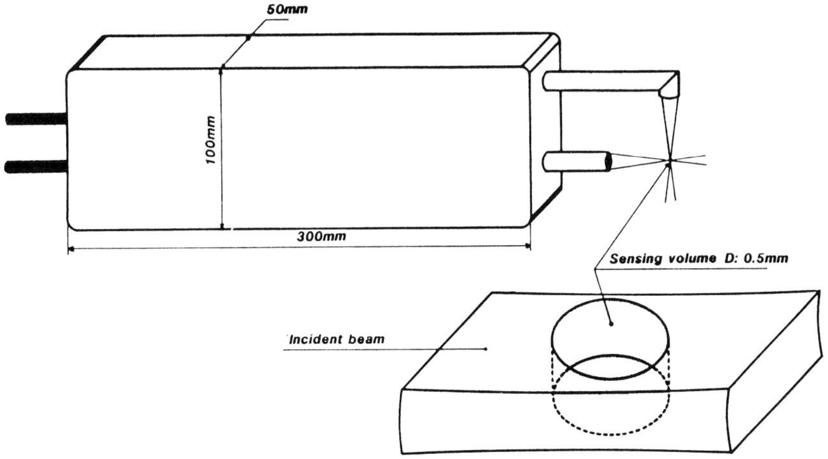

Figure 4. Optical probe and sensing volume.

sensing volume was initially selected for the prototype to prevent the possibility of having two bubbles at the same time in this volume. It rapidly became apparent that this volume was far too small - 0.06 mm^3 - to detect enough bubbles (500 bubbles in a 3 hour run, Avellan and Resch (1983)). It was therefore increased to its maximum value compatible with the geometrical configuration of the probe. The slit D1 was removed and diaphragm D2 was enlarged to give a maximum volume of 1 mm^3. The uniformity of the light intensity in the sensing volume was then provided by an optical diffuser. A few thousand air bubbles were detected per hour run leading to statistically stable results. To the extent that the bubble concentration is low enough, so that the probability of having more than one bubble at a time in the sensing volume is low enough, the larger sensing volume should be selected. In the next 'field type' generation of probe, this volume will be adjustable from 0.1 mm^3 to 2000 mm^3, corresponding to extreme values of concentration of 10 bubbles/mm^3 and 10^{-5} bubbles/mm^3. A larger volume has several advantages: - it decreases the side effect error: a bubble crossing the boundaries of the sensing volume and therefore being only partially detected, -it also increases the residence time of the bubble in the sensing volume. For a given sampling rate this makes it possible to obtain many more data points to identify a particular bubble.

ii) Data acquisition system

If we consider the two characteristic times, the time interval between bubble arrivals 't_a' (elapsed time between two successive detected bubbles) and bubble residence time 't_r' (time spent by the bubble in the sensing volume) it rapidly turns out that $t_a \gg t_r$. As it is desirable to use the highest sampling rate, here 40,000 Hz, to detect and analyse the signal of a bubble crossing the sensing volume in time t_r, it is impossible to continuously sample and record (time $t_a + t_r \cong t_a$) the signal. A conditional sampling system was developed using an adjustable threshold. Although only data corresponding to a detected bubble are recorded, it is nevertheless possible to keep track of the successive values of 't_a' and 't_r', which is desirable

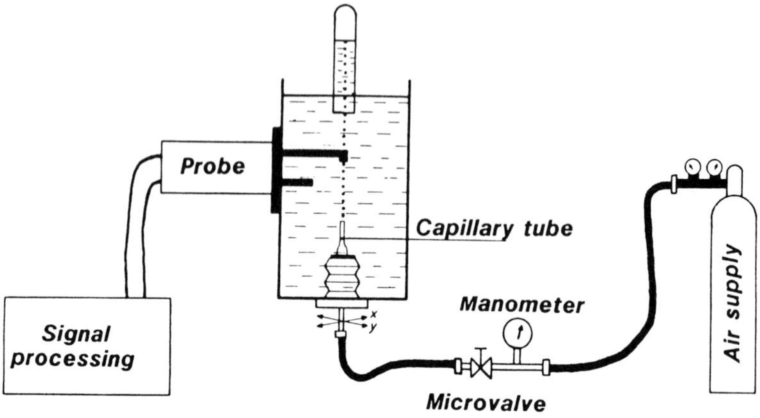

Figure 5. Calibration unit.

since their distribution leads to interesting physical insights into bubble configuration (patterns of isolated or clustered bubbles).

This conditional sampling was performed with a hardware system in the prototype version. The present software system has the advantages of a) decreasing the signal-to-noise ratio, b) allowing the threshold of detection to be varied easily, and c) saving the information on characteristic times 't_a' and 't_r'.

iii) Data processing system

A sophisticated data processing system was developed and used to extract various physical information from the signal.

The 'side effect error' was minimized and in most cases corrected. The highest values of the signal, directly related to the bubble diameter, are no longer recognized by eye (as was the case with the prototype) but by the method of maximum likelihood. Digital filtering is applied to the signal which allows for an improvement in the signal-to-noise ratio and therefore makes possible the use of the smallest detection threshold value.

of the smallest detection threshold value.

By this processing one has access to the histograms of bubble size, concentrations, and 'stirring' velocities as a function of bubble diameter as well as histograms of arrival times and residence times. The corresponding averages for various size bandwidths are available.

II-3 Calibration of the probe

Calibration of the probe is a very important step in getting reliable results. A special calibration unit was built for this purpose (figure 5).

Bubble diameters ranging between 50μm and 1mm were obtained with various capillary glass-tubes. For each bubble diameter approximately 5000 bubbles were detected giving one calibration point, i.e. one voltage versus diameter. Special attention was paid to the data processing of the calibration signal, normalized probability density skewness and flatness factors were duly considered.

For each bubble diameter considered, several parameters were changed, i.e. sampling frequency, filtering frequency, number of bub-

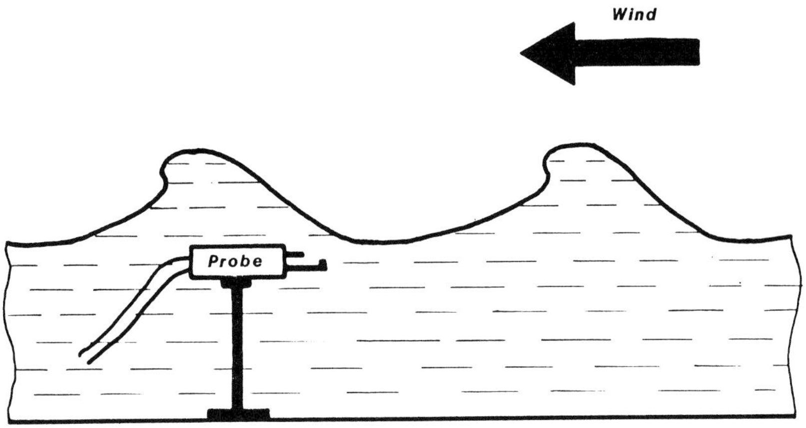

Figure 6. Experimental set-up.

bles, and the distance between the sensing volume and the tip of the capillary tube.

The calibration curve showing the square of the bubble diameter versus the output voltage is found to be a straight line passing through the origin (as predicted by the theory), at least for the considered diameter range (50μm–1mm).

II-4 Experimental results

Measurements were carried out with the probe in an air-sea simulating facility, 40 meters long with a fetch of 26 meters and a maximum wind velocity of 14 m/s. Measurements were carried out at 6 stations between the trough and the crest (figure 6). For this last situation, a special signal conditioning was used. For a fixed depth under the wave trough, measurements were carried out as the wind velocity was varied: 11, 12, 13 and 14 m/s.

Other series of measurements were performed in which a wave-maker was used.

Wave height measurements were simultaneously recorded so that surface displacement can be directly correlated to the various bubble measurement positions. With the second generation probe a few thousand bubbles were recorded per hour for the highest concentrations. The smallest bubbles detected had a diameter of 50 μm.

Qualitatively the main results are as follows:

a) Bubble concentrations increase rapidly when the wind velocity increases.

b) The variations of bubble concentrations or bubble frequencies of occurrence with increasing bubble diameters show a strong decrease: slopes range from -3 to -4 on a log - log plot.

c) In the diameter range considered (50 μm to 3000 μm) no maximum concentration or frequency of occurrence was recorded.

d) Both concentrations and frequencies decrease rapidly (exponentially) from the surface to lower depths.

e) Bubble stirring velocities are quite uniform close to the surface where the turbulence intensity is strongest and much larger than the rising terminal velocity.

f) Probability densities of 't_a' show a peak for measurements close to the surface thus indicating the occurrence of packs of clustered bubbles. This is not the case at greater depths where

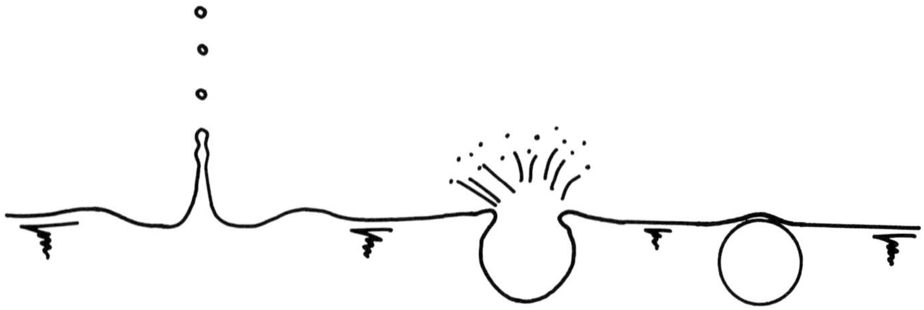

Figure 7. Film-drop and jet-drop production.

times 't_a' are more uniformly distributed.

g) Very close to the surface, there is a strong correlation between the peak values of 't_a' densities and wave frequency showing clearly the generation of bubbles by the waves.

II–5 Future developments

Detailed quantitative results will be published in the near future.

A third generation probe is presently being developed. It should fit all the requirements needed to perform direct measurements in the sea under storm conditions.

This last version will be tested in the laboratory and results will be compared with results of the modified (second) version.

A field station will be equipped in order to conduct experimental field studies.

III. Film Drop-Jet Drop Partition

When the oceanic air bubbles reach the air-sea interface they burst producing two families of droplets, the jet-drops and the film-drops. This bubble bursting phenonenon is described by numerous authors, among them Blanchard and Woodcock (1957) Blanchard (1963) and McIntyre (1972) (see figure 7).

i) Jet-drops: a vertical jet of water is produced by the collapse of the internal cavity of the bubble. This water jet becomes unstable and produces 1 to 10 jet drops. These droplets have a diameter d ranging from 0.1 to 0.15 times the generating bubble diameter ϕ_b. Their number is a *decreasing known* function of the bubble diameter. Their ejection height is approximately 100d for $\phi_b < 2$ mm and decreases when $\phi_b > 2$ mm.

ii) Film-drops: when bubbles break, the interfacial film between the atmosphere and the bubble is shattered, producing film-drops. Contrary to the case with jet drops, the number of film-drops is an *increasing unknown* function of the bubble diameter ϕ_b. The film drop diameter usually ranges from 1 μm to 30 μm. The ejection height is approximately 1 cm.

Which family of droplets plays the most important role in air-sea particulate exchange? This has long been a matter of controversy. For a long time it was felt that the jet-drops were the predominant contributors (Blanchard and Woodcock, 1957). More recently, evidence was found to prove that film drops also play a very important role in the ocean to atmosphere transfers. In this regard see MacIntyre (1972), Cipriano and Blanchard (1981), Blanchard (1982), and Cipriano et al. (1983).

In fact, we should ask, with Blanchard (1982), the question: Does the sea produce more film drops than jet drops? This question may be answered only when we know how many film-drops are produced by a bubble bursting at the air-water interface. Until now this uncertainty was of two orders of magni-

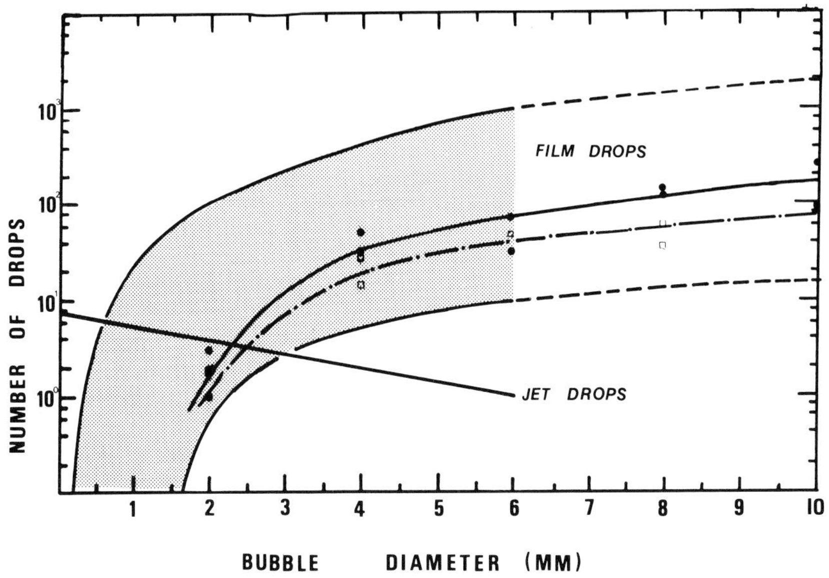

Figure 8. Number of film drops and jet drops as a function of bubble size. ● fresh water □ sea water. Shaded region and jet-drop line from Blanchard (1982).

tude, as is shown on figure 8 (from Blanchard, 1982). A photographic method was selected for use in a study undertaken to obtain an estimate of the number of film-drops produced as a function of the bubble diameter.

III-1 Holographic method

A photographic method has been successfully used since 1954 (Kientzler et al.), to describe jet-drop formation.

As far as we know, no attempt using a similar method has been tried to get the same kind of information about film-drops. This is not surprising if one realizes that the conditions of film-drop formation are much more drastic than those of jet-drop formation. Droplet sizes are smaller (a few microns to tens of microns), they form within a few hundred microseconds after the bubble bursting starts, their ejection velocity may reach 10m/s, their number varies between 1 and a few thousand, and they are distributed in a volume of a few cubic centimeters. By way of comparison, jet-drops measure hundreds of microns in size, they form a few milliseconds after the film drops, their number varies between 1 and 10, and they are directed in a single (vertical) direction.

In the case of film-drops it is therefore necessary to have a high light intensity during a very short period of time to get pictures of these very small particles moving at a high velocity. Microphotography is not suitable as the depth of field is too small. The voluminous spatial distribution of the droplets led us to choose the holographic method.

The simplest holographic technique, referred to as the Gabor technique, was used with a single direct laser beam (Bretheau, 1983).

A powerful laser source was needed: a YAG (Quantel company) was used. Its energy is 105 mJ for the wavelength of 532 nm. The duration of the pulses is 13 ns; its power is therefore:

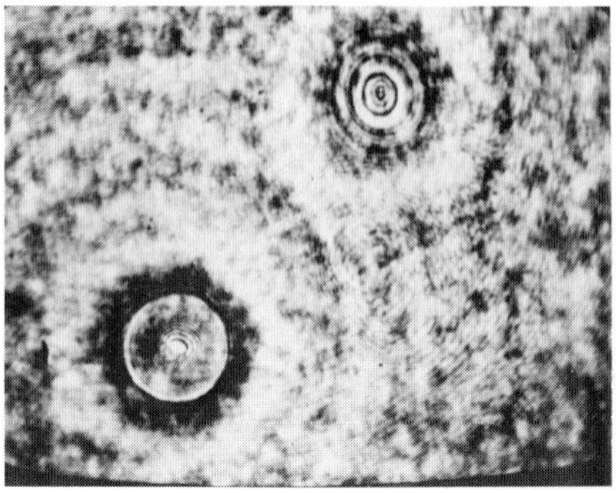

Figure 9. Example of hologram of film-drop; distance between the two drops is 600 μm.

105×10^{-3} J/13 $\times 10^{-9}$ s = 8 megawatts. The beam has a 7mm diameter. Holographic plates Holotest Agfa 8 E 75 HD were used. Images of the film-drops are obtained with a 1 mw He–Ne laser and observed with a microscope coupled with a camera. A typical image is depicted in figure 9.

To get proper holograms of film-drops, the most delicate operation is to coordinate the laser pulse with the bubble bursting. To obtain this fundametal control, an optico-electronic system was especially designed. When a bubble arrives at the air-water interface a He–Ne laser beam is directed at it in such a way that when the bubble bursts, the beam is received by a photoelectrical cell. The subsequent signal is then amplified to command the Pockels cell opening of the YAG laser. This time response of the Pockels cell was set up at 135 μs. The electronic time response of the over-all system can be varied between 15 μs and 30 ms. Therefore the total time response can be varied continuously from 150 μs to 30 ms after the time of bubble bursting.

III–2 First experimental results

It turns out that the minimum time delay of 150 μs is small enough to see the onset of the bubble bursting as can be seen in spectacular fashion in figure 10 for a bubble of 10mm diameter. At 300 μs all of the film-drops have been ejected, as is shown in figure 11. After 1ms the film-drops have disappeared by gravity or evaporation. The jet-drops appear between 3 and 20 milliseconds after the bursting as is shown in figure 12.

As far as we know, these pictures, which are simply the photographs of the hologram plates, are the first ones showing clearly film-drops and their ejection from a bursting bubble. To count all the film-drops present in the volume and to measure their diameter a complete analysis of the holograms is undertaken.

This is achieved with the help of a TV camera and microscope. Only drop diameters of 2 μm and larger can be detected and only drops of 10μm and larger can be measured. Data analysis yields results concerning bubbles of 2, 4, 6, 8 and 10 mm.

Experiments are performed with deionized fresh water and with sea water. Bubbles are produced with a capillary tube in a 20μm diameter cylindrical container.

Although the final results of this research program are not yet available, preliminary results concerning the number of drops produced by 2, 4, 6, 8 and 10 mm diameter bubbles, both in fresh and sea-water, can be shown. These results are plotted here on figure 8 along with the findings of Blanchard (1982). These results represent a first answer to the question posed at the beginning of this section.

First, it seems that fewer film-drops are produced by bubbles in sea water than in fresh water, an observation which is in agreement with the interpretation already given by several authors.

Although not shown here, histograms of film-drops (numbers of drops versus drop diameters) show that film-drops larger than those usually reported are present. There is a continuous spectrum of these drops up to 150 μm diameter.

Although less crucial, a detailed analysis can be made for the jet drops as regards size, ejec-

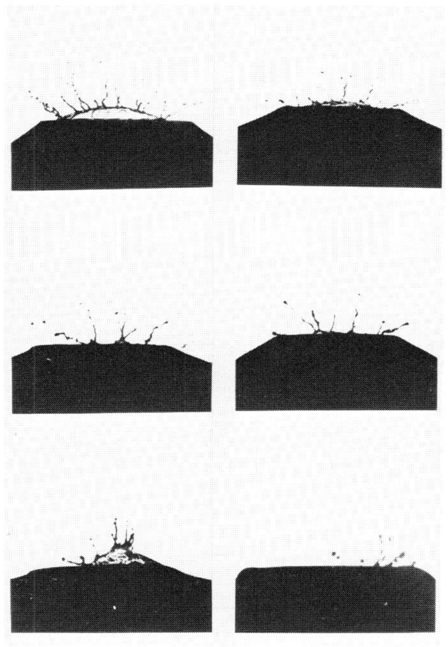

Figure 10. Bubble bursting. Photograph of a hologram. Bubble diameter is 10mm. Picture was taken 150μs after bursting commenced.

Figure 11. Film-drops produced 300μs after bursting commenced. From left to right and top to bottom, bubble diameter was 4mm, 6mm, 8mm, and 10mm.

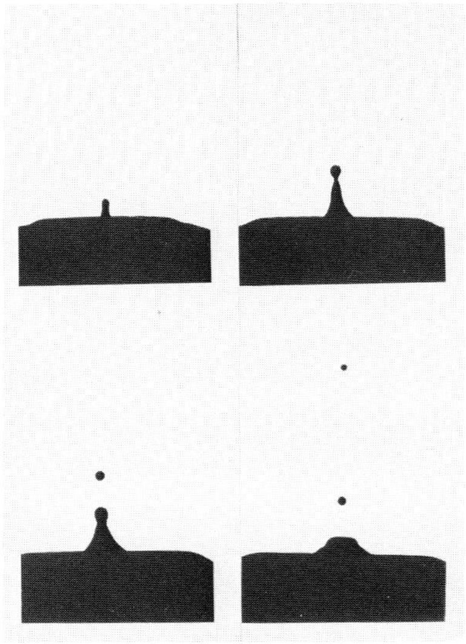

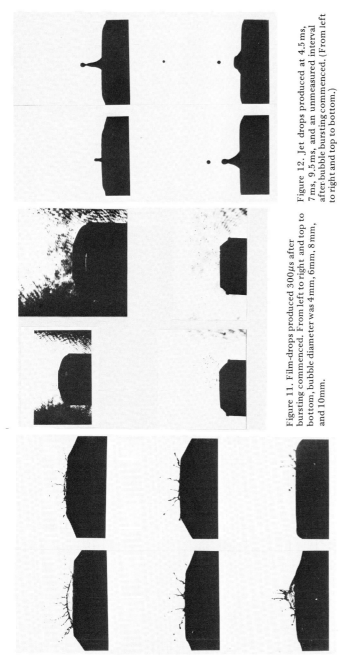

Figure 12. Jet drops produced at 4.5ms, 7ms, 9.5ms, and an unmeasured interval after bubble bursting commenced. (From left to right and top to bottom.)

tion height and speed. This analysis should be even more accurate than the present analysis of the film drops.

IV-Conclusions

Our main goal is to be able to predict the aerosol flux from the sea into the atmosphere. For this purpose, we need to get a good approximation of oceanic bubble spectra: estimates of bubble size, concentration, and injection depth.

The first part of this paper describes how we are approaching this problem. A new kind of optical probe is proposed. The prototype performances have been markedly improved to give initial valuable results in simulating laboratory experiments. The first results are qualitatively reported. The complete quantitative analysis will soon be published.

The next step is to define a transfer function relaying bubble characteristics and droplet production through the physical phenomenon of bursting. For this purpose, the generation of jet-drops is now quite adequately described. But little is known about film drop production. This is where we have tried to make a significant contribution by taking instantaneous holograms of the bursting bubbles producing film-drops. Preliminary results show that our optical technique can provide a very relevant answer to the question of how many film-drops are produced by the bursting of a bubble of a particular size.

REFERENCES

Avellan, F. & F. Resch, 1983: A scattering lightprobe for the measurement of oceanic air bubble sizes. *Int. Journal of Multiphase Flow* 9, No. 6, 649-663.

Blanchard, D. C. & A. H. Woodcock, 1957: Bubble formation and modification in the sea and its meteorological significance. *Tellus*, 9, no. 2, 145-158.

Blanchard, D. C., 1963: The electrification of the atmosphere by particles from bubbles in the sea. *Prog. oceanog.*, 1, 71-202.

Blanchard, D. C., 1982: The production, distribution, and bacterial enrichment of sea-salt aerosol. Proc. of the NATO advanced institute on the air-sea exchange of gases and particles, Durham, New Hampshire, 19-30 July 1982 - D. Reidel Pub. Co.

Bretheau, D., 1983: Etude de la visualisation par holographie de l'aerosol crée lors de l'éclatement d'une bulle d'air à une interface air-liquide. Rapport de stage, Ecole Supérieur d'Optique, Orsay, France.

Cipriano, R. J. & D. C. Blanchard, 1981: Bubble and aerosol spectra produced by a laboratory 'breaking wave'. *J. Geophys. Res.*, 86, C9, 8085-8090.

Cipriano, R. J., D. C. Blanchard, A. W. Hogan & G. G. Lala, 1983: On the production of Aitken nuclei from breaking waves and their role in the atmosphere. *J. Atmos. Sci.*, 40, 2, 469-479.

Johnson, B. D. & R. C. Cooke, 1979: Bubble populations and spectra in coastal waters: A photographic approach. *J. Geophys. Res.* 84, 3761-3766.

Kientzler, C. F., A. B. Arons, D. C. Blanchard & A. H. Woodcock, 1954: Photographic investigation of the projection of droplets by bubbles bursting at a water surface. *Tellus*, 6, 1, 1-7.

Kolovayev, D. A., 1976: Investigation of the concentration and statistical size distribution of wind-producing bubbles in the near-surface ocean. *Oceanology*, Engl. transl. 15, 659-661.

MacIntyre, F., 1972: Flow patterns in breaking bubles. *J. Geophys. Res.*, 77, 5211-5228.

Medwin, H., 1977: In situ acoustic measurements of micro-bubbles at sea. *J. Geophys. res.* 82, 971-975.

Resch, F., 1982: Air-sea particulate exchanges in coastal regions. Proc. of first int. conf. on meteorology and air/sea interaction of the coastal zone, May 10-14. The Hague, Published by the Am. Meteor. Soc., Boston, Mass., 54-57.

Wu, J., 1981: Bubble populations and spectra in near-surface ocean: summary and review of field measurements. *J. Geophys. Res.*, 86, *457-463*.

ACKNOWLEDGEMENTS

The study of film-drop generation by holography was supported by the Centre National de la Recherche Scientifique, via grants ATP no. 4144-4145 and 117. This research work was carried out under the scientific direction of F. Resch (University of Toulon) and J. S. Darrozes (University of Paris VI). The holograms were taken and analysed by G. Afeti and D. Bretheau.

WHITECAPS, BUBBLES, AND SPRAY

JIN WU
Air-Sea Interaction Laboratory,
College of Marine Studies,
University of Delaware,
Lewes, Delaware 19958

Results reviewed earlier are summarized along with those reported recently to further explore parameterizations of whitecap coverage of the sea surface, bubbles in the near-surface ocean, and spray in the atmospheric surface layer.

1. Introduction

Under the continuous influence of the wind, waves grow and eventually the water surface becomes unstable locally; the waves then break to dissipate the excess energy provided by the wind. The breaking is marked by whitecaps, which first appear near the wave crests. Field experiments have been conducted by Monahan (1971) in the Atlantic Ocean and by Toba and Chaen (1973) in the Pacific Ocean to measure whitecap coverage at various wind velocities. Their results were reanalyzed by Wu (1979a) to establish a power-law description of the whitecap coverage dependence on wind velocity, and to reveal effects of the air-sea temperature difference on whitecap coverage.

Air entrained at the sea surface by breaking waves forms bubbles in the near-surface ocean. Populations and size spectra of bubbles were measured by Kolovayev (1976) and Johnson and Cooke (1979). Both sets of results were found (Wu, 1981a) to have consistent trends. The bubble population decreases exponentially with depth and increases in a power-law form with wind velocity; the shape of the size spectrum appears to be invariant with either depth or wind velocity.

It has been suggested that sea spray is mainly produced through bubble bursting. Measurements of spray were conducted in the field by Monahan (1968) and Preobrazhenskii (1973); the former were performed at a single low elevation, while in the latter the results from various wind velocities were lumped together. Consequently, the size distributions of droplets and their fluxes were parameterized (Wu, 1979b) on the basis of laboratory results (Toba, 1961; Wu, 1973; Lai and Shemdin, 1974; and Wang and Street, 1978).

All three sets of results, describing whitecaps, bubbles, and spray, are summarized and reviewed along with those reported in more recent studies.

2. Whitecaps

(a) Summary of Earlier Results

Analytical considerations: Wave breaking is the result of the continuous supply of excessive energy by the wind. In the equilibrium state, we can consider that the energy lost by wave breaking is balanced by the energy gained from the wind, and that the pattern of wave breaking is similar at all wind velocities. The percentage of the sea surface covered by breaking waves under these equilibrium conditions can be related to the energy flux from the wind as shown below (Wu, 1979a),

$$W \sim \dot{E} = \tau V \sim \tau u_* = (C_{10} U_{10}^2)(C_{10}^{1/2} U_{10}) \sim U_{10}^{3.75} \quad (1)$$

E. C. Monahan and G. Mac Niocaill (eds.), Oceanic Whitecaps, 113–124.

where W is the fraction of the sea surface with whitecap coverage, $\dot{E}$ the rate at which energy is supplied by the wind to a unit area of the sea surface, τ the wind stress, u_* the wind-friction velocity, U_{10} the wind velocity measured at 10m above the mean sea surface, C_{10} the wind-stress coefficient considered to be proportional to $U_{10}^{1/2}$ (Wu, 1969), and V the sea-surface drift current considered to be proportional to u_* (Wu, 1975).

Observational Results: Photographic observations of whitecaps were made by Monahan (1971) in the Atlantic Ocean and adjacent salt-water bodies, and by Toba and Chaen (1973) in the East China Sea and in the southern Pacific Ocean. Both sets of data, regardless of the stability conditions (stable, neutral, or unstable) of the atmospheric surface layer were compiled and replotted (Wu, 1979a) to illustrate the general variation of whitecap coverage with wind velocity; see Fig. 1. The results are seen to follow rather closely the power-law variation shown in Eq. (1), especially the main body of the data enclosed by dashed lines. It was shown (Wu, 1979a), however, that the whitecap coverages reported by Monahan and Toba and Chaen were influenced by stability conditions.

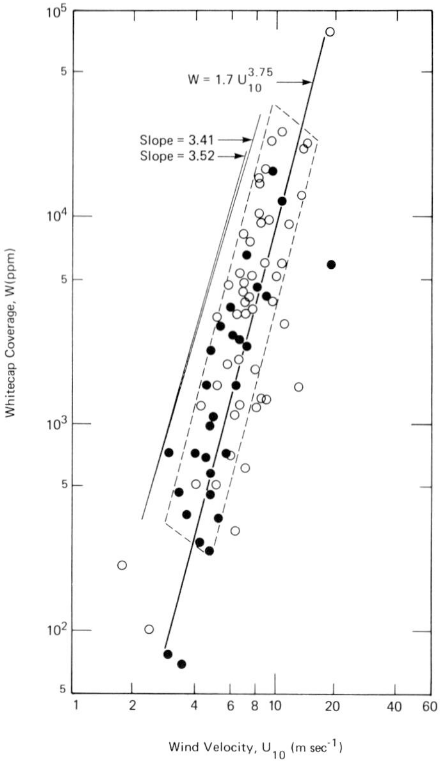

Figure 1. Whitecap coverages under various sea states. Monahan's (1971) results are indicated by o, Toba and Chaen's (1973) results by •.

(b) Recent Results

Power-law variation with wind velocity: For the same data presented in Fig. 1, Monahan and O'Muircheartaigh (1980) suggested that the exponent for the best curve fitting should have a value of either 3.41 or 3.52. Two linens corresponding to these two slopes were added (Wu, 1982a) to the figure, showing, however, no improvement of the fit.

More recently, Monahan et al. (1983) published a new set of data; their results were replotted in Fig. 2. The power-law formula proposed by them, $W \sim U_{10}^{3.3}$, is shown as the dashed line in the figure, while the solid line corresponds to the form of the variation indicated by Eq. (1). It appears then that even Eq. (1) underrepresents the increase of the whitecap coverage with wind velocity, with the data following more closely $W \sim U_{10}^{4.5}$. Adopting the same argument that was used in deducing Eq. (1), the last variation corresponds to a linear increase of C_{10} with U_{10}, as reported in Garratt (1977).

Effects of Air and Sea Temperatures: Effects of the air-sea temperature difference on the whitecap coverage were already discerned (Wu, 1979a) in the results of Monahan (1971) and of Toba and Chaen (1973), and these effects have been further verified recently by Smith (1981). The effects of stability conditions are twofold; their influence on the wind-stress coefficient is well known (Garratt, 1977), while the life of foaming is either affected by the air-sea temperature difference or by the water temperature (Miyake and Abe, 1948). Studies are needed to parameterize the life time of foaming, and the sea-surface whitecap coverage should be related to the wind-friction velocity.

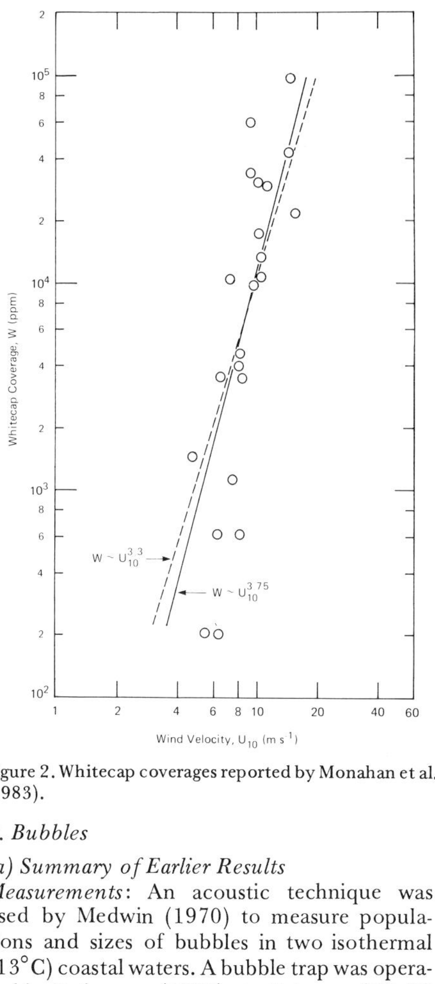

Figure 2. Whitecap coverages reported by Monahan et al. (1983).

3. *Bubbles*

(a) Summary of Earlier Results

Measurements: An acoustic technique was used by Medwin (1970) to measure populations and sizes of bubbles in two isothermal (13°C) coastal waters. A bubble trap was operated by Kolovayev (1976) at a distance of 60-70 m from a ship in the subtropical part of the Atlantic Ocean, where the water temperature was uniform from the sea surface to a depth of 25 m and averaged 14°C. A camera suspended from a float was used by Johnson and Cooke (1979) in coastal waters during a winter season with the water temperature between 2-3°C. The mechanism of bubble production in Medwin's investigation may have been mainly subsurface biological processes, while that in other investigations was mainly air entrainment at the sea surface. Consequently, only the results of Kolovayev and Johnson and Cooke were discussed by Wu (1981a).

Variation of Bubble Populations with Depth: The total number of bubbles per unit volume of water, N, expressed in m^{-3}, obtained at different depths by Kolovayev and by Johnson and Cooke were replotted by Wu (1981a). The populations measured in both investigations were found to have a similar variation with depth, but many more bubbles were measured by Johnson and Cooke than by Kolovayev. The bubble population at the sea surface was then obtained by extending the line fitted through each set of data to zero depth. The results were then normalized with respect to this surface value, N_0. Rather consistent variations of relative bubble population, N/N_0, with depth are found in both sets of data. This consistency indicates that the discrepancies in absolute bubble populations between the two investigations are likely due to either differences in the experimental conditions or systematic errors in the measurements, while the processes of bubble entrainment by turbulent water are likely to be the same.

Near the sea surface, the bubble population was found to decrease exponentially with depth, or

$$N/N_0 = \exp(-z), \quad z < 3\,\mathrm{m} \tag{2}$$

where z, expressed in meters, is the depth below the sea surface.

Variation of Bubble Populations with Wind Velocity: The main portion of Kolovayev's and Johnson and Cooke's results were obtained for the same wind-velocity range, U_{10} = 11-13 m s^{-1}. In addition, Kolovayev (1976) measured the bubble population for a lower wind velocity, U_{10} = 6-8 m s^{-1}; Johnson and Cooke (1979) for U_{10} = 8-10 m s^{-1}. All these results were first corrected to obtain the bubble population at the 1.5-m depth; subsequently, the overall variation of bubble population at this depth with wind velocity was found (Wu, 1981a) to be

$$N \sim U_{10}^{4.5} \tag{3}$$

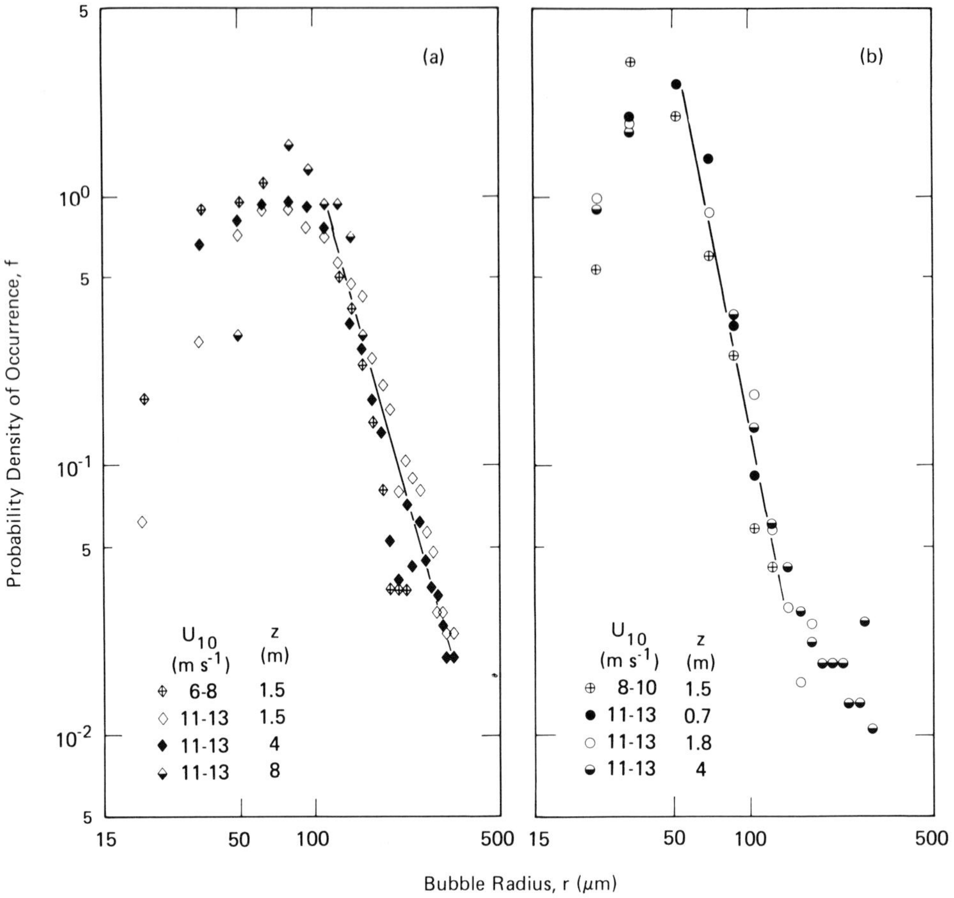

Figure 3. Bubble spectra at various depths and under different wind velocities. The results in (a) were obtained by Kolovayev (1976) and in (b) by Johnson and Cooke (1979).

This expression is not too much different from the variation of sea-surface whitecap coverage with wind velocity discussed earlier.

Equilibrium Shape of Size Spectra: The size spectra of bubbles obtained by Kolovayev and by Johnson and Cooke were normalized with the corresponding total bubble populations to obtain the probability density distributions of bubbles of various sizes. The normalized results are presented in Fig. 3a, b, where different symbols are used for results obtained at various depths under different wind velocities. It is seen that the size spectra for each investigation are invariant with either depth or wind velocity. Comparing the two sets, the results obtained by Johnson and Cooke are seen, however, to consist of relatively many more smaller bubbles than those by Kolovayev.

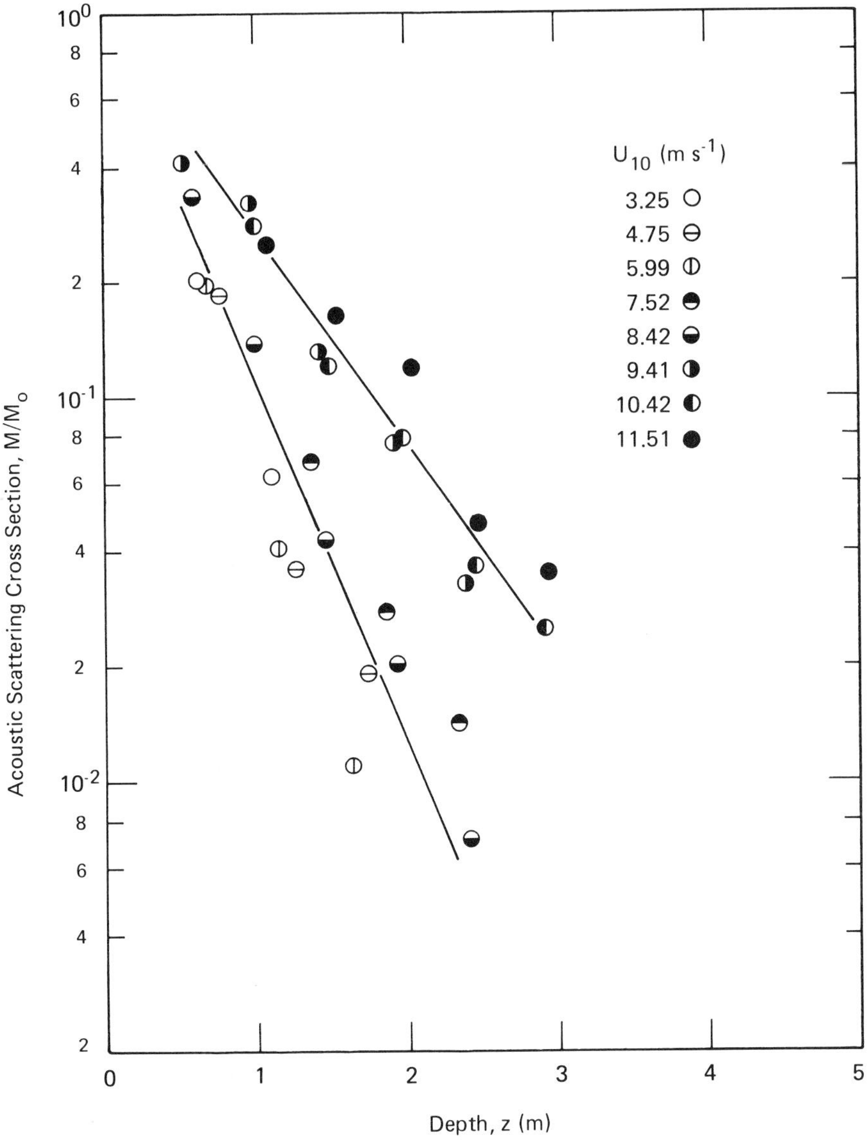

Figure 4. Variations of acoustic scattering cross section with depth at various wind velocities. The data were obtained from Thorpe (1983).

(b) Recent Results

Field Measurements: Observations have been made by Thorpe (1983) with a sonar in the sea near Oban. The transducer was mounted on a quadpod resting on the sea bed, and the beam was directed upward towards the water surface. The acoustic scattering cross section (M) per unit volume of a bubble cloud with N(r) bubbles per unit radius is given by

(4) $$M = \int_0^\infty N(r)\sigma dr$$

where $\sigma = 4\pi r^2$ is the surface area of the bubble with radius r. Results in terms of scattering cross sections at different depths under various wind velocities were reported by Thorpe. At each wind velocity, the logarithm of the cross section was found to vary linearly with depth. Following the same procedure discussed earlier, we extended the linear variation shown by Thorpe to the water surface to obtain the surface value of the scattering cross section M_0; subsequently, his data were normalized with the corresponding surface value and replotted in Fig. 4. It is very interesting to note that each set of data follows the description shown in Eq. (2). The variation of the bubble concentration with depth becomes more gradual as the wind velocity increases. Roughly, the data are divided at $U_{10} = 9$ m s^{-1} into two groups; two straight lines are drawn to illustrate this trend. The dividing wind velocity is very likely associated with the transition of the regimes of the aqueous boundary layer. The atmospheric surface layer becomes aerodynamically rough when the wind velocity reaches 7 m s^{-1}; the transition of the aqueous boundary layer is known (Wu, 1975) to occur at a higher wind velocity.

The average values of the scattering cross section for various wind velocities were also presented by Thorpe. His data are replotted in Fig. 5. A straight line corresponding to Eq. (3) is drawn in the figure and is seen to represent closely his results.

Laboratory Measurements: The measurements cited in this section are often considered to relate to the background bubbles (Blanchard and Woodcock, 1957). A bubble trap was used by Blanchard and Woodcock to collect 'fresh' bubbles produced by breaking waves on a beach. The trap was exposed for a period of two seconds at a depth of 10 cm a few seconds after the passing of the breaker. Although the bubble concentration measured by them directly under the breaker should differ from the 'background', i.e. the temporally averaged concentration measured by others, the slope of the Blanchard and Woodcock spectrum actually falls in between those of Kolovayev, and of Johnson and Cooke, shown in Fig. 3. In order to further explore the distribution of bubbles

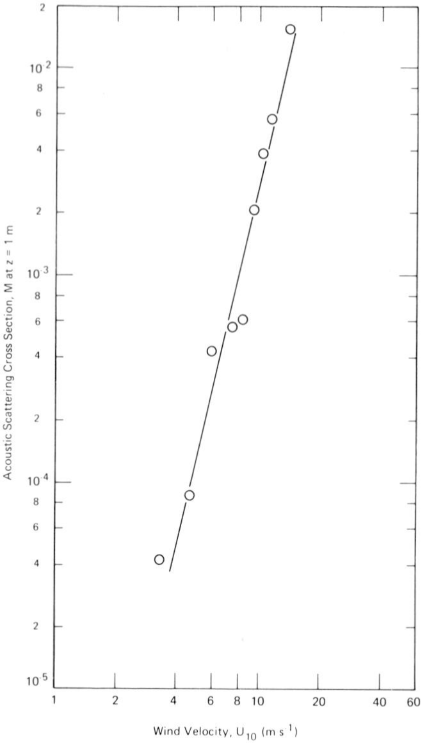

Figure 5. Acoustic scattering cross sections at various wind velocities. The data were obtained from Thorpe (1983).

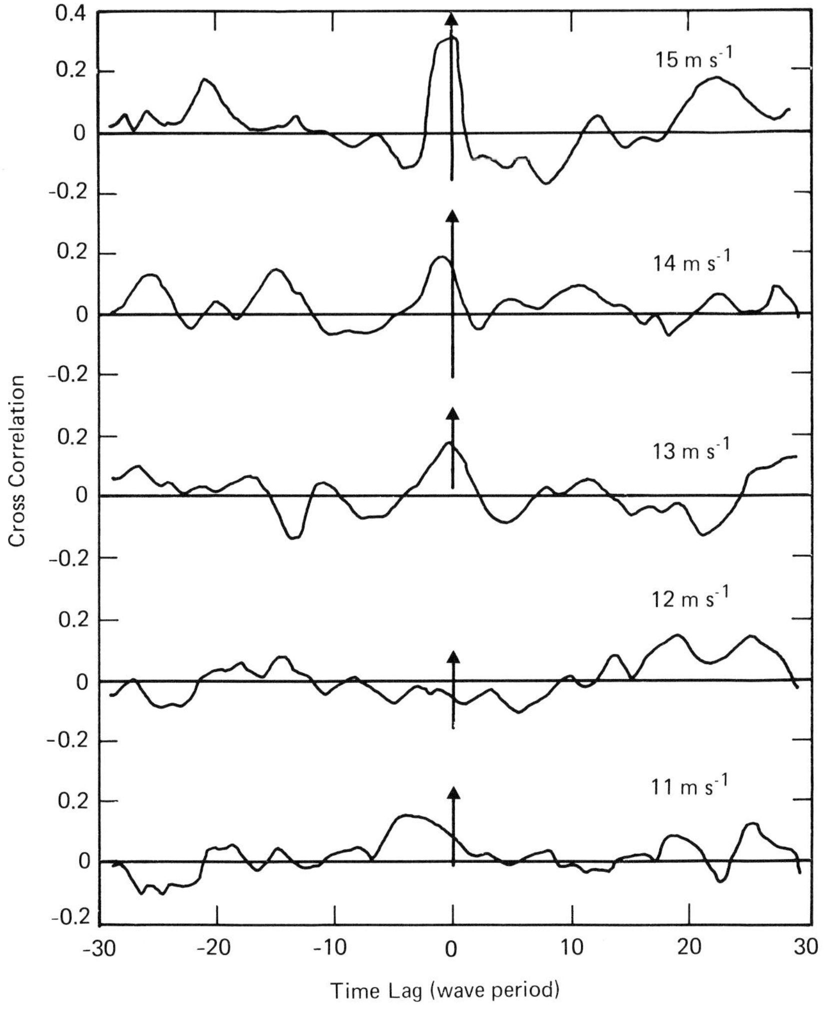

Figure 6. Correlation between bubble occurrence and wave height.

directly under the breaker, we have performed a series of experiments in the Wind-Wave-Current Research Facility (Hsu et al., 1983). The cross correlation between the occurrence of bubbles with diameters larger than 160 μm and the wave profile is shown in Fig. 6. No correlation is seen at low wind velocities, and the correlation remains small even at the highest wind

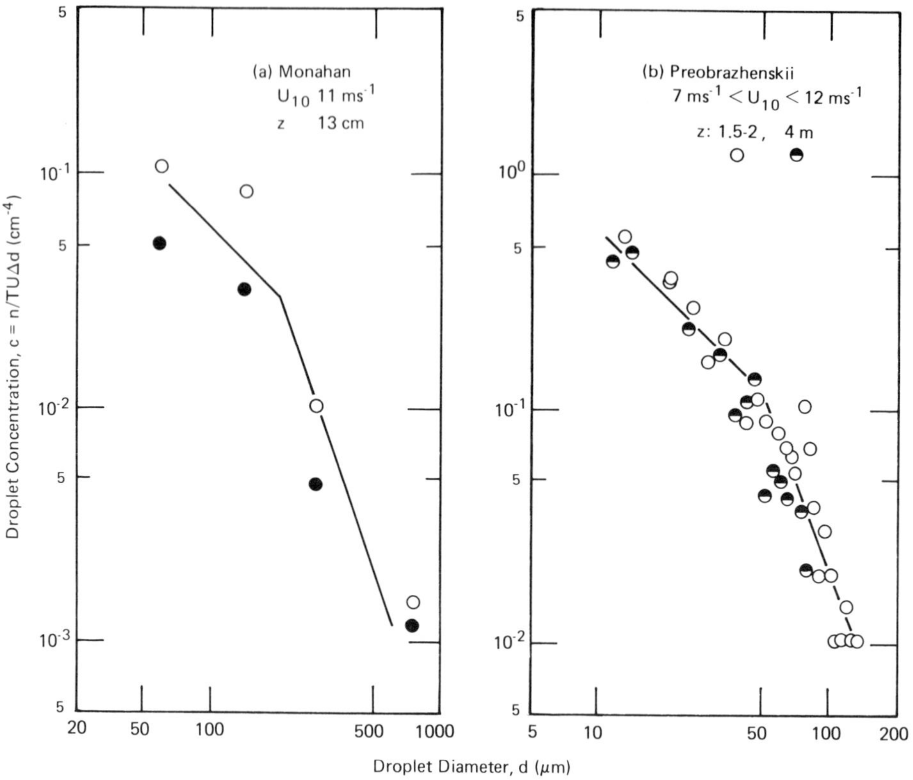

Figure 7. Distributions of droplet sizes measured by Monahan (1968) and Preobrazhenskii (1973).

velocity when every wave breaks. It is concluded that the relatively long residence time of bubbles in water in comparison with the wave period makes the laboratory tank inappropriate for this kind of study.

4. Spray

(a) Summary of Earlier Results

A photographic device mounted on a raft was used by Monahan (1968) to measure spray at a single elevation, 13 cm, above the sea surface under several wind velocities. Later, Preobrazhenskii (1973) made measurements with oil-coated plates attached to a boom extended from a ship; the plates were supported at three elevations, 1.5–2, 4, and 7m, above the sea surface. His results, however, were grouped into two wind speed categories: moderate winds ranging between 7 and 12 m s^{-1}, and strong winds ranging between 15 and 25 m s^{-1}. The size distributions of droplets were reported by both investigators and are reproduced in Fig. 7; we included here only Preobrazhenskii's results under moderate winds, the same range under which Monahan's experiments were conducted. Furthermore, we have omitted Preobrazhenskii's data obtained at the highest elevation which are distinctly different from those at the two lower elevations. In the figure, c is the droplet concentration, n the droplet count,

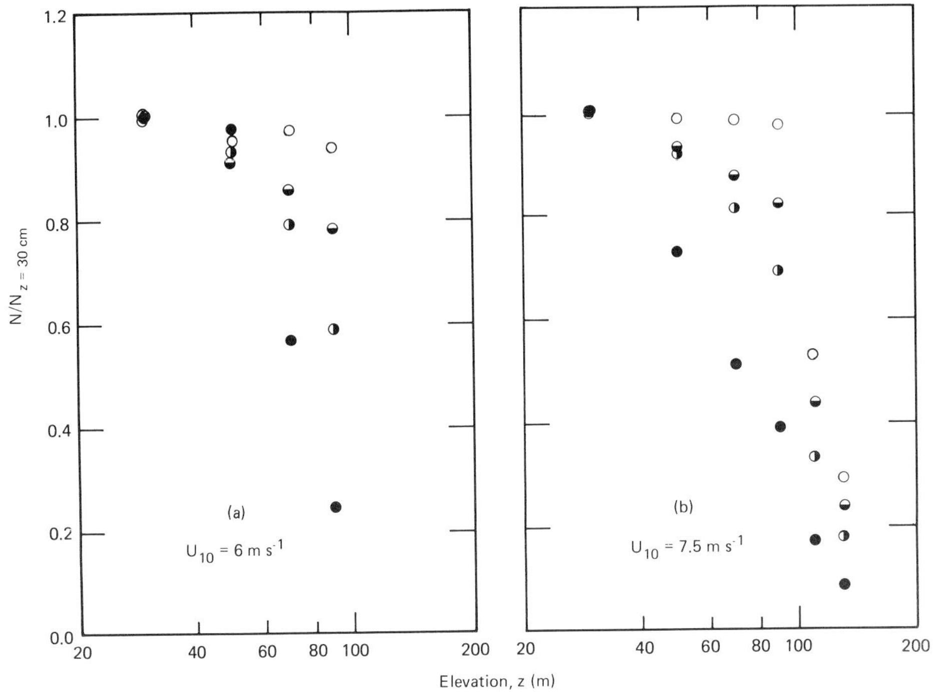

Figure 8. Vertical distributions of droplets of various diameters (μm): o (30-50), ◒ (50-100), ◑ (100-200), • (200-400).

T the sampling area, U the wind velocity at the droplet sampling level, and Δd the diameter-band of sampling. Both Monahan's and Preobrazhenskii's results are seen in the figure to have the same trend. In light of these earlier results, when taken together with the recent results (Wu et al., 1984), a two-segment size distribution, as illustrated in the figure, was proposed (Wu et al., 1984).

$$c \sim d^{-1} \quad d < 75\,\mu m$$
(5)
$$c \sim d^{-3} \quad d > 75\,\mu m$$

where d is the diameter of the droplet.

(b) Recent Results

Sea spray within the lowest meter of the atmospheric surface layer was measured by Wu et al. (1984) with an optical instrument supported on a raft. Two series of experiments were performed, each with a different raft. In addition to the droplet spectrum, two new sets of results are discussed in the following.

Vertical Distributions of Droplets: The vertical distributions of droplets of various size ranges are shown in Fig. 8, where the data for each size group are normalized with respect to the frequency of occurrence measured at the lowest height. A clear dependence of the vertical distribution on the droplet size is seen; smaller droplets are more homogeneously distributed than larger droplets. For the smallest size group, the droplets appear to be rather homogeneously distributed within the lowest meter, and the droplet concentration falls off abruptly above this elevation. Vertical distributions of the other size groups follow the same

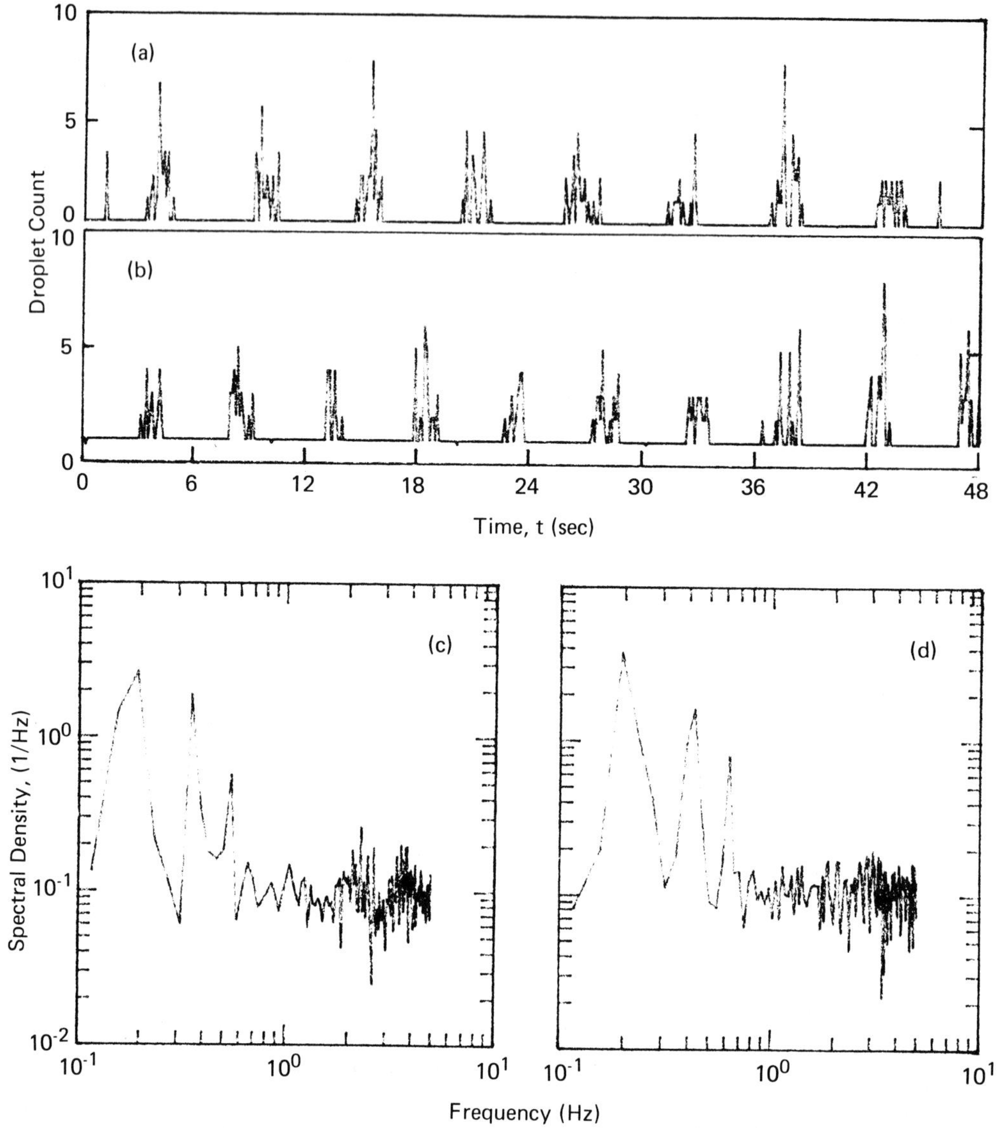

Figure 9. Patchy distributions of droplets. The droplet counts per 1/10-sec increment are shown in (a) and (b) and the spectral densities of droplet occurrences are shown in (c) and (d). The data were obtained at $U_{10} = 7.5\,\mathrm{ms}^{-1}$; (a) and (c) with $z = 30\,\mathrm{cm}$, (b) and (d) with $z = 50\,\mathrm{cm}$.

general trend, but their sudden falloff occurs at lower elevations for the larger sizes.

As discussed earlier (Wu, 1979a), the vertical distribution of large droplets is probably governed by the ejection height (up to about 20 cm), as their fall velocity is greater than the turbulence intensity. The boundary layer thickness becomes important for the distribution of small droplets when their suspension is governed primarily by the convective action of turbulence. The results shown in Fig. 8 indicate more detailed effects of the droplet size on the vertical distribution for the wind-velocity range 6-8 ms^{-1}. There appear to be two distinct patterns (Wu, 1982b) divided by $d = 50 \mu m$. Droplets with diameters smaller than this, as discussed earlier (Wu, 1982b), are more evenly distributed vertically; the distribution is probably governed by the boundary-layer thickness. Droplets with diameters larger than 50 μm failed to be lifted off; their vertical distribution is probably governed by the ejection height.

Horizontal Distribution of Droplets: An accelerometer was mounted on the raft to provide simultaneous records of waves along with those of droplets; see Fig. 9. In Fig. 9a, b, the data records have been divided into 1/10-s increments corresponding approximately to one fortieth of the wave period. It is evident from these time series records that the spatial distribution of droplets is nearly periodic in the wind direction. Spectral analyses were then performed on the time series of droplet occurrences along with those of waves; the results are shown in Fig. 9c, d, in which the main peak coincides with the peak of the wave spectrum.

As discussed earlier (Wu, 1979a, 1981b, 1982b), the droplets detected were probably produced mainly by bursting air bubbles, which are entrained by breaking waves. In laboratory tanks, the time needed for most air bubbles to reach the water surface is comparable to, or longer than, the period of the dominant (breaking) waves. In addition, the mean traveling distance of droplets, reported (Wu, 1979a) to be about 4m, is also much longer than the length of the dominant (breaking) waves in laboratory tanks, generally about 0.5 m or less. Consequently, the spatial distribution of droplets in laboratory tanks tends to be rather uniform.

Contrary to laboratory conditions, the period and the length of dominant, or breaking, waves in the field are much longer than both the rise time of bubbles and the mean traveling distance of droplets. Therefore, both temporal and spatial distributions of droplets in the field, as shown in Fig. 9a, b, are patchy rather than uniform. In other words, as the length of the dominant waves increases and with the whitecaps thus being separated farther apart, the droplets are concentrated near the wave crest. Therefore, the distribution of droplets varies greatly along a wave profile, deviating from the temporally averaged number of droplets. This is clear from the results shown in Fig. 9, where the occurrence of patches at the peak frequency of the wave spectrum is apparent.

ACKNOWLEDGEMENTS

I am very grateful for the sponsorship of my work provided by the Office of Naval Research (N00014-83-K-0316) and the National Science Foundation (OCE-8214998).

REFERENCES

Blanchard, D. C. and A. H. Woodcock, 1957: Bubble formation and modification in the sea and its meteorological significance. *Tellus*, **9**, 145-158.

Garratt, J. R., 1977: Review of drag coefficient over ocean and continents. *Monthly Weather Rev.*, **105**, 915-929.

Hsu, Y.-H. L., P. A. Hwang and Jin Wu, 1983: Bubbles produced by breaking waves. *Proc. Int. Sym. on Gas Transf. at Water Surface*. Cornell University.

Johnson, B. D. and R. C. Cooke, 1979: Bubble populations and spectra in coastal waters: a photographic approach. *J. Geophys. Res.*, **84**, 3761-3766.

Kolovayev, P. A., 1976: Investigation of the concentration and statistical size distribution of wind-produced bubbles in the near-surface ocean layer. *Oceanology*, (Engl. Transl.), **15**, 659-661.

Lai, R. J. and O. H. Shemdin, 1974: Laboratory study of the generation of spray over water. *J. Geophys. Res.*, **79**, 3055-3063.

Medwin, H., 1970: *In situ* acoustic measurements of bubble populations in coastal ocean waters. *J. Geophys. Res.*, **75**, 599-611.

Miyake, Y. and T. Abe, 1948: A study on the foaming of sea water, Part I. *J. Mar. Res.*, **7**, 67-73.

Monahan, E. C., 1968: Sea spray as a function of low elevation wind speed. *J. Geophys. Res.*, **73**, 1127-

1137.

Monahan, E. C., 1971: Oceanic whitecaps. *J. Phys. Oceanogr.*, **1**, 139-144.

Monahan, E. C., C. W. Fairall, K. L. Davidson and P. J. Boyle, 1983: Observed inter-relation between 10-m winds, ocean whitecaps and marine aerosols. *Quart. J. Roy. Met. Soc.*, **109**, 379-392.

Monahan, E. C. and I. O'Muircheartaigh, 1980: Optimal power law description of oceanic whitecap coverage dependence on wind speed. *J. Phys. Oceanogr.*, **10**, 2094-2099.

Preobrazhenskii, L. Yu., 1973: Estimate of the content of spray-drops in the near-water layer of the atmosphere. *Fluid Mech. (Soviet Res.)*, 2/2. 95-100.

Smith, P. M., 1981: Measurements of whitecap coverage and surface winds over the Gulf of Mexico loop current, Rep. 43, Naval Ocean R & D Activity.

Thorpe, S. A., 1983: On the clouds of bubbles formed by breaking wind-waves in deep water, and their role in air-sea gas transfer. *Phil. Trans. Roy. Soc.*, **A304**, 155-210.

Toba, Y. 1961: Drop production by bursting of air bubbles on the sea surface (III), Study by use of a wind flume. *Mem. Coll. Sci. Univ. Kyoto, Ser. A*, **29**, 313-344.

Toba, Y. and M. Chaen, 1973: Quantitative expression of the breaking of wind waves on the sea surface. *Rec. Oceanogr. Works Japan*, **12**, 1-11.

Wang, C. S. and R. L. Street, 1978: Measurements of spray at an air-water interface. *Dyn. Oceans Atmos.*, **2**, 141-152.

Wu, Jin, 1969: Wind stress and surface roughness at air-sea interface. *J. Geophys. Res.*, **74**, 444-555.

Wu, Jin, 1973: Spray in the atmospheric surface layer: Laboratory study. *J. Geophys. Res.*, **78**, 511-519.

Wu, Jin, 1975: Wind-induced drift currents. *J. Fluid Mech.*, **68**, 49-70.

Wu, Jin, 1979a: Oceanic whitecaps and sea state. *J. Phys. Oceanogr.*, **7**, 1064-1068.

Wu, Jin, 1979b: Spray in the atmospheric surface layer: review and analysis of laboratory and oceanic results. *J. Geophys. Res.* **84**, 1693-1704.

Wu, Jin, 1981a: Bubble populations and spectra in near-surface ocean: Summary and review of field measurements. *J. Geophys. Res.* **86**, 457-463.

Wu, Jin, 1981b: Evidence of sea spray produced by bursting bubbles. *Science*, **212**, 324-326.

Wu, Jin, 1982a: Comments on optimal power-law description of oceanic whitecap coverage dependence on wind speed. *J. Phys. Oceanogr.*, **12**, 750-751.

Wu, Jin, 1982b: Sea spray: A further look. *J. Geophys. Res.*, **87**, 8095-8912.

Wu, Jin, J. J. Murray and R. J. Lai, 1984: Production and distributions of sea spray. *J. Geophys. Res.* **89**, 8163-8169.

STATISTICAL ASPECTS OF THE RELATIONSHIP BETWEEN OCEANIC WHITECAP COVERAGE WIND SPEED AND OTHER ENVIRONMENTAL FACTORS

I. G. Ó MUIRCHEARTAIGH & E. C. MONAHAN
University College, Galway

1. Introduction

The major purpose of this paper is to review the various statistical methods which have been applied by the authors to evaluate the nature of the dependence of oceanic whitecap coverage (W) on 10 m elevation wind speed (U), and on a number of other environmental factors.

These methods have been applied to a number of different data sets, and to various combinations of these sets. The data sets involved were those described in:

A: Monahan (1971)
B: Toba and Chaen (1973)
C: Monahan et al. (1981) - The JASIN data set and
D: The STREX data set.

Most of the effort involved has been directed towards the derivation of the optimal W(U) expression from the several available data sets. We describe in the following sections the different methods used and present in each case the results of the application of the method to the STREX data set, and to the combined data sets. We also, in some cases, for reasons which are given, present the results of the application of the methods to the combined data sets *excluding* the STREX data.

2. Method 1 - Power-law fit by least squares

The most widely used model relating W to U (Wu, 1979; Monahan and O'Muircheartaigh, 1980) has been:

$$W = \alpha U^{\lambda} \qquad (1)$$

i.e. a power-law. This model can be fitted to the data in a number of ways, viz.,

(a) By taking a log transformation, i.e.

$$\log W = \log\alpha + \lambda \log U, \qquad (2)$$

and estimating the parameters log α and λ by a simple linear regression of log W on log U

(b) By finding the values of α, λ which minimise

$$R(\alpha, \lambda) = \Sigma_i (W_i - \alpha U_i^{\lambda})^2 \qquad (3)$$

i.e. by non-linear least squares.

Each of these methods has its advantages and disadvantages. In applying method (a), the main difficulty is presented by the existence of zero values (particularly relating to W) in the data. Also the question of the validity of the underlying assumptions arises. Method (b) suffers from the same disability in relation to the validity of assumptions, but zero values do not pose a problem. Accordingly method (b) is preferred. We present, in Table 1, the results for method (b) as applied to the STREX data set and also for the combined data sets.

TABLE 1. Parameter estimates, MSE, and SE's (in brackets) for eq. 1 fitted by least squares.

PARAMETER	Data Set: STREX	COMBINED
α	9.56×10^{-5}	2.20×10^{-5}
	(1.15×10^{-5})	(0.16×10^{-5})
λ	2.10 2	2.71
	(0.35)	(0.35)
MSE	10.05×10^{-5}	7.59×10^{-5}

E. C. Monahan and G. Mac Niocaill (eds.), Oceanic Whitecaps, 125–128.

The quantity MSE included in Table 1 is the mean squared error of the fitted model and is defined:

$$\text{MSE}\,(\hat{\alpha}, \hat{\lambda}) = R(\hat{\alpha}, \hat{\lambda})/n \tag{4}$$

It is presented because it is an intuitively reasonable measure of goodness of fit, and because it facilitates comparisons with other models. In Table 2, we present, for each of the data sets, the optimal models fitted by this method and the corresponding MSE.

TABLE 2. Optimal LS Power-Law and corresponding MSE

DATA SET	OPTIMAL λ	MSE $\times 10^{-5}$
A	3.7	3.47
B	3.3	10.06
C	2.5	8.43
D	2.1	10.05

This table indicates clearly that the latter data sets do not yield the same precision of relationship as do the earlier sets, although in all cases we have statistically significant values of α, λ.

3. Method 2 – Power-law fit by biweight fitting
Following Gaver (1979), we now fit the model of equation 1 to the data via the technique of (Robust) biweight fitting (RBF). This method utilises the least squares computation iteratively. In the course of these computations weights are automatically developed which reduce the influence of any outlying points permitting the curve to more closely approximate the main body of the data. In Table 3 we present the results for the RBF fitting of equation 1 to the STREX and the combined data sets.

TABLE 3. Parameter estimates, and MSE's for equation (1) using RBF

	Data set	
PARAMETER	STREX	COMBINED
α	1.389×10^{-4}	1.309×10^{-5}
λ	1.92	2.89
MSE	10.99×10^{-5}	7.65×10^{-5}

4. Method 3 – Generalised power-law model
In an earlier paper (O'Muircheartaigh and Monahan (1983a), we have described the fitting of a modified power-law expression

$$W = \alpha U^{\lambda} - \beta \tag{5}$$

to data set C, and to data sets A, B and C combined. This model was fitted primarily in an attempt to estimate U_B, the Beaufort velocity, the wind speed below which whitecapping does not occur. This model was fitted via a process of non-linear least squares, i.e. by minimising

$$R(\alpha, \lambda, \beta) = \Sigma (W_i - \alpha U_i^{\lambda} - \beta)^2 \tag{6}$$

This estimation implicitly involved the assumption of homoscedasticity in the data i.e. the assumption that in essence all the observations were equally reliable. It was shown that this assumption did not hold in this particular case and the necessary modifications were made to the model. We describe here the fitting of the same modified model to (a) the STREX data and (b) data sets A, B, C and D combined. Essentially, having established that there was evidence of heteroscedasticity in the data, the problem is to determine the form of the heteroscedasticity. When no assumptions are made about the nature of the heteroscedasticity, we have to rely entirely on the sample information and estimate the variances of the residuals from the data. Since, for each specific value of U, the value of the variance of the residuals may be different, we need several observations on the dependent variable W for each U_i in order to estimate this variance.

Since our data consisted of single observations at different values of U, it was not possible to determine the form of the dependence of the variance of the residuals on U from the data. However, an analysis of the dependence of the absolute values of these residuals on the independent variable U revealed that their absolute value was strongly dependent on U, and that this dependence had a particular form.

In this paper, we present the results of the application of this methodology to the STREX data, an application which produced results similar to, though somewhat different from, those obtained for the other data sets.

In fitting the model described in equation 5, it was deemed appropriate to scale the wind speed by dividing the U-values by 10. Hence the model actually fitted was

(7) $$W = \alpha(U/10)^{\lambda} - \beta$$

Table 4 gives initial estimates of the parameters and their standard errors for the STREX data.

TABLE 4. Initial estimates, SE's for eq. 7 fitted to STREX data

PARAMETER	ESTIMATE	SE
α	6.87×10^{-3} 3	3.28×10^{-3}
λ	3.12 0	0.93
β	-4.43×10^{-3}	2.95×10^{-3}

An analysis of the residuals (r_i) for this data set revealed that there was a positive dependence on U. Exploratory analysis indicated that a good description of this dependence was given by

(8) $$E(|r_i|) = \alpha^* + \beta^* (U/10)$$

Table 5 gives the parameter estimates and SE's for this model fitted to the STREX data.

TABLE 5. Parameter estimates, SE's for eq. (8) fitted to STREX data

PARAMETER	ESTIMATE	SE
α^*	-6.36×10^{-4}	2.08×10^{-3}
β^*	7.68×10^{-3}	1.89×10^{-3}

It is clear from Table that E (|r|) is essentially proportional to U. It should perhaps be noted here that for the other data sets, as reported in O'Muircheartaigh and Monahan (1983a), the form of this dependence was found to be quadratic rather than linear.

Given that E(|r|) is proportional to U, it follows that the variance of the residuals is proportional to U^2; we now perform a weighted least squares by minimising

$$R^* (\alpha, \lambda, \beta) = \Sigma (W_i - \alpha U_i^{\lambda} - \beta)^2 / U_i^2 \qquad (9)$$

The results are presented in Table 6.

TABLE 6. Weighted estimates of parameters, SE's for eq. 5 fitted to STREX data

PARAMETER	ESTIMATE	SE
α	1.36×10^{-2}	3.3×10^{-3}
λ	1.73 0	0.53
β	6.42×10^{-4}	2.43×10^{-3}

Hence the final model for STREX is

$$W = 0.0136 (U/10)^{1.73} - 0.000642 \qquad (10)$$

The corresponding estimate U_B of U_B is obtained by putting W = 0 in equation (10) which yields

$$U_B = 1.71 \text{ ms}^{-1}$$

5. Method 4 – The Box-Cox transformation

In an earlier paper (O'Muircheartaigh and Monahan (1983b)), we have applied the Box-Cox (1964) family of transformations to data set C and data sets A, B, C combined to determine an optimal functional form (in a sense defined) relating W to U. The technique has also been applied to the same data sets augmented by the inclusion of two additional variables, air temperature (T_a) and water surface temperature T_w. It was shown that, using this technique, a dependence of W on T_a and T_w is detected which more straightforward methods fail to reveal.

In this paper we describe the method briefly and present the results of its application to the STREX data. The method essentially involves the application of the transformation

$$Z(\lambda) = (Z^{\lambda} - 1)/\lambda \qquad (11)$$

to some or all of the variables in a regression relationship. The choice of the power transformation parameter (λ) is made based on the maximisation of a likelihood, which in turn is obtained by making certain assumptions about the data. A full description may be found in either O'Muircheartaigh and Monahan (1983) or in the seminal Box and Cox (1964).

An advantage of this method over those previously outlined, is that it facilitates the construction of confidence intervals for the power-law or power transformation parameters. The relevant theory is given in the papers referred to above. Furthermore, the values of any of the λ's may be subjected to constraints (such as equality among a subset of them, or a particular λ being fixed etc.) so that any specific subset of functional forms may be fitted.

In our case we fitted the model (using the notation of equation (11)

$$W(1) = \alpha + \beta U(\lambda) \tag{12}$$

to the STREX data. The results are presented in Table 7.

TABLE 7. Box-Cox results, STREX data

Maximum likelihood estimate, λ	3.2
95% confidence interval for λ (0.9,	(0.9, 5.1)

From Table 7, we see that the optimal value of λ is not very different from that obtained by other methods or for other data sets. However, the additional information given here is the confidence interval, which indicates that no value of λ between approximately 1 and 5 inclusive can be rejected for this data set. This is due to the inherently large scatter of this data set, compared to the other data sets considered here. A reference to Table 2 confirms this point.

For information, we present in Table 8 the corresponding results for other data sets.

TABLE 8. Optimal, 95% confidence intervals for eq. (12) fitted to other data sets.

Data Set	Optimal λ	95% confidence interval
A	2.14	(1.5, 3.1)
ABC	3.03	(2.1, 3.9)
ABCD	3.08	(2.2, 4.0)

ACKNOWLEDGEMENT

This work was supported by ONR grant No. N00014-82-G-0024.

REFERENCES

Box, G. E. P. and Cox, D. R. (1964). An analysis of transformations. *J. R. Statist. Soc.* B25, 211-252.

Gaver, D. P. (1979). Statistical Methods of Probable Use for understanding remote sensing data. Technical Report NPS55-79-020 Naval Postgraduate School, pp. 1-35.

Monahan, E. C., 1971. Oceanic Whitecaps, *Journ. of Physical Oceanography*, **1**, pp. 139-144.

Monahan E. C., and I. G. O'Muircheartaigh, 1980. Optimal Power-Law Description of Oceanic Whitecap Coverage Dependence on Wind Speed. *Jour. of Physical Oceanography*, **10**, pp. 2094-2099.

Monahan, E. C., I. G. O'Muircheartaigh, and M. P. Fitz-Gerald, 1981. Determination of surface wind speed from remotely measured whitecap coverage, a feasibility assessment, *Proceed. of an EARSel-ESA Symp., Applic of Remote Sensing Data on the Continental Shelf, Voss, Norway, 19-20 May 1981*, European Space Agency SP-167, pp. 103-109.

O'Muircheartaigh, I. G., and E. C. Monahan, 1983a. Aspects of oceanic whitecap coverage dependence on wind speed: heteroscedasticity in the data and the estimation of the Beaufort velocity, *Proceedings, Second International Conference on Statistical Climatology, Lisbon* (in press).

O'Muircheartaigh, I. G. and E. C. Monahan (1983b). Use of the Box-Cox transformation in determining the functional form of the dependence of oceanic whitecap coverage on several environmental factors. *Proceedings, Eighth Conference on Probability and Statistics in Atmospheric Sciences*. Hot Springs, USA (in press).

Toba, Y., and M. Chaen, 1973. Quantitative expression of the breaking of wind waves on the sea surface, *Records of Oceanographic Works in Japan*, **12**, pp. 1.11.

Wu, J., 1979. Oceanic whitecaps and sea state, *Journal of Physical Oceanography*, **9**, pp. 1064-1068.

CHARACTERISTIC FEATURES OF A WIND WAVE FIELD WITH OCCASIONAL BREAKING, AND SPLASHING DROPLETS AT HIGH WINDS

MOMOKI KOGA
Department of Geophysics, Faculty of Science,
Hokkaido University, Sapporo 060 Japan

The characteristics of intermittent and sporadic wind wave breaking and the splashing of droplets along a wave are investigated through wind-wave tank experiments (wind-wave tank: 20m long, 0.6m wide, 1.2m high, with fresh water 0.6m deep). For investigating the characteristics of wind wave breaking, the intensity of breaking of individual wind-wave crests has been related to the parameters of wave form (e.g. height, period, steepness and skewness), and their process of change has been studied (reference wind speed $15 ms^{-1}$, fetch 16m). Distributions of the wave form parameters are different for breaking and nonbreaking waves. Fully breaking waves seem to be ones for which the relation $H \propto T^2$, between the individual wave height H and period T, holds. The condition of breaking is more complicated than a simple criterion of Stokes' limit. The quantities wave height and steepness for a breaking wave are not always larger than these quantities for a nonbreaking wave. This may suggest that there is an overshooting nature to the breaking wave. The wave form parameters change cyclically during wave propagation. The period of the cycle in the present case is longer than 4 wave periods. An intermittency in the wave breaking is associated with this cyclic process. Roughly speaking, two or three successive breaking-waves sporadically exist among a series of non-breaking waves moving downwind. The direct production of droplets from breaking individual waves is described, based on experimental results. The spreading of the droplets (diameter >0.8mm) under the direct action of the wind field along a breaking wind wave is also described, based on the comparison of the actual distribution of droplet velocities with the wind field (reference wind speed: $16 ms^{-1}$, fetch 16m). Both the movement of the droplets and that of styrofoam flakes used as tracers for the air flow visualization are measured by a simple photographic technique involving the use of multi-colored overlapping exposures. This comparison gives a concrete picture of the acceleration of droplets by the wind.

I. Introduction

Wind wave breaking occurs intermittently and is distributed sporadically. It results in unsteady and nonuniform distributions of whitecaps, and hence of bubbles and spray droplets (e.g. Donelan et al., 1972; Thorpe and Humphries, 1980). Therefore, the effects of whitecaps on air-sea interaction processes, such as the exchange of momentum and heat, and on the scattering of light and microwave radiation, cannot be precisely estimated without taking account of this unsteady and nonuniform nature of whitecaps. However, in most cases, only the overall whitecap coverage or the overall rate of wave breaking has been the subject of measurement.

Since whitecapping itself is a phenomenon which is associated with individual wave crests, the unsteady and nonuniform nature mentioned above should be investigated in relation to the characteristics of individual waves. There are a few measurements of the breaking of individual waves, and of the associated bubble entrainment and direct splashing of droplets (e.g. Toba, 1961; Toba et al., 1975; Koga, 1981, 1982). The life time of individual whitecaps was also measured by some investigators (e.g. Monahan, 1969, 1971; Kondo et al., 1973; Thorpe and Humphries, 1980), and the mean characteristics of individual wind waves were studied by Toba (1978) and Tokuda and Toba (1981). The internal flow pattern of individual waves in nonbreaking conditions was investigated by Okuda (1982a, b, c). However, the re-

E. C. Monahan and G. Mac Niocaill (eds.), Oceanic Whitecaps, 129–145.

lationship between the individual whitecap and the characteristics of the individual wave, including its unsteady nature, is still not clear.

The splashing of droplets is one of the most distinctive phenomena accompanying the breaking of individual waves. As for the velocity of the droplets; its mean vertical distribution for the droplet size of about 0.2 mm diameter was measured by Wu (1973), and the distribution of droplet velocity along the wave profile for droplets of $d > 0.81$ mm by Koga (1981) (hereafter referred to as KG). However, there have been no studies which measured both the droplet velocity and the wind field along a wave profile. Therefore, the effect of the wind field on the droplet spreading process along a wave profile is still not clear.

The present paper, firstly, introduces wave form parameters, which describe the individual wave characteristics such as its time and space scales and shape, and experimentally investigates the relationship between these parameters and the whitecapping phenomena, including the unsteadiness and sporadic distribution of whitecaps, in a wind-wave tank (section 2).

Secondly, in section 3, the process of direct droplet production from breaking wind waves is described as is the distribution of the droplet velocities along a wave profile, which had been already measured by KG. The wind field along a wave profile is measured by a flow-visualization technique in the same wind condition as pertained in KG. By comparing the distribution of the droplet velocities and the wind field along a characteristic wave, the effect of the air flow on the droplet spreading process is elucidated.

2. *Characteristics of the wind wave field with occasional breaking*

2.1. *Experiment and analysis*

The wind-wave tank used in this work was 20m long, 0.6m wide and 1.2 m high, with fresh water of 0.6 m depth, and belongs to the Department of Geophysics, Tohoku University. The relationship among the wind wave conditions at various fetches F and the reference wind speed at the center of the flume U_r, has been described in detail by KG (see Table 1). For the case of a fixed fetch of 16m, the critical wind speed for the occurrence of wave breaking with bubble entrainment was about U_r = 12m s^{-1}. The present experiment has been conducted in the high-wind condition $U_r = 15$m s^{-1}.

A schematic picture of the experimental apparatus is shown in Fig. 1. The intensity of breaking of an individual wind wave is classified

Table 1. Wind wave conditions at various fetches and wind speeds (from Koga, 1981)

Reference wind speed U_r (m s^{-1})	Fetch F (m)	Friction velocity of air u_* (cm s^{-1})	Roughness length Z_0 (cm)	Significant wave: Height $H_{1/3}$ (cm)	Significant wave: Period $T_{1/3}$ (s)	Significant wave: Mean phase speed of wave (cm s^{-1})
14.0	16	124	0.26	7.5	0.61	110
15.0	16	148	0.40	8.2	0.68	116
16.0	16	197	0.80	9.0	0.73	125
16.0	12	–	–	7.7	0.62	119
16.0	9	–	–	6.6	0.57	101
16.0	5	–	–	4.3	0.44	94

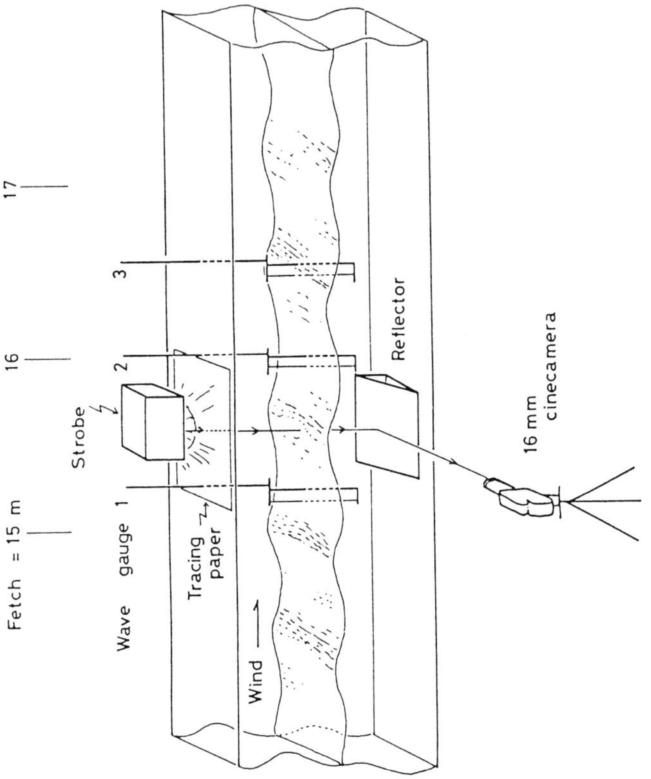

Fig. 1. Schematic picture of experimental arrangements.

based on the amount of entrained bubbles. For the precise measurement of this quantity, the wave surface was photographed from below by the use of a 16-mm cinecamera and a mirror placed at the bottom of the tank. A strobotron illuminates the tank from the top through a sheet of tracing paper, and distinct outlines of the entrained bubbles and the characteristic pattern of the water surface were obtained as shadowgraphs. The elevation of the wave surface was simultaneously measured by three wave gauges installed at the three fetches before and behind the photographed area (40 cm X 30 cm). The fetches for the wave gauges 1, 2 and 3 are 15.16 m, 15.92 m and 16.45 m, respectively. The times of the photographs were recorded simultaneously on the wave records.

Wave form parameters

The definition of an individual wind wave was made from wave records as illustrated in Fig. 2 (a) (the zero-crossing trough to trough method after Tokuda and Toba, 1981), and its wave form parameters were determined by the use of the wave records according to Tokuda and Toba (1981). These parameters are defined in the following list:

Wave height	$H \equiv (h_1 + h_2)/2$
Wave period	$T \equiv t_1 + t_2$
Phase speed of wave	$C \equiv l/\Delta t$
Wave length	$L \equiv TC$
Steepness	$\delta \equiv H/L$

Here h_1 is the vertical distance between the crest and the trough on the downwind side and h_2 is the vertical distance between the crest and the trough on the upwind side, t_1 is the time lag between the passing of the trough of the downwind side and that of the crest and t_2 the time lag between the crest and the trough of the upwind side. Δt is the time lag between the passage of a wave peak past two wave gauges whose fetches differ by a distance l. These quantities are illustrated in Fig. 2 (b).

In addition, the present study has introduced the following nondimensional quantities,

$$R_1 \equiv t_1/T, \qquad R_2 \equiv (t_3 + t_4)/T$$
$$R_3 \equiv t_3/t_1, \qquad R_4 \equiv t_4/t_2$$

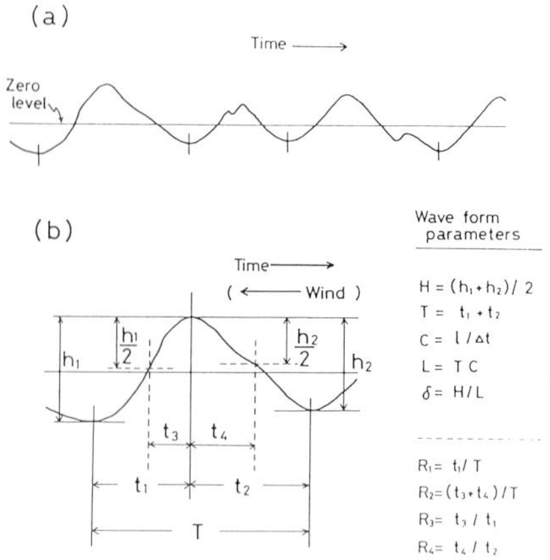

Fig. 2. Definition of individual waves and wave form parameters on a wave record.
(a) A wave crest between adjacent vertical segments is the unit of the present analysis and is called an individual wave (zero-crossing trough-to-trough method).
(b) Definition of wave form parameters.

where t_3 is the time lag between the passing of the surface point of height $h_1/2$ from the trough of the downwind side and the passing of the crest, and t_4 the time lag between the passing of the crest and that of the surface point of height $h_1/2$ from the trough of the upwind side. These are also illustrated in Fig. 2 (b). R_1 is an indication of the asymmetry or the skewness of the wave shape in the horizontal direction and R_2 is an indication of the kurtosis. R_3 and R_4 indicate the inclinations of downwind slope and upwind slope, respectively.

Wave types classified by intensity of breaking

Waves, which were continuously photographed by a 16-mm cinecamera, were classified into four types by the amount of entrained bubbles as follows:

Type 1: Fully breaking
Type 2: Partly breaking
Type 3: Nonbreaking (1)
Type 4: Nonbreaking (2)

A clear definition of each of the four types is given in the following. Type 1 is a wave in which bubble entrainment is seen in the whole

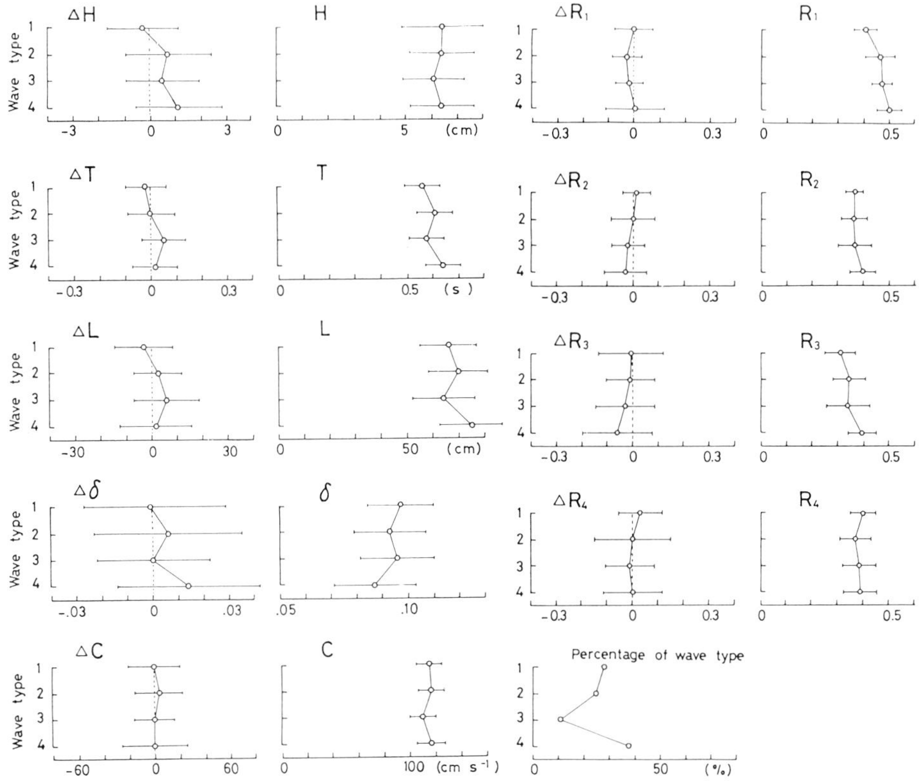

Fig. 3. Wave form parameters and their difference between the 1st and 2nd wave gauges for the four wave types. Each of the parameters with the mark Δ indicates the difference obtained by subtracting the value at the 1st wave gauge from the value at the 2nd wave gauge. A parameter without the mark Δ indicates the mean of the two values. Phase speed of wave C was calculated from the phase lag between the two wave records at the 1st and 2nd wave gauges, and the difference ΔC was given by $\Delta C = C' - C$, where C' was calculated from the phase lag between the records at the 2nd and 3rd wave gauges.

width of the tank for the whole time when the wave is passing the photographed area. Type 2 is a wave in which bubble entrainment is seen over part of the tank width or for part of the time interval, during which the wave was photographed. Types 3 and 4 are both waves without bubble entrainment, but they are distinguished from one another by the criterion of whether or not they have a clear convergent line on the downwind wave slope (e.g. Koga, 1982). Type 1 and Type 4 are the two extreme stages with and without breaking, respectively. Type 2 and Type 3 represent the intermediate states.

In the case when an individual wave crest recorded at the 1st wave gauge cannot be identified in the record of the 2nd wave gauge, or, even if the same crest is identified, when at least one of the two crests on the records is so distorted that the form parameters may not be clearly measured, the individual waves were excluded from the analysis. The ratio of the excluded waves to the total ones (164 waves) was about 25%. The relative occurrence of each of the four wave types is 27.6%, 24.4%, 10.6% and 37.4%, for the types 1, 2, 3 and 4, respectively.

2.2. *Results*

Relation between wave types and wave form parameters

Fig. 3 shows the wave form parameters and their difference(Δ) with fetch for each of the four wave types. For the parameter R_1 the difference between the two extreme types (Type 1 and Type 4) is distinctive. For the other parameters, even the differences of these parameters between the two extreme types (Type 1 and Type 4) are somewhat obscured by the overlap of the standard deviations (expressed by the horizontal line segments extending from the mean values). This is due to the intrinsic irregularity of the individual wind waves, rather than to the scantiness of the wave data (123 waves of all four types). However, the differences among the mean values for different types will be significant.

For Type 1, H is the same as that for Type 4, and T and L are a little smaller than those for Type 4. These facts result in δ for Type 1 being larger by about 0.01 than δ for type 4. R_1 for Type 1 is 0.4 which is smaller than the corresponding value for Type 4, and R_2 and R_3 are also a little smaller. These values indicate that individual waves of Type 1 are skewed forwards and have a somewhat sharp crest. On the other hand, for Type 4, the wave shape is not skewed forwards as $R_1 \approx 0.5$. Therefore, the shape of individual waves is different in the two extreme cases of Types 1 and 4.

Phase speeds of individual waves are not so much different in the two wave types and their changes with fetch are nearly zero for both of the wave types. However, if we note the fact that L and T for Type 1 are smaller than those for Type 4, then we can say that waves of Type 1 have a relatively large phase speed for their smaller L and T compared with the case of Type 4.

Figs. 4(a) and (b) are the correlation plots of these parameters where the wave types are indicated with various symbols. Fig. 4(a) shows

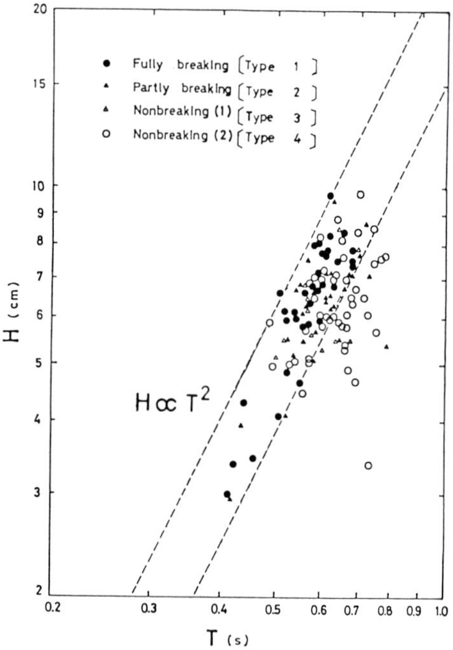

Fig. 4. (a) Correlations between the wave form parameters: H-T correlation. Dashed lines indicate a line defined by $H \propto T^2$. Four wave types are indicated by four respective symbols.

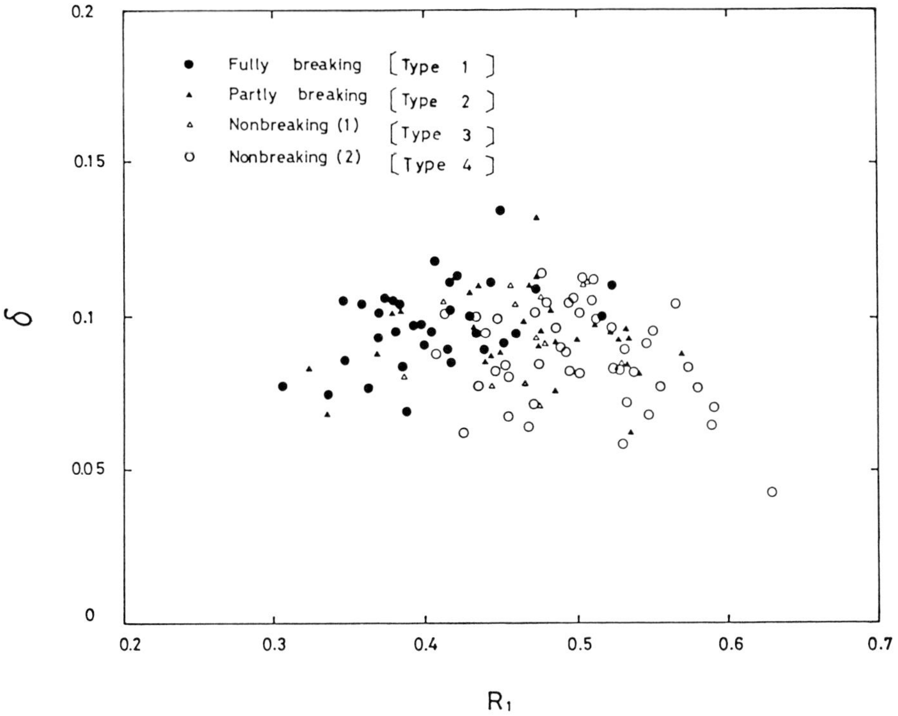

Fig. 4. (b) Correlations between the wave form parameters: δ-R_1 correlation.

the H-T correlation. Individual waves of Type 1 (solid circles) are distributed approximately within the area enclosed by two dashed lines defined by $H \propto T^2$. Fig. 4 (b) shows the δ-R_1 correlation. A positive correlation is seen for Type 1 and a weak negative correlation for Type 4.

Change of wave form parameters during propagation and the relation between wave types

Let us examine the change of the wave form parameters using the data obtained at the 1st and 2nd wave gauges (parameters with the mark Δ in Fig. 3). For Type 1, H, T and L decrease on average, keeping δ, R_1, R_2, and R_3 nearly constant. This indicates that waves of Type 1 (breaking waves) are in a decreasing stage. For Type 4, on the other hand, H, T and L increase, on average, as these waves propagate. Especially, the increase of H is distinctive. This increase is the main cause of the marked increase of δ. At the same time, the decreases of R_2 and R_3 indicate a sharpening of the crest. However, waves of Type 4 seem not to be skewed forwards and have no clear tendency to become skewed, as can be seen from the R_1 of about 0.5 and its small change.

Fig. 5 shows a change of mean values of the representative parameters H, T, δ and R_1 with increasing fetch, by adding the data from the 3rd wave gauge. The changes are shown for wave Types 1 and 4, where the wave type was judged from the photographs taken between the 1st and 2nd wave gauges. For Type 4, T scarcely changes with fetch, but H and δ are increasing with fetch up to near an upper limit. R_1 begins to decrease at the fetch of the 2nd wave gauge. On the other hand, for Type 1, T decreases and then increases a little from its lowest value. H and δ are decreasing with fetch

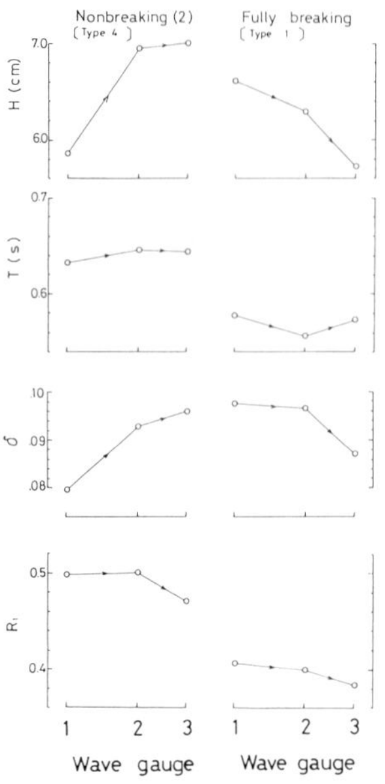

Fig. 5. Change of wave form parameters during wave propagation. The changes of the mean values of wave form parameters H, T, δ, and R_1 with increasing fetch (for wave gauges 1-3) are shown for the waves of Types 1 and 4.

from their maximum values. R_1 is nearly at the lower limit. Therefore, the four wave types can be interpreted as the stages in a cyclically changing process involving an individual wind wave, where Types 1 and 4 will be the two extreme stages and Types 2 and 3 will be the transient stages between them. Distinct changes of the parameters T and R_1 will take place in the stages represented by Types 2 and 3.

The cyclic process results in an intermittency in the wave breaking. Fig. 6 schematically shows the process on H–T and δ–R_1 correlation plots. Actually, since waves in a wind-wave tank develop with increasing fetch on average, the cyclic process may not be a closed one. Each of the individual waves does not necessarily undergo the whole process without losing its identity because of the intrinsic irregularity of these waves. Therefore, the cycle shown in Fig. 6 is a statistical one.

The period of the cyclic process will be much longer than two times the time interval, t_{13}, in which the mean individual wave passes through from the 1st wave gauge to the 3rd wave gauge, as can be inferred from the change of H or δ in Fig. 5. The quantity t_{13} is roughly estimated using l_{13} (the distance between the 1st and the 3rd wave gauges) and C (the mean phase speed), as follows:

$$t_{13} \approx l_{13}/C = 129\ \mathrm{cm}/115\ \mathrm{cm\ s^{-1}} = 1.1\ \mathrm{s}$$

Therefore, the period of the cyclic process will be much longer than 2.2 s in the present case. This corresponds to a time interval much longer than 4 wave periods.

Variation of wave types with fetch

The variation with time of wave types has been determined in the present study through the use of a 16-mm cinecamera whose lateral field of view extends to about a half of the wavelength. The transition of a wave type hardly occurs in one wave period, as mentioned above. Therefore, the spatial variation of wave types between two consecutive individual waves will not be different from the timewise variation of type of a single wave. We can examine the variation of wave types with fetch using the present experimental data.

Table 2 shows the percentage frequency of occurrence of a wave type for a wave just before (downwind side) or behind (upwind side) a remarked wave of wave Type 1 or 4. For the remarked wave of Type 1, the adjoining wave of its downwind side has a large probability of being Type 1 compared to the overall mean occurrence of Type 1 waves and has a relatively small probability of being of type 4. The same tendency is seen in the frequency of occurrence of the wave types for the wave of its upwind side. Therefore, waves of Type 1 have a wave of the same type before or behind them more frequently than expected from the overall mean frequency of occurence of this type wave. Similarly, waves of type 4 have a wave of type 4 before or behind them more frequently than would be expected from the mean frequency of occurrence of these waves.

These results imply that some number of con-

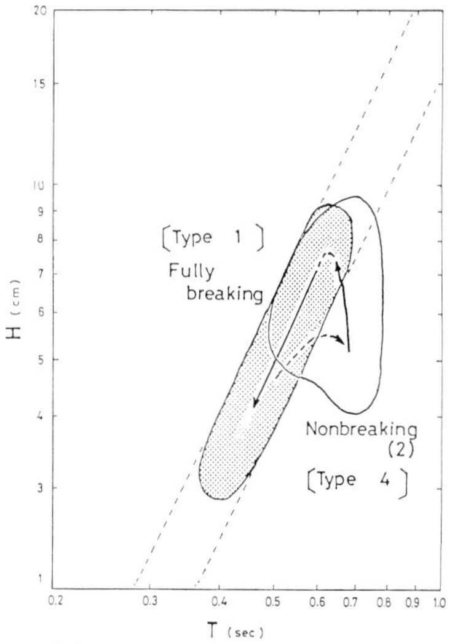

(a)

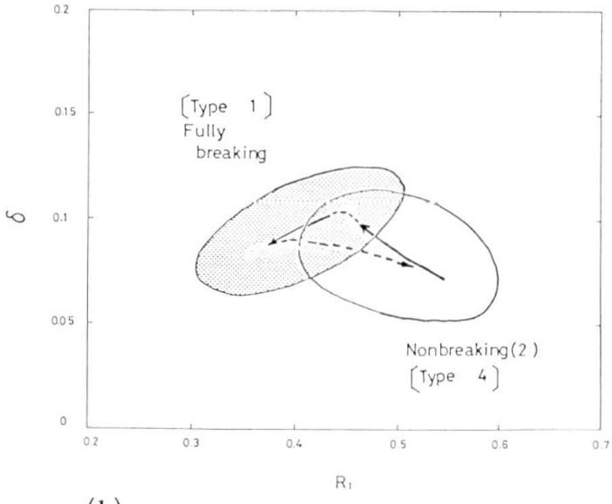

(b)

Fig. 6. Cyclical change of individual waves schematically represented in correlation plots. Arrows indicate the direction of the cycle.
(a) Cycle in the H–T correlation plot. Dashed lines indicate the relation given by $H \propto T^2$.
(b) Cycle in the δ–R_1 correlation plot.

secutive breaking waves (as represented characteristically by Type 1) sporadically exist among a series of nonbreaking ones (as represented characteristically by Type 4). Intermediate wave types, Types 2 and 3, will modulate the above pattern. At the same time, it should be remembered that each individual wave in the pattern changes its wave type with time according to the cyclic process. Therefore, the actual arrangement of wave types shifts relative to the propagating waves, keeping the sporadic arrangement of breaking waves.

2.3. Discussion

The wave form parameters are useful indicators for investigating the characteristic differences between breaking and nonbreaking waves and the cyclically changing character of individual waves. The breaking of individual waves occurs intermittently and the breaking waves exist sporadically in accord with the cyclically changing process.

The present results show through a detailed inspection that the conditions of the breaking are more complicated than a simple criterion of Stokes' limit. The values of H and δ of a break-

Table 2. Percentage frequency of occurrence of a wave type before or behind the remarked wave whose type is 1 or 4

		Frequency of occurrence of a wave type (%) 1	2	3	4
overall mean		27.6	24.4	10.6	37.4
Remarked wave type 1	down-wind wave type	36.1	30.6	11.1	22.2
	up-wind wave type	36.1	22.2	11.1	30.6
Remarked wave type 4	down-wind wave type	17.5	28.1	10.5	43.9
	up-wind wave type	14.0	31.6	14.0	40.4

ing wave are not always a larger than those of a nonbreaking wave, and their distributions for both the breaking waves and the nonbreaking waves rather overlap each other. This may suggest that there exists an overshooting nature to the breaking process or recovering process as it relates to an individual wave.

Such a complexity may be caused partly by a direct coupling of the individual waves with the air flow over them. Okuda (1982a, b) has shown that a vorticity layer caused by the shearing stress is concentrated near the crest and that growth and decay of the layer are closely related to the cyclically changing character of individual waves. The coupling will be very effective in strong wind conditions such as pertained in the present study ($u_* = 148$cm s^{-1}). It is an indication of the coupling of the waves with the air flow that the overall mean, profile of the individual waves is skewed forward, as can be seen in the present experiment. The steepness of the fully breaking waves ranges rather widely from 0.07 to 0.13 and its mean value (about 0.1) is much smaller than the Stokes' limit 0.142, as can be seen in Fig. 6(b). This may be attributed to the fact that the condition of breaking (especially when it involves the formation of bubbles) is strongly controlled by the wind-induced local surface current as pointed out by Koga (1982). For short wind waves like those encountered in the present case, most of the bubbles are formed by entrainment caused by the convergence of the surface current on the downwind slope near the crest.

In this paper, we have treated the individual waves as units of a wave field. The similarity structure of these waves will be the result of their direct coupling with the air flow over them as discussed by Toba (1978) and Tokuda and Toba (1981). The present study has shown that each of the wave types (especially Types 1 and 4) has associated with it a characteristic range of each of the wave form parameters. This suggests that the similarity structure is a little different for the various wave types in relation to the cyclically changing character of the individual waves. As a matter of fact, in Fig. 4(a), only the waves of Type 1 (solid circles) are distributed approximately within the area enclosed by two dashed lines defined by $H \propto T^2$. The similarity relation $H \propto T^2$ for fully breaking (Type 1) waves can be derived dimensionally, if H is not dependent on the friction velocity u_*, but on g (the acceleration of gravity) and T only. Namely,

$$H = AgT^2$$

where A is a nondimensional universal constant. A is about 2.0×10^{-2} for the mean of the two dashed lines in Fig. 6 (a).

3. Splashing of droplets along a wave

3.1. Direct production of droplets and the droplet velocity distribution along a wave

The mechanism of the direct production of droplets from breaking wind waves was investigated by KG using the same wind-wave tank as that used in Section 2, for the wind condition $U_r = 16$m s^{-1} and at a fetch of 16m. The micro-scale configurations of the surface of breaking wind waves and their process of change were investigated with the use of the multi-colored overlapping exposures technique (MOET). The MOET (two-color case) shadowgraph photographic system is shown in Fig. 7. The most conspicuous micro-scale phenomenon was the appearance of small projections, mainly on the

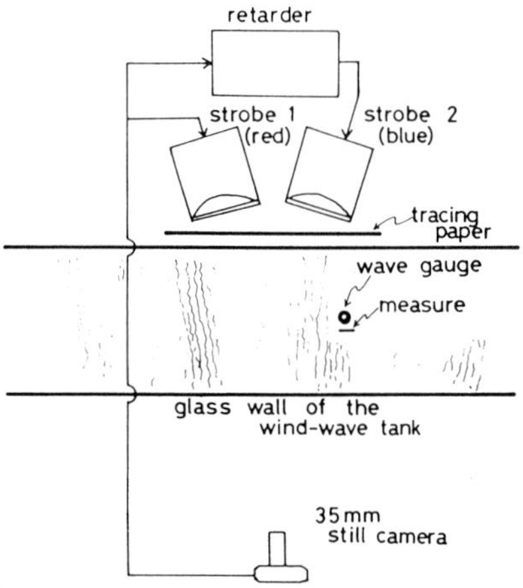

Fig. 7. Schematic representation of the photographic arrangements (top view) for the multi-colored overlapping exposure technique (MOET). (From Koga, 1981).

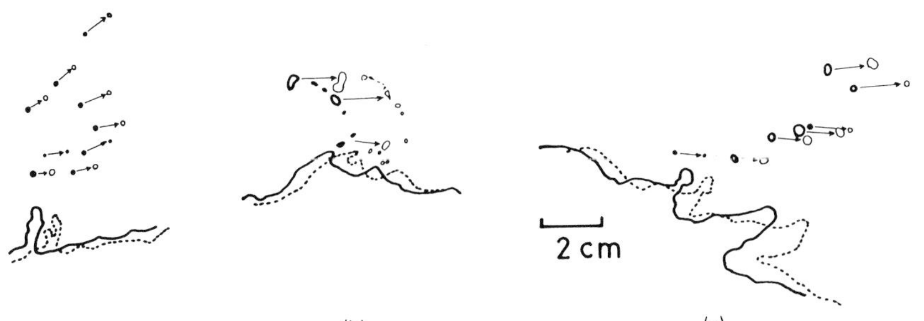

Fig. 8. Pictures of direct production of droplets traced from MOET (two-color) photographs. Thick solid line indicates first-time images, and dotted line or thin solid line indicates second time images. (a) Upwind slope near the crest, fetch F = 5 m, reference wind speed U_r = 16 m s^{-1}, flashing time interval t = 6 ms. (b) At the crest, F = 16 m, U_r = 16 m s^{-1}, t = 5 ms. (c) Downwind slope near the crest, F = 9 m, U_r = 16 m s^{-1}, t = 5 ms. (From Koga, 1981).

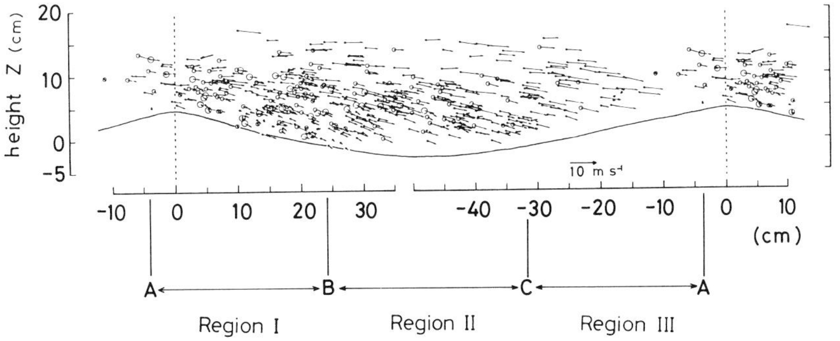

Fig. 9. The droplet velocity distribution along a representative wave (F = 16m, U_r = 16m s^{-1}) obtained by Koga (1981), with slight modifications. Three kinds of circle are used to indicate droplet size; size 1 (droplet diameter d = 0.81mm–1.74mm), size 2 (1.74mm–3.76mm), size 3 (3.76mm–8.10mm). Arrows indicate the droplet velocity vectors in a coordinate system moving with the phase speed of the wave. The phase of the wave is divided into three regions as shown at the bottom of the figure; from crest to downwind slope (A–B), trough (B–C) and upwind slope (C–A).

crest of waves, and the subsequent stretching and breaking of these projections into small droplets. This sequence of events represents the process of direct production of droplets by breaking waves. Fig. 8 shows examples of the direct production of droplets traced from MOET photographs. These droplets then spread along the wave.

The experimentally determined velocity distribution of the spreading droplets first described in KG is summarized briefly here. The droplet velocity distribution along a representative wave profile is shown in Fig. 9, which is a slightly modified version of the original figure (Fig. 6 in KG). Fig. 9 and the other results described in KG indicate that the mean movements of splashing droplets have the following features: (1) Most of the droplets larger than 0.81 mm in diameter are produced on the downwind slope near the crest, with initial

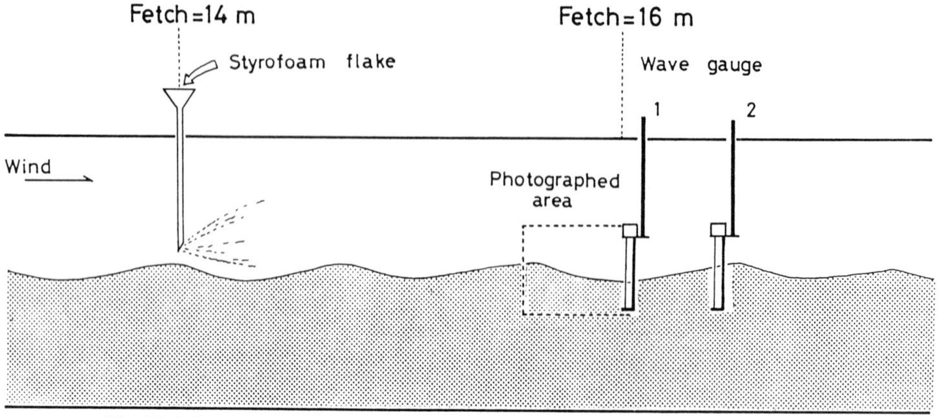

Fig. 10. Schematic figure of the experimental arrangement for the air flow visualization in the wind-wave tank.

speeds of the same order of magnitude as, or somewhat larger than, the phase speed of the wave. (2) They spread upward and forward acclerating due to the drag of the mean wind, but most of them fall on the upwind slope near the crest of the next wave as a consequence of the acceleration of gravity. (3) The short life span of these droplets may be due to their insufficient acceleration by the mean wind.

3.2. Experiment for air flow visualization over breaking wind waves

Air flow visualization over breaking waves in the wind-wave tank was realised by the use of styrofoam flakes (SF flakes). The movement of flakes was photographed by use of MOET. In the present experiment, two colors, red and blue, were used for MOET. The flashing time interval was selected in the range from 0.5 ms to 1.0 ms. The velocity of an SF flake was calculated by the positions of the pair of images, blue and red, and the wind velocity was estimated by subtracting the terminal velocity of a falling SF flake in stagnant air from the calculated velocity of the SF flake. These flakes were prepared by grating a lump of styrofoam, whose specific gravity is 0.013. Their shape is far from that of a sphere and rather like sawdust. Flakes larger than about 2.7 mm in their maximum length L were eliminated by the use of a sieve. For the precise identification of flakes without any confusion with water droplets in the photographs, only flakes greater than about 0.8 mm in L were used. The terminal velocity of falling SF flakes in stagnant air measured by the MOET ranges from 35 cm s^{-1} to 50 cm s^{-1}, with a mean value of about 40 cm s^{-1}.

The experimental arrangement for the air flow visualization is shown in Fig. 10. SF flakes were continuously scattered into the flume at a point 2.0 m upwind from the field of view of the camera which was at a fetch of 16 m. Most of the scattered flakes reached the photographed region without falling onto the wave surface. Each photograph covered a region 35 cm (vertical) × 52 cm (horizontal). Wave gauges and a scale were set up at the downwind end of the photographed area (fetch 16 m) as shown in Fig. 10, and a mark designating the time when the photographs were taken was also recorded simultaneously on the wave records.

The equality of the flake velocity with the air velocity in the photographed region will be examined by comparing the flake velocities with the wind profile. In Fig. 11 the mean velocity profile of the flakes obtained by averaging at every height all the flake velocity data is compared with a wind profile measured by a pitot-static tube in the case where the reference wind speed, U_r, was 16 m s^{-1}. The data on the

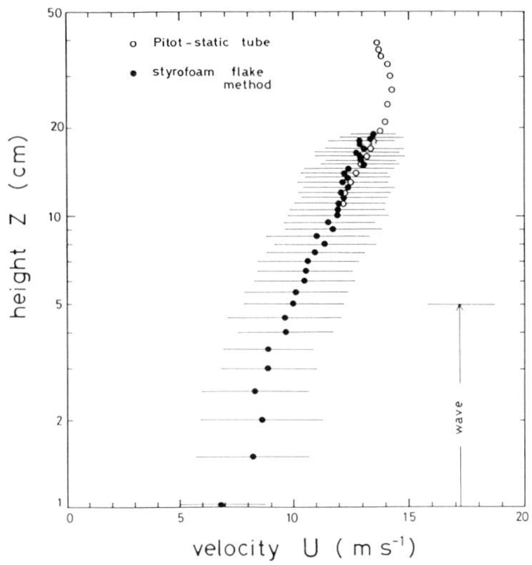

Fig. 11. Overall mean wind profile over wind waves obtained by the flow visualization technique and that measured by a Pitot-static tube (F = 16m, U_r = $16 m s^{-1}$). The horizontal bars indicate the standard deviation of the wind velocity as determined by the flow visualization technique. A vertical arrow at the right of the figure indicates the vertical region lower than the wave crest.

flake velocities was obtained by using all of the analysed 90 photographs (35 cm X 52 cm) which were taken randomly with time. The profile obtained by use of the pitot-static tube is the profile averaged in the cross wind direction for a reasonable comparison with the mean velocity profile of the flakes. Above 11 cm, these two profiles almost agree with each other. This means that SF flakes move with nearly the same velocity as that of the local wind at the experimental site. However, at heights lower than 11 cm, the wind profile could not be measured by the pitot-static tube due to the clogging of its head with droplets.

Examples of the instantaneously visualized air flow over representative waves are shown in Figs. 12(a) and (b). Fig. 12(a) covers the region from the upwind slope to the crest, and indicates that the air flow is approximately following the upwind slope. Fig. 12(b) covers the region from the downwind slope to the trough. A separation area is found above the downwind slope. This is an indication that the air moves over the downwind slope avoiding the separation area near the water surface.

3.3. Comparison of the distribution of droplet velocities with the wind field

The profiles of the mean horizontal component of the droplet velocity were calculated from the measured droplet velocities shown in Fig. 9 for various droplet sizes (size 1: diameter d = 0.81 mm - 1.74 mm, size 2: d = 1.74 mm - 3.76 mm, size 3: d = 3.76 mm - 8.10 mm). The results are shown in Fig. 13(a), together with the wind velocity profile. At heights lower than the crest, the dependence of the droplet velocity on size is not clear. However, at heights higher than the crest, the droplet velocity increases with decreasing droplet size. Droplets of size 3 have a nearly constant horizontal velocity of $3\text{-}4 m s^{-1}$. On the other hand, for the droplets of size 1, the velocity increases from $4 m s^{-1}$ to $7 m s^{-1}$ with increasing height. However, for every size the droplet velocity is considerably smaller than the wind speed.

Fig. 13(b) shows the ratio of the droplet velocity to the local wind speed. For droplets of size 1 the ratio is about constant at 0.5. For droplets of size 2 or 3, the ratios are about 0.45 and 0.35, respectively, though they decrease gradually with increasing height. In the figure, the ratio for droplets of about 200 μm in diameter, as measured by Wu (1973), is also shown. In his case, the ratio is also nearly constant at 0.85. This large value means that the acceleration of the droplets by the local wind is more effective for these smaller size droplets.

In order to investigate the acceleration of the droplets by the wind in more detail, the variation of the droplet velocity along the wave was examined in the case of size 1 droplets. Figs. 14 (a), (b) and (c) show the profiles of droplet velocity in region I (from the crest to the downwind slope: A-B), region II (the trough: B-C) and region III (the upwind slope: C-A). These regions are marked along the bottom of Fig. 9. In each figure, the mean wind profiles calculated from the visualized wind data of the corresponding region are also shown. In region I,

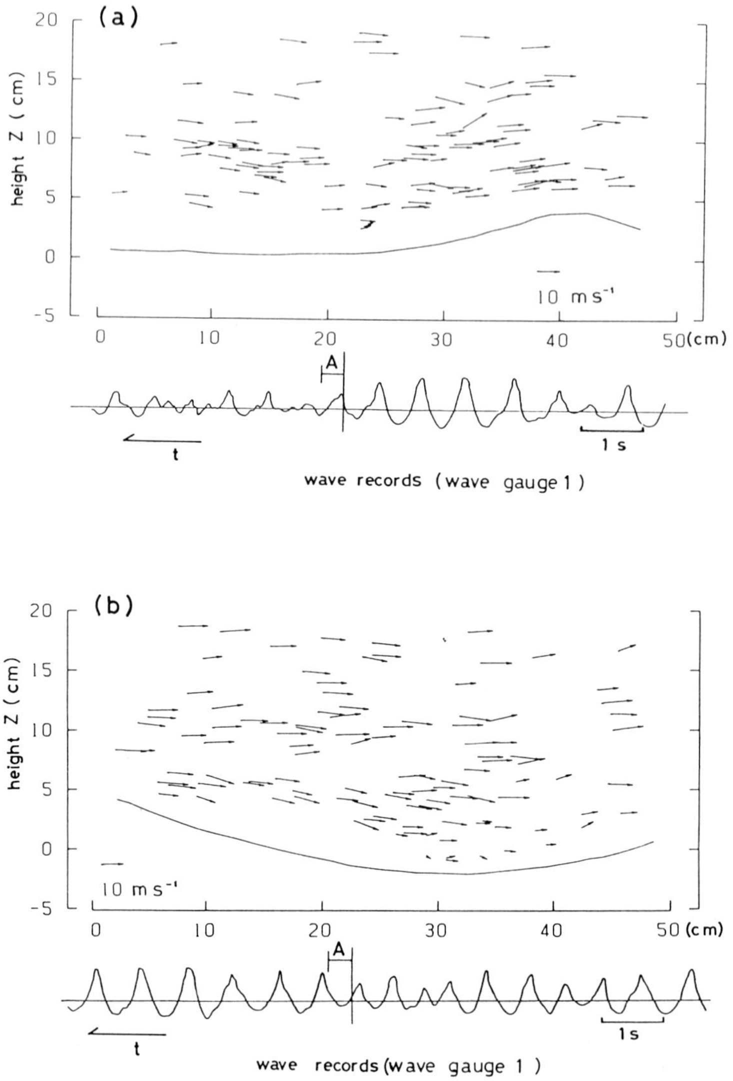

Fig. 12. Examples of the air flow pattern visualized over breaking wind waves, in the condition $F = 16$m and $U_r = 16\,ms^{-1}$, (a) From upwind slope to crest, (b) From downwind slope to trough.
Each arrow in the figure shows a local wind velocity vector in a coordinate system moving with the phase speed of the wave. The time of the photograph is marked with a vertical line on the wave record shown at the bottom of each figure. The horizontal line marked A represents the extent of the wave surface photographed.

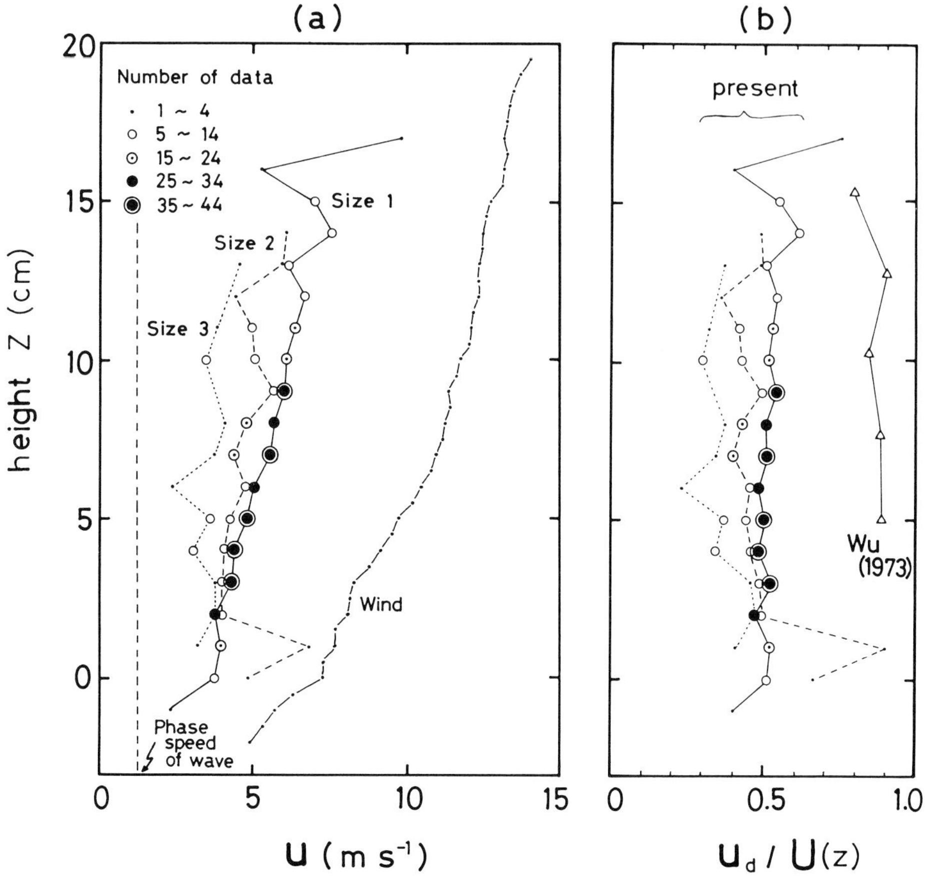

Fig. 13. Mean vertical distributions of horizontal droplet velocity. Size 1 (solid line), size 2 (dashed line) and size 3 (dotted line). Each symbol indicates number of data, as defined in the top left of the figure. The vertical dashed line indicates the phase speed of the wave.

(a) Droplet velocity distributions in the absolute coordinate system. The mean wind profile over the representative wave and the phase speed of the wave are also shown.

(b) Droplet velocity distributions normalized by the wind velocity at each elevation. Experimental results obtained by Wu (1973) for a droplet of about 200μm diameter are shown by triangles.

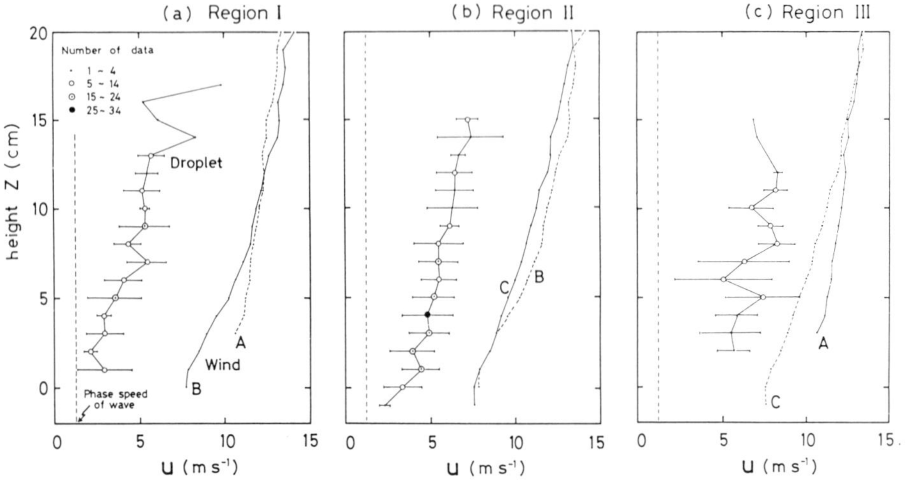

Fig. 14. The horizontal velocity of size 1 droplets compared with the wind profile at three regions along the wave defined in Fig. 9. Wind profile at left end of each region and that at right end are shown by dashed and solid lines respectively. A horizontal line from each mark indicates the standard deviation.
(a) From crest to downwind slope (Region I of Fig. 9); (b) Trough (Region II); (c) Upwind slope (Region III).

the droplet velocity is 2-3m s^{-1}, about two times the phase speed of the wave, at heights lower than the crest. At the height of 10 cm, it is about 5m s^{-1} and its ratio to the local wind speed is smaller than the general mean value of 0.5 shown in Fig. 13 (b). In region III, the horizontal droplet velocity attains a value of about 8m s^{-1}, and its ratio to the local wind is about 0.6-0.7, which is much larger than the general mean of 0.5. The horizontal droplet velocity thus increases gradually due to the acceleration by the wind as the droplet spreads forward along the wave. From a similar comparison for the vertical component of droplet velocity, it can be inferred that the effect of the wind on the vertical motion of the droplets is small compared with the vertical acceleration of gravity, at least for droplets larger than 0.81 mm in diameter. The initial velocity and the acceleration of gravity will thus dominate the vertical movement of these droplets, as inferred by Koga and Toba (1981) from the character of the vertical distribution of droplet concentration.

4. *Concluding remarks*

Wind-wave tank experiments like the present study are suitable for the detailed investigation of the characteristics of breaking wind waves at high winds. At high wind conditions, individual waves directly couple with the air flow over them and the surface drift current (vorticity layer) strongly controls the breaking condition. This coupling is a strongly nonlinear process. The concept of individual wind waves will be essential and useful for studying such a complex and nonlinear process.

The photographic technique is rather troublesome, although it is simple and more reliable than techniques incorporating various sophisticated electronic devices. Each of the present first and second experiments was conducted under only one wind condition. For a more quantitative discussion, further experiments under various conditions will be necessary.

REFERENCES

Donelan, M., M. S. Longuet-Higgins and J. S. Turner (1972): Periodicity in whitecaps. *Nature*, **239**, 449-451.

Koga, M. (1981): Direct production of droplets from breaking wind waves - its observation by a multicolored overlapping exposure photographing technique. *Tellus*, **33**, 552-563.

Koga, M. (1982): Bubble entrainment in breaking wind waves. *Tellus*, **34**, 481-489.

Koga, M. and Y. Toba (1981): Droplet distribution and dispersion process on breaking wind waves. *Sci. Rep. Tohoku Univ., Ser. 5* (Tôhoku Geophys. Journ.), **28**, 1-25.

Kondo, J., Y. Fujinawa and G. Naito (1973): High-frequency components of ocean waves and their relation to the aerodynamic roughness. *J. Phys. Oceanogr.*, **3**, 197-202.

Monahan, E. C. (1969): Fresh water whitecaps. *J. Atmos. Sci.*, **26**, 1026-1029.

Monahan, E. C. (1971): Oceanic whitecaps. *J. Phys. Oceanogr.*, **1**, 139-144.

Okuda, K. (1982a): Internal flow structure of short wind waves. Part I. On the internal vorticity structure. *J. Oceanogr. Soc. Japan*, **38**, 28-42.

Okuda, K. (1982b): Internal flow structure of short wind waves. Part II. On the streamline pattern. *J. Oceanogr. Soc. Japan*, **38**, 313-322.

Okuda, K. (1982c): Internal flow structure of short wind waves. Part III. On the pressure distributions. *J. Oceanogr. Soc. Japan*, **38**, 331-338.

Thorpe, S. A. and P. N. Humphries (1980): Bubbles and breaking waves. *Nature*, **283**, 463-465.

Toba, Y. (1961): Drop production by bursting of air bubbles on the sea surface (III). Study by use of a wind flume. *Memoirs Coll. Sci. Univ. Kyoto, Ser. A*, **29**, 313-344.

Toba, Y. (1978): Stochastic form of the growth of wind waves in a single-parameter representation with physical implications. *J. Phys. Oceanogr.*, **8**, 494-507.

Toba, Y., M. Tokuda, K. Okuda and S. Kawai (1975): Forced convection accompanying wind waves. *J. Oceanogr. Soc. Japan*, **31**, 192-198.

Tokuda, M. and Y. Toba (1981): Statistical characteristics of individual waves in laboratory wind waves. I. Individual wave spectra and similarity structure. *J. Oceanogr. Soc. Japan*, **37**, 243-258.

Wu, J. (1973): Spray in the atmospheric surface layer: Laboratory study. *J. Geophys. Res.*, **78**, 511-519.

ACKNOWLEDGEMENTS

The author wishes to express his sincere thanks to Professor Y. Toba of Tohoku University for his valuable advice and encouragement. Grateful thanks are also due to Dr K. Okuda of Tokohu University and the late Dr S. Kawai for their fruitful discussions and encouragement. He is also indebted to Dr M. Tokuda of the National Research Center for Disaster Prevention for his helpful discussions and encouragement.

SURFACE TENSION EFFECTS IN NONLINEAR WAVES

S. J. HOGAN
Department of Applied Mathematics
and Theoretical Physics, University of Cambridge,
Silver Street, Cambridge CB3 9EW

1. Introduction

This contribution deals with the effects of surface tension on nonlinear waves. Throughout this paper, the emphasis will be on nonlinearity. We shall review older work, present some recent developments and offer some problems for future study.

In Section 2 we present a brief outline of theories for the generation of capillary waves. This subject has already been considered by the present author in another context (Hogan 1980b).

In Section 3 we review work on *steady* nonlinear waves influenced by surface tension. Both centimetre-long and metre-long waves will be considered. Some very recent work on the stability of nonlinear capillary waves is outlined in Section 4. This is of relevance to very short steep waves which may play a role in the generation of bubbles. One aspect of the role of surface tension in longer waves is considered in Section 5. This is an account of work by Miller (1972), who showed that a reduction in surface tension can sometimes correspond to an increase in the height at which waves break. It follows that such waves would break nearer to a given coastline. The possibility of increased coastal erosion in the presence of severe pollution can be inferred from these results.

In Section 6, we highlight the lack of an adequate criterion for determining the highest wave of metre length in water.

Finally in Section 7 we present the governing equations for the propagation of waves in the presence of a surface-active monolayer. Again, the emphasis is on nonlinearity. The crucial difference here is the presence of a surface tension gradient which allows a tangential stress difference to be maintained across the interface. This in turn leads to the possibility of another form of wave, the elastic longitudinal wave. The crucial quantity here is not the surface tension, T, but the surface dilational modulus, $\epsilon \equiv dT/d(\ln A)$, where A is the surface area.

2. On the generation of capillary waves

This subject has been reviewed in another context by the present author (Hogan 1980b). The main points in that discussion are summarised here.

Casual observation of a wind-wave field indicates that small capillary waves are present on the *front* face of longer gravity waves. This is consistent with the 'fishing-line' problem and can be associated with some form of disturbance at the gravity wave crest. Cox (1958) performed a series of experiments on plunger-generated waves (length 4.7cm) and found capillaries even in the absence of wind.

Longuet-Higgins (1963) put forward a physical mechanism for these observations. In a steep wave, surface tension is important at the crest. This implies a travelling disturbance, which gives rise to capillary waves. He therefore considered a basic flow which was a pure gravity wave near maximum steepness, with surface tension considered as a perturbation. He concluded that the ripple steepness S_c near the crest was given by

E. C. Monahan and G. Mac Niocaill (eds.), Oceanic Whitecaps, 147–158.

$$S_C = \frac{2\pi}{3} \exp\left(\frac{-g}{6T\Lambda^2}\right) \tag{2.1}$$

where T is the surface tension and Λ is the curvature of the basic gravity wave. This result was subject to the restriction

$$T\Lambda \ll \tfrac{1}{2} q_0^2 \tag{2.2}$$

where q_0 = particle speed of the gravity wave. Equation (2.1) is a good approximation for the first two capillaries, despite the restriction in equation (2.2) not being strictly adhered to in Cox's data. Also no ripples are allowed behind the gravity wave crest, a constraint not in accord with Cox's observations. This theory was modified by Crapper (1970), Vanden-Broeck (1974) and Ruvinski & Freydman (1981), with slight improvements in agreement with Cox's work.

McGoldrick (1972) solved a time-dependent model equation because his own experiments had shown that a *steady* profile was not possible. He concluded that the generation mechanism was a nonlinear resonant interaction. Benney (1977) considered the interaction of a nonlinear capillary wave train with a small amplitude gravity wave. He found that when the group velocity of the capillary waves equals the phase velocity of the gravity wave, the capillary waves have a maximum amplitude at the gravity wave crest.

Ferguson et al. (1978) solved a more sophisticated model equation than McGoldrick (1972). They concluded that if viscosity is neglected, capillary waves could cover the entire wave. But if it is included, then they were a transient phenomenon.

Chang et al. (1978) performed an experiment to verify Longuet-Higgins' (1963) theory. They found rather good agreement if the main wave was long but experienced difficulty in attempting to measure experimentally the quantity Λ.

In summary, the most advanced theory turns out to be almost impossible to verify. Another candidate, that of resonant interaction, has yet to be tested experimentally. In Section 4, more attention will be given to this latter mechanism.

3. Steady nonlinear capillary-gravity waves

The computation of steady wave properties as a function of amplitude, in the presence of uniform surface tension, has become more tractable with the advent of computers. In the absence of gravity, however, an exact solution has already been found by Crapper (1957). This is given in Figure 3.1, where H is the wave height and λ is the wave length. Above H/λ = 0.73, the solution is unphysical, in that the surface crosses itself and a detached bubble does not arise. The possibility of these very short waves being able to produce bubbles has aroused considerable interest. It is natural to enquire if similar results hold in the presence of gravity.

We consider two-dimensional progressive waves on the surface of an irrotational, incompressible, inviscid fluid of infinite depth. At the surface $y = \eta(x, t)$ the boundary condition on the pressure is

$$\tfrac{1}{2}(u^2 + v^2) - g\eta - \frac{T\eta''}{(1+\eta'^2)^{3/2}} = K, \quad \text{on } y = \eta \tag{3.1}$$

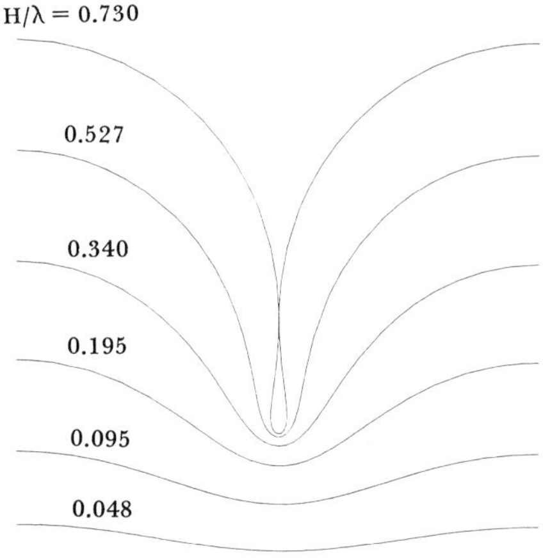

Fig. 3.1. Wave profiles of nonlinear capillar waves, after Crapper (1957).

where (x, y) are Cartesian coordinates with the line y = 0 in the undisturbed surface and the y-axis pointing upwards. The velocity in a frame of reference moving with the wave is (u,v), g is gravity, $' \equiv \partial/\partial x$ and K is a constant. In addition we have the kinematic condition

$$u \frac{\partial \eta}{\partial x} = v \text{ , on } y = \eta \tag{3.2}$$

This problem has been considered by Hogan (1979a, 1980a, 1981), Chen & Saffman (1979, 1980), Rottman & Olfe (1979), and Schwartz & Vanden-Broeck (1979).

Typical results are shown in Figures 3.2–3.6. In Figure 3.2 the highest waves for different wavelengths are given (all scaled so that $\lambda = 2\pi$). The parameter K is defined by

$$K = \frac{4\pi^2 T}{\lambda^2 g} \tag{3.3}$$

A bubble is clearly visible in each trough. The wave K = ∞ corresponds to Crapper's solution. For a given value of K, the wave profiles as a function of H are given in Figure 3.3 (K = 1.0).

Chen & Saffman's work led them to discover several bifurcations at both finite and infinitesimal amplitudes. An example is given in Figure 3.4, where h = ½H and c is the phase speed of the wave. It is difficult to imagine that such a profile could be stable.

Several properties of these waves are given in the cited work of the present author.

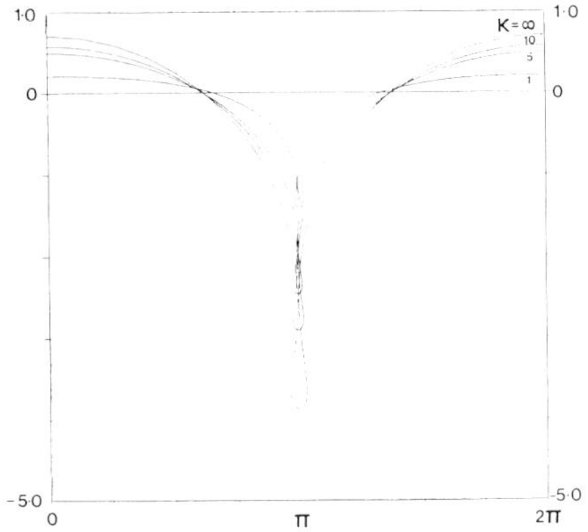

Fig. 3.2. Highest wave profiles of nonlinear capillary-gravity waves for K = 1, 5, 10 and ∞, after Hogan (1980a).

From all the work done on this problem, it is clear that bubbles are present in many limiting profiles of short waves. Two examples from Hogan (1980a) for longer waves (20cm and 2m respectively) show that this may not generally be so. Figure 3.5 illustrates half wavelength profiles for K = 0.0075 for various values of h and Figure 3.6 is for K = 0.000075.

All the profiles in this section are of steady waves. It is pertinent to enquire if they are stable. That question is considered in the next section.

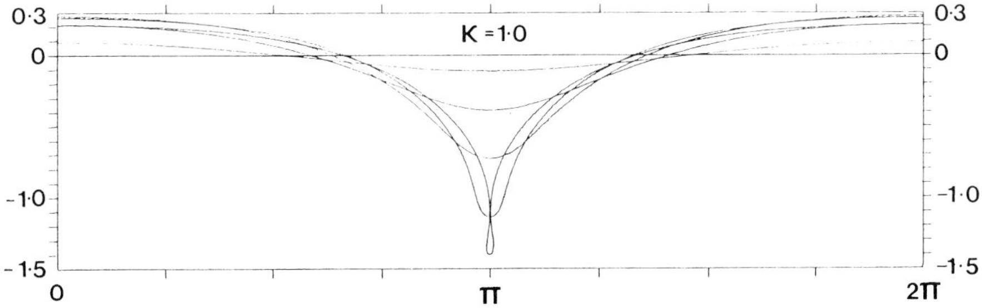

Fig. 3.3. Profiles of various wave heights for K = 1, after Hogan (1980a).

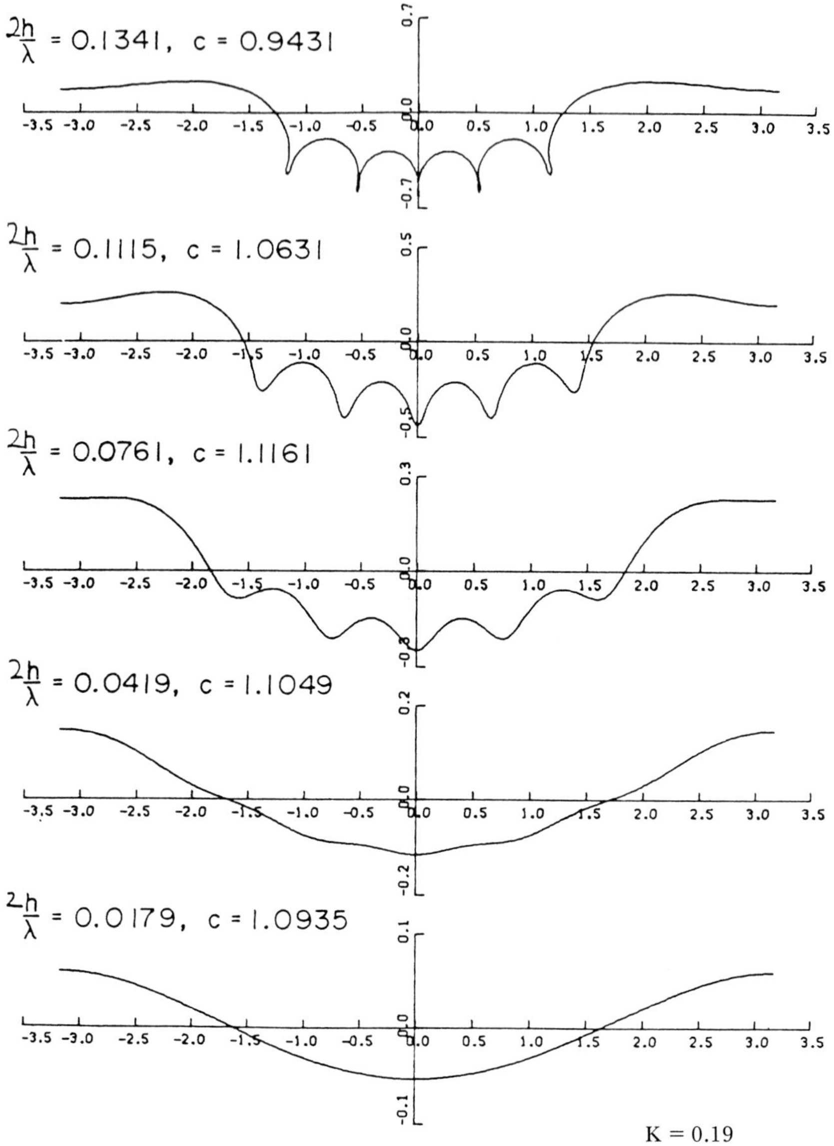

Fig. 3.4. Profiles for various wave heights for a bifurcated solution at K = 0.19, after Chen & Saffman (1980).

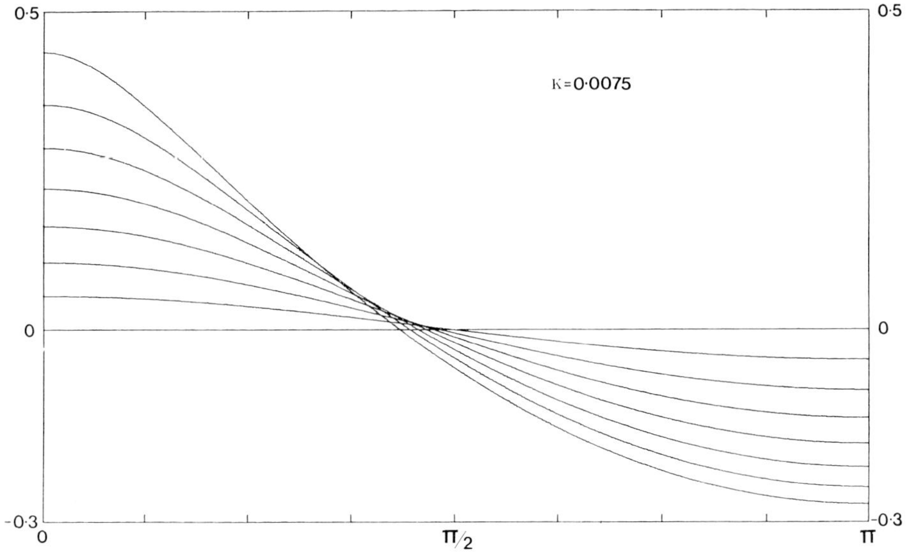

Fig. 3.5. As Fig. 3.3, with K = 0.0075.

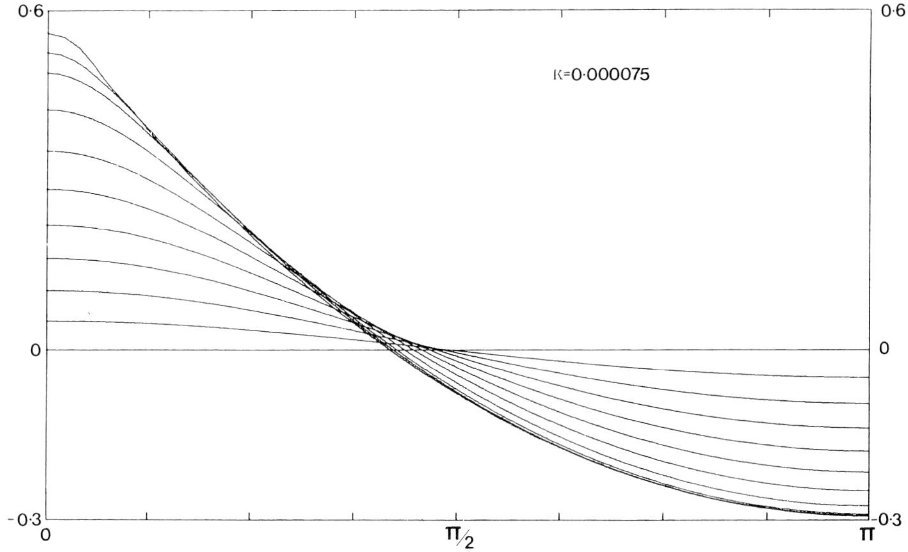

Fig. 3.6. As Fig. 3.3, with K = 0.000075.

4. The stability of nonlinear capillary waves

Some work has recently been completed (Hogan 1983) on the stability to normal mode perturbations of the nonlinear pure capillary wave of Crapper (1957). These waves are those where only uniform surface tension acts as the restoring force and gravity is absent.

The purpose of this study is to determine whether the enclosed bubble in the trough of these waves could ever possibly form. The waves were found to be very unstable. This must cast doubt on the bubble-forming ability of these *short* waves. Further calculations on *longer* waves (length in metres), have yet to be performed. The analysis involves considered linear perturbations of a nonlinear wavetrain. The full calculation of a nonlinear perturbation is now known to require novel numerical techniques (Srokosz, private communication) because the methods of Longuet-Higgins & Cokelet (1976) and others break down very quickly owing to *spurious* instabilities.

The equations to be solved are the time-dependent versions of equations (3.1) and (3.2). Bernoulli's condition becomes

$$\tfrac{1}{2}q^2 + g\eta - \frac{T}{R} + \frac{\partial\phi}{\partial t} = B(t)\ , \text{ on } y = \eta(x,t) \tag{4.1}$$

where ϕ (x, y, t) is the velocity potential and R is the radius of curvature. The kinematic condition becomes

$$\frac{D}{Dt}(y-\eta) = 0\ , \text{ on } y = \eta(x,t) \tag{4.2}$$

The free surface derived in Section 3 are given by $\psi = 0$, where ψ (x, y, t) is the stream function. To include perturbations, we must consider that $y = \eta$ (x, t) is given by

$$\psi = F(\phi,t) \tag{4.3}$$

where F is a small quantity of order δ.

Then we take

$$\begin{aligned} x(\phi,\psi,t) &= X(\phi,\psi) + \xi(\phi,\psi,t) \\ y(\phi,\psi,t) &= Y(\phi,\psi) + \eta(\phi,\psi,t) \end{aligned} \tag{4.4}$$

where (X, Y) is the steady solution derived in Section 3 and (ξ, η) are perturbations of the same order as F.

We now substitute equations (4.3) and (4.4) into equations (4.1) and (4.2) and Taylor expand each term about $\psi = 0$. We retain the 0(1) and 0(δ) terms only. The details are omitted here, since they are extremely lengthy. The 0 (1) term of equation (4.1) is merely equation (3.1). There is no 0 (1) term in equation (4.2).

We seek perturbations with time-dependence $e^{-i\sigma t}$ where $\mathrm{Im}(\sigma) > 0$. The equations derived are true for general X and Y. We checked our calculations for (X, Y) = (ϕ, ψ), that is, a uniform flow. The perturbation has the phase speed of capillary-gravity waves, as expected.

The next stage is to take (X, Y) as the solution derived by Crapper (1957). This requires that g = 0. The results for the real part of the radian frequency σ of *normal mode* perturbations, plotted against the steepness H/λ of the wave, are depicted in Figure 4.1. The modes are designated $n = \pm1, \pm2, \pm3$ etc. according to their character for small H/λ. Thus mode n contains $|n|$ crests (and troughs) per wavelength of the unperturbed wave. A plus sign means the perturbation is travelling with the wave, a minus sign that it is moving against the wave. As $H/\lambda \to 0$,

$$\sigma = r \mp r^{3/2} \tag{4.5}$$

for $n = \pm r$. A continuous line represents a stable mode, a dotted line an unstable mode.

Clearly, the wave is unstable in many ways to these superharmonic perturbations. The cascade to lower frequencies with increased H/λ (for the unstable modes) corresponds to the appearance of ripples on the surfaces of the main wave. The growth rates of some of the modes are given in Figure 4.2.

For subharmonic perturbations, Re (σ) and Im (σ) are given in Figures 4.3 and 4.4 for typical values of n. Again the wave can be seen to be very unstable.

The source of these finite-amplitude instabilities is the fact that capillary waves can undergo triad and quartet resonances. The stronger $O(H/\lambda)$ growth rates correspond to the former case (Zacharov 1968), the weaker $O((H/\lambda)^2)$ rates to the latter. See Hogan (1983) for full de-

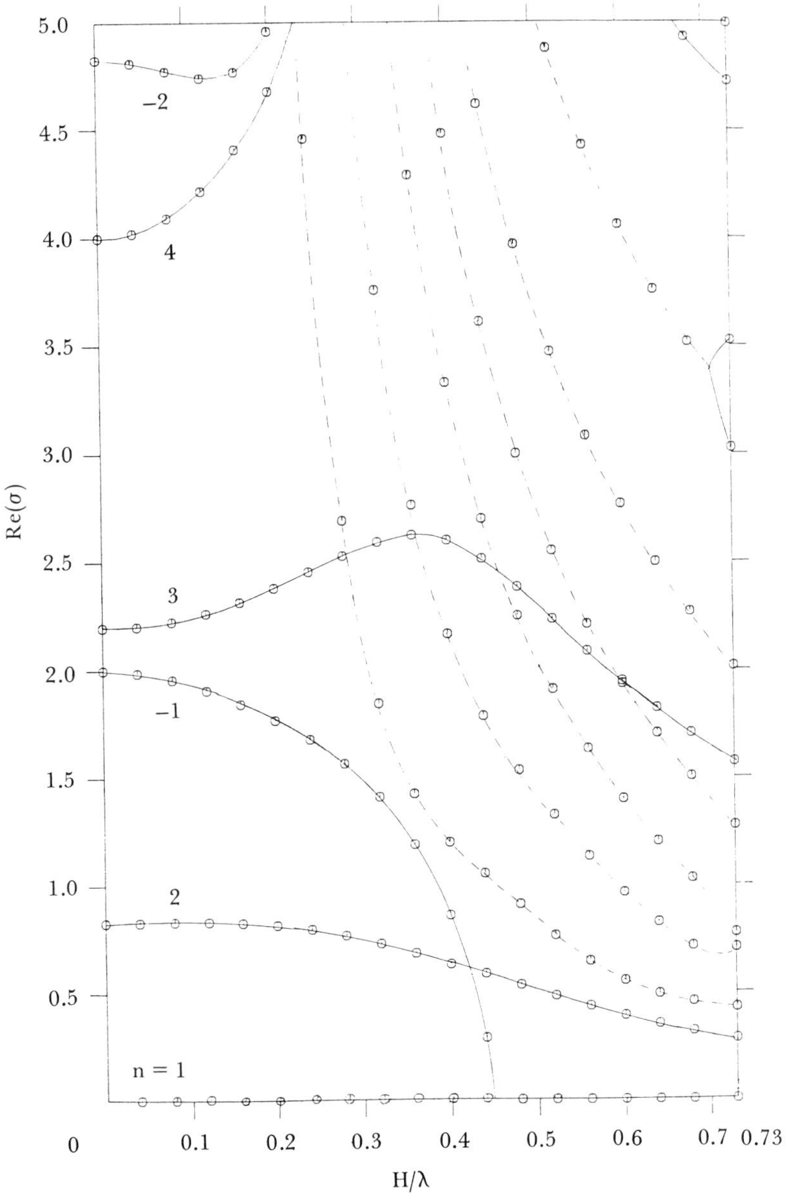

Fig. 4.1. Re(σ) of the normal mode superharmonic perturbations as a function of H/λ.

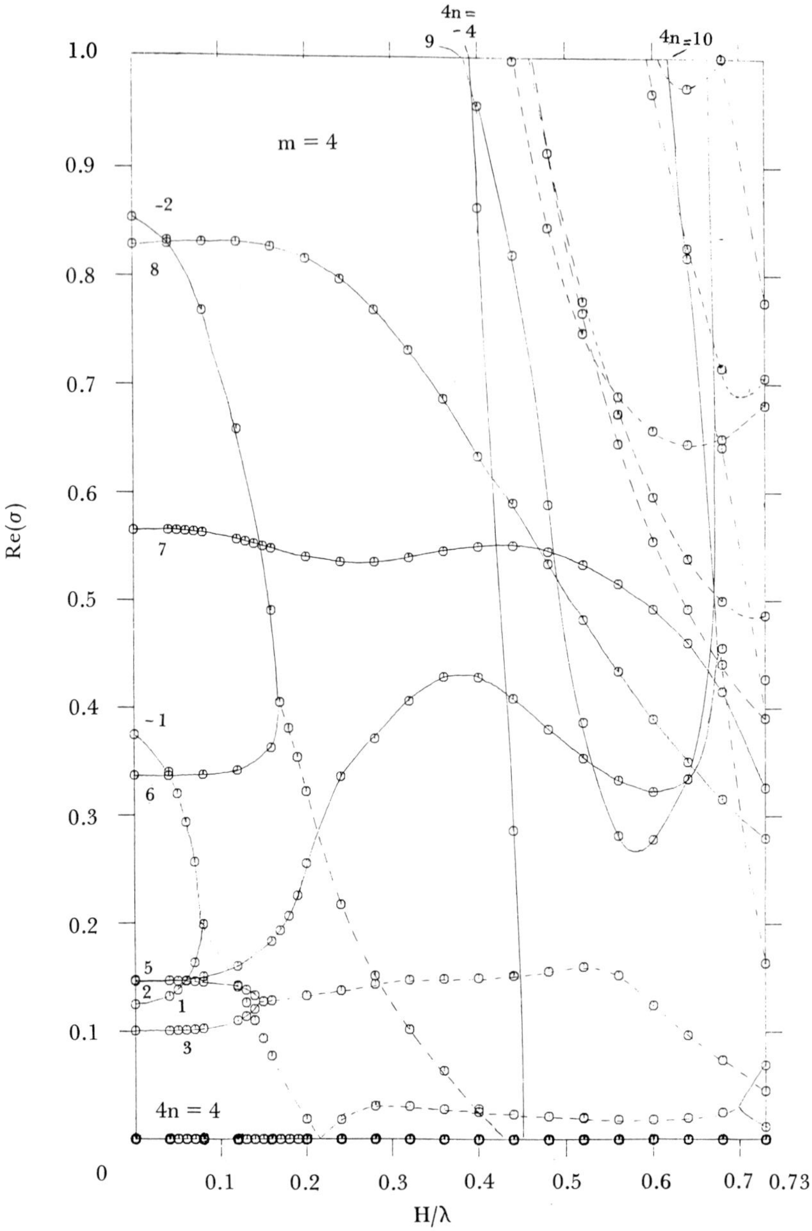

Fig. 4.3. Re(σ) of the normal mode subharmonic perturbations (based on four unperturbed waves) as a function of H/λ.

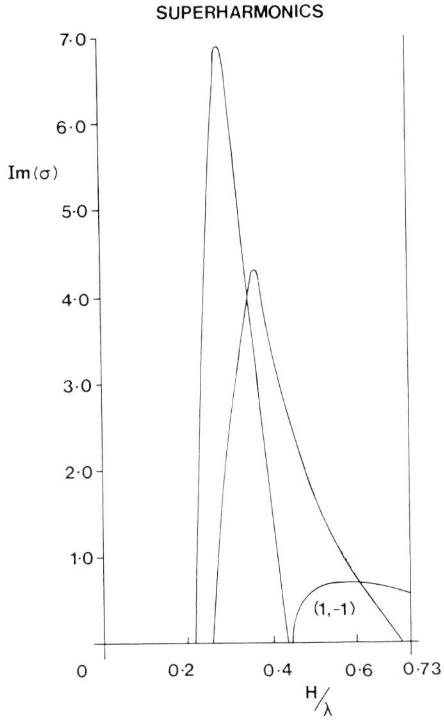

Fig. 4.2. As Fig. 4.1, Im(σ).

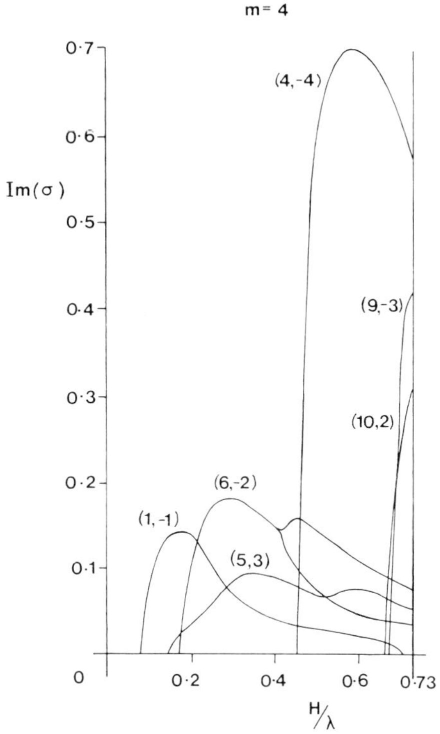

Fig. 4.4. As Fig. 4.3, Im(σ).

tails.

5. *The role of surface tension in breaking waves*

One of the first experimental studies of the role of surface tension in breaking waves was made by Miller (1972). He performed a series of experiments on many types of waves with varying surface tension. By adding chemicals to the water, he was able to obtain lower values for the surface tension coefficient. We shall consider his results for progressive waves only.

The experiments were performed in a wave tank of length 30m, filled with water to an undisturbed depth, d, of approximately 22 cm. A typical wavelength, λ, used was around 69 cm. Thus in these experiments the water was of intermediate depth with respect to the wavelength. With the aid of various chemical additives, Miller produced water-additive mixtures with surface tension coefficients, T, of 40 and 53 dynes cm^{-1}, to supplement the standard water-only value of 74 dynes cm^{-1}. This was accomplished with only moderate change in the value of the viscosity. He measured the limiting crest angle, θ, as well as the limiting wave steepness, H/λ, at the point of breaking (H is the wave height). His results are included in Table 5.1. At values of H/λ greater than those given,

Table 5.1. Values of the limiting crest angle, θ, and limiting wave steepness, H/λ, as function of the surface tension coefficient, T, after Miller (1972).

T (dynes, cm^{-1})	θ	H/λ
40	116°	0.120
53	123°	0.093
74	129°	0.088

the wave broke. In the absence of surface tension the limiting value for θ is 120°. It should be noted that the work of Cokelet (1977) can be used to predict a limiting value for H/λ of 0.135, for this value of d.

A clear trend is apparent from these results, despite difficulties in measuring θ, a trend which was confirmed by experiments on waves propagating over a uniform sloping beach. In these sloping beach experiments Miller found that, as the surface tension decreased, the waves got higher before they broke and hence did so closer to the shoreline. Surface tension can thus influence the breaking process in waves, once they are of sufficient height.

The present author's theoretical work on inviscid gravity-capillary waves (Hogan 1980a) tends to confirm this result. In particular, it was found that the highest wave, in conditions of steady propagation on water of infinite depth, increased as the surface tension decreased.

It should be pointed out that the experimental work involved a water-additive mixture and *not* just the presence of a layer of surfactant on the surface of normal water. Miller's conclusions are not therefore in contradiction with the observations of increased damping in the presence of monolayers (Lucassen 1982), but rather they are of relevance in the case of major pollution in shallow waters, where the entire water column may be occupied with water-additive mixtures.

Finally, the density of these mixtures was not measured. The extra wave height attained in these experiments may well be partly due to the reduction in density difference across the interface, as well as to the decreased surface tension.

6. *The highest gravity-capillary wave*

In the previous section, we discussed the breaking of waves where surface tension is present, but not dominant over gravity. This is of course what occurs for all the large waves in the ocean. But it has never been made clear in the literature precisely what is the limiting form of these waves.

If the steady wave has a wavelength of the order of a few centimetres, then the highest wave is geometrically determined by the formation of a bubble in the trough. But for steady waves of length a few metres or more, numerical calculations (Hogan 1980a) and casual observations indicate that this form of limit is unlikely. In the absence of surface tension, Stokes (1880) conjectured that the highest wave in water would have a sharp corner at its crest and enclose an angle of 120°. In the presence of surface tension, a sharp corner cannot form, because the jump in the wave slope would give an undefined value for the radius of curvature, R, at that point and hence a zero value for the surface tension. This is needed in order to maintain a finite pressure difference across the interface.

A numerical investigation of other possible limiting criteria has been carried out by Hogan (1979b). It was found that as the wave grew in height, the radius of curvature approached a limit which was dependent only on the value of the surface tension and the position on the wave. No physical reason has been attached to this observation.

The presence of two criteria (radius of curvature limiting and bubble formation) can be reconciled at intermediate wavelengths. Here there is a competition between the two limits with the possibility of the wave profile intersecting itself before the radius of curvature limit is reached.

7. *The influence of viscosity and surface tension gradients*

Throughout this talk, the emphasis has been placed on nonlinear waves on an *inviscid* fluid with *constant* surface tension coefficient. We shall now briefly consider some consequences of relaxing both these assumptions.

Let us suppose that that is a surface active

material on the interface between two fluids. If changes in the surface area now occur, for example due to the passage of a wave, the surface tension will also change. It will be increased at the wave crests and decreased in the troughs. This leads to an elastic property of the interface associated with the system's tendency to return to the state of uniform surface tension. The precise nature of this process is dependent on the adsorbed material and the frequency of the wave. For example, a compression or extension of the surface may reorientate the adsorbed molecules and produce a different value of the surface tension from that of a similar area of surface at rest. A more tractable relaxation mechanism occurs when there is a diffusional interchange between the surface and the bulk of the fluid.

If either of these effects is significant during the passage of a wave, then the surface is said to be viscoelastic, otherwise elastic. The parameter describing the resistance to deformation is called the surface dilational modulus defined by $\epsilon \equiv dT/d(\ln A)$ where A is the surface area.

The main mathematical difference between this situation and the case of uniform surface tension occurs in the boundary conditions for the flow. With the introduction of surface tension gradients, a tangential stress difference can now be maintained across the interface. That is

$$p_{nr} = \frac{\partial T}{\partial r} \quad \text{on } y = \eta(x,t) \tag{7.1}$$

where p_{nr} is the component of stress in a system of coordinates where n is normal to the surface and r is tangential. η is the surface elevation and t is time. For a derivation of this relationship, see Lucassen-Reynders & Lucassen (1969), p. 361. This is to be compared with the normal stress balance (Bernoulli) given by

$$p_{nn} = \frac{T}{R} - p_0 \quad \text{on } y = \eta(x,t) \tag{7.2}$$

where R is the radius of curvature.

Solution of equation (7.2) gives capillary waves as we have already seen. Equation (7.1) however gives rise to elastic longitudinal waves (Lucassen 1968a, b; Lucassen & van den Tempel 1972). These in turn result in additional, significant damping of the wave motion. Both the wavelength and the damping are strongly dependent on the surface dilational modulus ϵ.

It should be pointed out that most of the analysis to date has involved small-amplitude waves. Steep capillary waves with uniform surface tension can possess bubbles in their troughs. It is not clear if these viscous counterparts must also behave in the same way. (We note that whilst a fluid can be viscous and have uniform surface tension, it can not be inviscid and have a surface tension gradient. This follows directly from equation (7.1) since Pnr is proportional to μ, the vicosity.)

The fully nonlinear equations were given by van den Tempel & van de Riet (1965) and are included here for reference.

By continuity

$$\frac{\partial u}{\partial x} + \frac{\partial v}{\partial y} = 0$$

The momentum equations are

$$\frac{\partial u}{\partial t} + u\frac{\partial u}{\partial x} + v\frac{\partial u}{\partial y} = -\frac{1}{\rho}\frac{\partial p}{\partial x} + \frac{\mu\nabla^2 u}{\rho}$$

$$\frac{\partial v}{\partial t} + u\frac{\partial v}{\partial x} + v\frac{\partial v}{\partial y} = -\frac{1}{\rho}\frac{\partial p}{\partial y} + \frac{\mu\nabla^2 v}{\rho} - g$$

The kinematic condition is

$$\frac{\partial \eta}{\partial t} = v - u\frac{\partial \eta}{\partial x} \quad \text{on } y = \eta$$

The normal stress balance, equation (7.2), becomes

$$-p + \frac{2\mu}{[1+\eta'^2]}\left\{\frac{\partial v}{\partial y} - \frac{\partial \eta}{\partial x}\left(\frac{\partial u}{\partial y} + \frac{\partial v}{\partial x}\right) + \left(\frac{\partial \eta}{\partial x}\right)^2 \frac{\partial u}{\partial x}\right\}$$

$$= \frac{T}{R} - p_0 \quad \text{on } y = \eta$$

and the tangential stress balance, equation (7.1), becomes

$$2\mu\frac{\partial\eta}{\partial x}\left(\frac{\partial v}{\partial y}-\frac{\partial u}{\partial x}\right)+\mu\left(1-\left(\frac{\partial\eta}{\partial x}\right)^2\right)\left(\frac{\partial u}{\partial y}+\frac{\partial v}{\partial x}\right)$$

$$=\left(1+\left(\frac{\partial\eta}{\partial x}\right)^2\right)^{1/2}\left(\frac{\partial T}{\partial x}+\frac{\partial\eta}{\partial x}\frac{\partial T}{\partial y}\right)\quad \text{on } y=\eta$$

The actual surface under consideration will determine further equations. For example, if the surfactant can diffuse rapidly enough during the passage of a wave, then the surfactant concentration, c, in the subsurface layer is governed by the equation

$$\frac{\partial c}{\partial t}+u\frac{\partial c}{\partial x}+v\frac{\partial c}{\partial y}=D\nabla^2 c$$

where D is the diffusion constant of the surfactant. At a surface element of area α, the amount of adsorption of surfactant is given by $\Gamma\alpha$ where Γ is the surfactant adsorption. This varies with time due to diffusion. It is then possible to show that, at the surface, $y = \eta(x, t)$, the following equation holds,

$$\frac{\partial\Gamma}{\partial t}+u\frac{\partial\Gamma}{\partial x}+v\frac{\partial\Gamma}{\partial y}=$$

$$-\Gamma\left[\frac{\partial u}{\partial x}+\frac{\partial\eta}{\partial x}\left(\frac{\partial u}{\partial y}+\frac{\partial v}{\partial x}\right)+\left(\frac{\partial\eta}{\partial x}\right)^2\frac{\partial v}{\partial y}\right]\times$$

$$\left[1+\left(\frac{\partial\eta}{\partial x}\right)^2\right]-\frac{D}{[1+(\partial\eta/\partial x)^2]^{1/2}}\left[\frac{\partial c}{\partial y}-\frac{\partial\eta}{\partial x}\frac{\partial c}{\partial x}\right]$$

The solution procedure takes the form of assuming that each quantity u, v, p, η, T, c and Γ take the form of a Fourier series in time, with the coefficients of each harmonic depending on x and y.

One further assumption which could be made is that there is always equilibrium between the surface and the sub-surface. The T and Γ are completely determined by the surface concentration C_s (by, for example, the Gibbs' or Langmuir adsorption isotherm).

There has been considerable progress in the use of algebraic computing languages since 1965 to make these problems now tractable.

REFERENCES

Chang, J. H., R. N. Wagner & H. C. Yuen (1978). *J. Fluid Mech.* **86**, 401-413.

Chen, B. & P. G. Saffman (1979). *Studies in Appl. Math.* **60**, 183-210.

Chen, B. & P. G. Saffman (1980). *Studies in Appl. Math.* **62**, 95-111.

Cokelet, E. D. (1977). *Phil. Trans. Roy. Soc. A* **286**, 183-230.

Crapper, G. D. (1957). *J. Fluid Mech.* **2**, 532-540.

Crapper, G. D. (1970). *J. Fluid Mech.* **40**, 149-159.

Ferguson, W., P. G. Saffman & H. C. Yuen (1978). *Studies in Appl. Math.* **58**, 165-185.

Hogan, S. J. (1979a). *J. Fluid Mech.* **91**, 167-180.

Hogan, S. J. (1979b). PhD thesis, Cambridge University.

Hogan, S. J. (1980a). *J. Fluid Mech.* **96**, 417-445.

Hogan, S. J. (1980b). *Berichte aus dem SFB 94 Meeresforschung. Universität Hamburg.* **17**, 5-11. Trier Symposium on Capillary Waves and Gas Exchange.

Hogan, S. J. (1981). *J. Fluid Mech.* **110**, 381-410.

Hogan, S. J. (1983). Manuscript in preparation.

Longuet-Higgins, M. S. (1983). *J. Fluid Mech.* **16**, 138-159.

Longuet-Higgins, M. S. & E. D. Cokelet (1976). *Proc. Roy. Soc. Lond. A* **350**, 1-26.

Lucassen, J. (1968a). *Trans. Faraday Soc.* **64**, 2221-2229.

Lucassen, J. (1969b). *Trans. Faraday Soc.* **64**, 2230-2235.

Lucassen, J. (1982). *J. Colloid Interface Sci.* **85**, 52-58.

Lucassen, J. & van den Tempel (1972). *J. Colloid Interface Sci.* **41**, 491-498.

Lucassen-Reynders, E. H. & J. Lucassen (1969). *Adv. Colloid Interface Sci.* **2**, 347-395.

McGoldrick, L. F. (1972). *J. Fluid Mech.* **52**, 725-751.

Miller, R. L. (1972). University of Chicago, Fluid Dynamics and Sediment Transport Laboratory, Dept. of Geophysical Sciences. Technical Report No. 13.

Rottman, J. W. & D. B. Olfe (1979). *J. Fluid Mech.* **94**, 777-793.

Ruvinskii, K. D. & G. I. Freydman (1981). *Isv. Atmos. Oceanic Phys.* **17**, 548-553.

Schwartz, L. W. & J.-M. Vanden-Broeck (1979). *J. Fluid Mech.* **95**, 119-139.

Stokes, G. G. (1880). *Math. & Phys. Papers* **1**, 225-228.

Vanden-Broeck, J.-M. (1974). Université de Liége. Mémoire d'Ingénieur physicien, Faculté des Sciences appliquées.

Zacharov, V. E. (1968). *J. Appl. Mech. Tech. Phys.* **9**, 190-194.

THE EFFECT OF ORGANIC FILMS ON WATER SURFACE MOTIONS

JOHN C. SCOTT*
Fluid Mechanics Research Institute,
University of Essex,
Colchester, Essex CO4 3SQ

Water motions near surfaces are predominantly determined by two surface chemical factors: by the surface tension itself, and also by the surface dilatational elasticity, which is (simply) how much the surface tension changes with variations in surface compression and extension. Taking the example of ripples on water, whilst their velocity is strongly influenced by the surface tension, their damping is determined mainly by the elasticity. For waves of larger wavelength, gravity waves, the surface tension loses most of its influence on the velocity, but it is found that the elasticity becomes even more important in determining the wave damping.

Another example, even more relevant in the context of whitecapping, concerns the influence of the wind on waves that are very close to breaking. It appears, in this case, that the stabilizing effect exerted on the surface by the dilatational elasticity of an organic film leads to a marked reduction in the energy input to the waves, and can lead to retardation or prevention of the breaking process. This is the effect of 'pouring oil on troubled water', well-known in maritime folklore.

Surface chemical effects appear again, more subtly, in the coalescence behaviour of air bubbles, vitally important in the dynamics of the bubble clouds caused by wave breaking. Significant differences observed in fresh water and salt water whitecaps may be ascribed to these effects, bubbles in salt water being (as a result) greater in number, smaller, more densely packed, carried deeper, and slower to rise to the surface than those formed in fresh water by a similar wave-breaking event.

Another factor concerning the behaviour of bubbles relates to their velocity of rise through water, and their ability to scavenge surface-captive material from the water. The two effects are related, because the scavenging is likely to be more efficient as the bubble velocity decreases, and a bubble that is contaminated will tend to rise more slowly.

1. Introduction

Water is unique among liquids (liquid metals excepted) in having a high surface tension and thus the consequent ability to attract contamination. Surface tension arises from the imbalance in molecular forces in the surface regions, there being a net inwards-directed force into the bulk of the liquid. When this imbalance is marked, as it is in water, it becomes possible for foreign molecules, such as those of organic materials, to be energetically acceptable in the surface region. This process is called *adsorption* at the surface, and adsorption is particularly strong for molecules whose structure, while being, in general, strongly inimical to that of water, includes one or more sections that are water-like, or at least quite acceptable to the water structure.

The requirements for quite strong adsorption at the water surface are easily met by whole classes of organic materials, particularly those associated with natural life processes. (This is hardly surprising in view of the probable dependence of most life processes on the properties of thin membranes arranged at the interfaces of essentially aqueous media.) Water is, in fact, so prone to surface contamination that the most common experience of the appearance of liquid water, and of how it behaves when stirred, shaken, or otherwise moved about, is likely to be of contaminated water rather than clean water.

The concept of clean water itself, in the fluid mechanical context, is essentially different

*Present address: Admiralty Research Establishment, Southwell, Portland, Dorset DT5 2JS

E. C. Monahan and G. Mac Niocaill (eds.), Oceanic Whitecaps, 159–165.

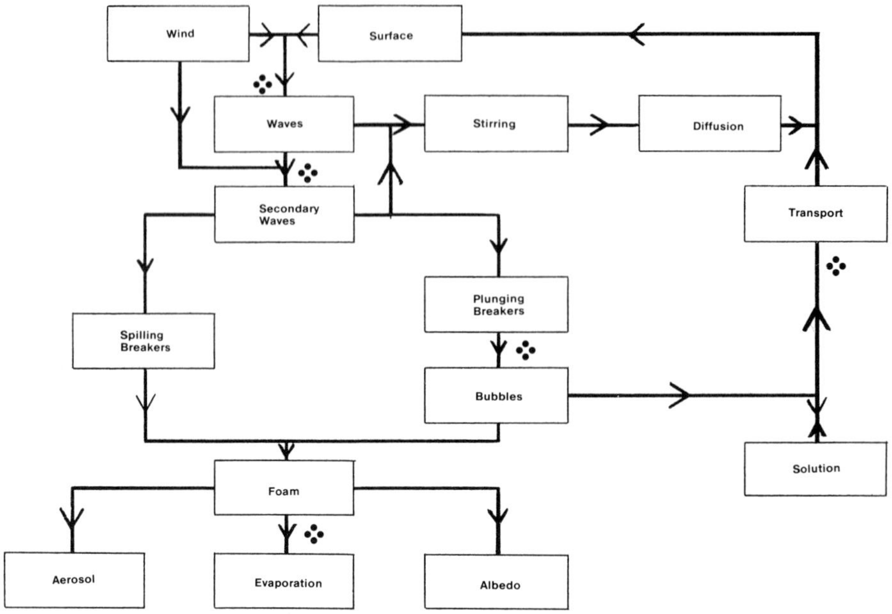

Figure 1. The links between the major stages in whitecap production, the asterisks showing the points at which surface-active effects may be important.

from other considerations of water purity. Whilst purity is essentially associated with volume or mass (bulk) proportions, surface cleanliness is restricted to two dimensions only, in the form of a surface concentration of contaminating material. If the surface is adjacent to a large volume of pure water (i.e. water containing only small fractions of a p.p.m. of contaminant) it is nevertheless possible that a large fraction of the bulk contaminant would eventually become adsorbed at the surface, making the bulk water more pure, but making the surface more dirty.

Figure 1 shows schematically the links between the major mechanisms involved in whitecap production, and indicates the areas in which surface chemical effects, particularly organic material effects, are likely to play an important role. In this context it is convenient to divide the wave generation process into two parts, the second being associated with the production of 'secondary waves', waves that begin at ripple wavelengths on the backs of well-developed waves which appear to play a role in determining the aerodynamic roughness of the large waves, and hence the energy flow between wind and waves in a white-capped ocean. The effect of surface-active materials on waves is thus two-fold: firstly an effect on retarding the growth of the wave field (particularly in the early stages); and secondly on the secondary process, where it gives rise to the well-known phenomenon of 'wave-oiling'.

2. *Effects of surface contamination on water surface motions*

A. *The effect of surface dilatational elasticity.*

The effect of organic surface contamination on water is primarily to reduce the surface tension. A typical variation of surface tension as a

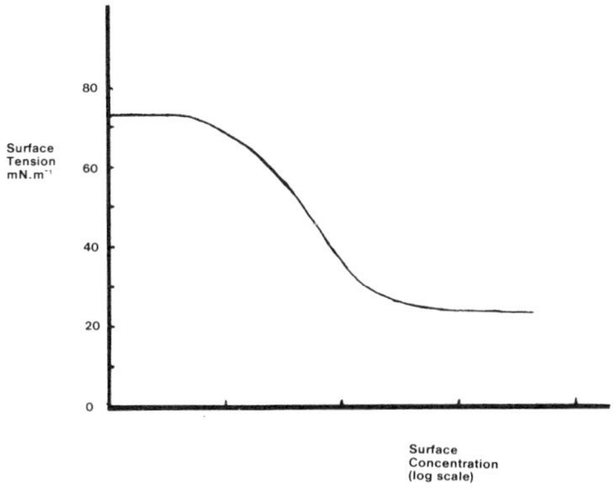

Figure 2. The variation of surface tension with surface concentration in a typical surface-active organic material (schematic).

function of surface concentration of contaminant would be similar to that shown in figure 2. Initially, at very small surface concentrations, for which the molecules are still far apart, the surface tension reduction would be small, and proportional to the concentration. As the contaminating molecules become sufficiently close packed to interact, the surface tension decreases more quickly with concentration, until, as the adsorbed layer approaches a packed state, one molecule thick, the surface tension tends to a lower limit.

The surface tension itself is important for the determination of some water surface motions. As is well known, the velocity of freely propagating ripples is controlled by the surface tension, the effect of gravity being predominant only at wavelengths greater than about 20mm. The surface tension also has an influence on the stability of certain fluid flows, such as the shear flows in the adjacent air that lead to the generation and growth of wind waves.

However, there is a second parameter characteristic of water surfaces that is also important; indeed, in the present context it is even more important than the surface tension itself. This parameter is the *surface dilatational elasticity*, the variation of surface tension with changes in surface concentration of contaminant:

$$\epsilon = d\gamma / d\,(\ln A),$$

where γ is the surface tension and A is the surface area per molecule.

The surface dilatational elasticity is essentially the (negative) slope of the curve shown in figure 2, and it therefore begins as zero for a clean (uncontaminated) surface, usually increasing steadily as the surface concentration increases. Because curves such as those shown in figure 2 are intrinsically reversible, this elasticity itself involves no energy loss when a surface expands and contracts. That is not to say that the expansions and contractions do not usually involve some viscous motion of the substrate water, but it is important to realise that this parameter expresses a concept quite different from other surface rheological considerations.

Surface rheology, in general, involves effects associated with the molecular interactions between the contaminating molecules in the surface. These concepts have been enumerated by Goodrich (1962), and most surface analogues of bulk rheological properties of liquids (and solids) are anticipated to arise in some contaminant layers. However, comparatively little is known about such effects, and it is certain that, even if they are present at a given surface, their effect can only be to modify the influence of the dilatational elasticity, which is known to exist in all adsorbed layers.

The surface compressibility parameter is sometimes used to describe surface contamination effects. This is essentially the inverse of the dilatational elasticity, zero elasticity corresponding to infinite compressibility. In practice it is a less useful parameter, however. The dilatational elasticity is zero for a clean water surface, increasing to values perhaps as high as 200mN.m^{-1} for some materials. Typical values, however, tend to go no higher than about 50mN.m^{-1}, and many of the biological polymers of interest in the oceanographic context have maximum values no greater than 15mN.m^{-1}.

The surface dilatational elasticity affects the near-surface flow behaviour of water and aqueous solutions through its effect on the fluid mechanical boundary condition at the surface. In contrast to the properties of the clean surface (zero elasticity), the surface is al-

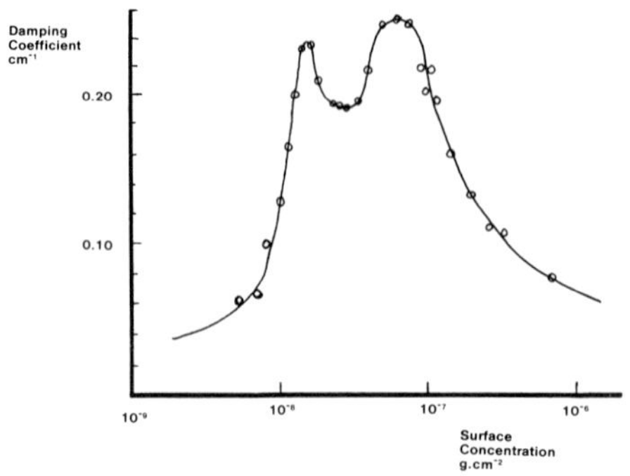

Figure 3. The surface concentration variation of the spatial damping rate of 30Hz ripples on a water surface covered by an adsorbed monolayer of poly(ethylene oxide) (molecular weight about 100 000). The two peaks can be explained on the basis of hydrodynamic theory, without recourse to mechanisms specific to the adsorbed molecules.

lowed to resist shear motion, and compensating surface tension gradients are set up within the film which oppose shear flows which tend to expand or compress the film.

B. The effect of dilatational elasticity on ripple propagation

The importance of the surface dilatational elasticity is illustrated in the results shown in figure 3, which describe the damping of 30Hz ripples on a still water surface as a function of the surface concentration of a high-molecular-weight water-soluble polymer, poly(ethylene oxide) (Scott and Stephens, 1972). This material is not typical of surface-active materials in general, but, as will be indicated below, it is much closer in its behaviour to the polymeric materials produced naturally in the oceans.

The region of rapid increase of the damping observed in figure 3 at low surface concentrations occurs while the surface tension is barely different from the clean water value, and the peak values of the damping, very much larger than the clean water value, occur for elasticities around 5-6mN.m^{-1}. In fact, this polymer material is only quite weakly surface active, and the surface tension of even a compressed monolayer is higher than 60mN.m^{-1}, compared with the clean water value of 72mN.m^{-1}.

The effect of surface tension itself on ripple damping is illustrated in the same results by the fact that the second damping peak is slightly higher than the first. Both peaks are associated with similar elasticity values, and the damping achieved in the second peak is greater because the surface tension is in the region of 8mN.m^{-1} less than it is in the first peak. It is thus clear that the surface dilatational elasticity is the dominant surface chemical parameter in the determination of ripple damping.

The damping effect of surface films on water is now well understood, and the results shown in figure 3 are adequately explained by hydrodynamic theory. The paper of Lucassen-Reynders and Lucassen (1969) gives a very good review of the work done in this area. In the case of insoluble surface films it is found that the damping can be accurately predicted from knowledge of the surface tension and the surface dilatational elasticity.

An examination of the hydrodynamic theory leads to some quite surprising conclusions, one of which is that, in the context of wave damping, a surface film is capable of a greater effect on gravity waves than on ripples. This is quite opposite to the popularly held view that 'surface effects' are important only for small-scale surface motions. While this is undoubtedly true for the surface tension itself, it is quite clear from hydrodynamic theory that surface elasticity is not so restricted. Of course, for a given elasticity, the damping decreases as the wavelength increases, as would be expected intuitively, but the clean-surface damping decreases more quickly, and the ratio of greatest possible damping on a film-covered surface to that on a clean surface increases with wavelength.

C. Water-soluble polymers at the water surface

The ripple damping results shown in figure 3 illustrate another factor which may be significant in some oceanographic contexts: that a water-soluble, weakly surface-active material, present in the water with a very low bulk concentration, can exert an appreciable influence on some surface motions. The idea of soluble molecules becoming adsorbed into films that behave in an insoluble manner is paradoxical, but it is explained by the fact that the monomer units can become individually adsorbed. Whilst one unit of a long chain being adsorbed will make the adsorption of its neighbour more likely, the desorption of a completely adsorbed chain will require the nearly simultaneous desorption of all of the units.

However, compression of the surface layer, or some other vigorous mixing action, will return the adsorbed molecules to a properly dissolved state, as found by Scott (1972).

D. Effects of surface films on wind waves

All studies reported so far on the effect of surface dilatational elasticity on the stability of shear flows at interfaces (Benjamin, 1964; Whitaker, 1964; Davies, 1966; and Gottifredi & Jameson, 1968) have concluded that the elasticity tends to increase the stability of the interface. The analysis of Gottifredi & Jameson, in fact, was on the wind-wave problem, using the Miles model. Although this model itself has considerable shortcomings, the message seems clear that a surface film, through its dilatational elasticity, is able to stabilize a water surface against wave generation by wind.

The age-old practice of using oil to prevent the breaking of dangerously large water waves (Scott, 1978) seems to be capable of explanation using this stabilizing tendency. It is apparent that a deep-water wave will only break if it exceeds the allowed maximum height, and that this can only happen if energy is being fed into the wave.

Now, the energy input to a large amplitude (close to breaking) wave is known to depend on the aerodynamic roughness of the exposed surface, and further it is known that the roughness of a wave comes from quite small-scale ripples (Wu, 1971), freshly generated on the backs of the waves by the wind. It is therefore likely that the effect of oil is in stabilizing the exposed wave surface against the formation of these ripples. The aerodynamic roughness would be thereby reduced, the energy input from the wind to the wave dramatically decreased, and the wave delayed from destructive breaking. Further evidence for the effect of surface films on wave damping and wave breaking is included in Scott (1979).

It should be noted that surface films are not expected to have an appreciable effect on the breaking behaviour of waves except when these waves are under the influence of wind. Waves breaking near beaches are one such case, and the results of Toba & Kunishi (1970) indicate that waves breaking as a result of interactions with other waves are unaffected by oil films.

3. The ocean as a clean water surface

Direct experimental evidence on the presence or absence of adsorbed material at the surface of the ocean is very scanty, and also questionable in its reliability. The surface concentrations involved appear to be very small, and it is very difficult to gather data while avoiding the possibility of adding contamination during the experiment. However, organic materials abound in the ocean, and experience would suggest that, given sufficient time, even a low overall bulk concentration of material will allow an accumulation at the surface by the processes of diffusion and adsorption. These processes would be aided by the presence of bubbles rising to the surface, scavenging material from greater depths.

It is likely that it is these naturally occurring materials, perhaps supplemented by civilization-related pollution, which give rise to the slick patterns observed in coastal waters.

However, the experiments of Scott (1972) showed that even moderate exposure to winds of 5–6 m/s was sufficient to mix even large surface concentrations of surface active material back into the bulk. It is likely therefore that under whitecapping conditions the ocean surface may be actually *clean* from the point of view adopted in this paper.

This point definitely needs further examination. On the one hand the wind will be continually cleaning the surface, preventing the adsorption of a monolayer which might stabilize the surface, and on the other hand the rising air bubbles created by breaking waves will undoubtedly have the effect of concentrating wave-damping materials in the surface region in general. Such a competition between the two processes may well play some part in explaining a 'hysteresis' that is sometimes reported in whitecap coverage as a function of wind speed.

In conditions of a sudden decrease of wind speed the concentration of surface-active material generated at the higher speed may be able to manifest a greater effect at the new (lower) wind speed than would have occurred if the wind speed had increased steadily up to the observed value.

4. Surface chemical effects on bubble transport of organic materials

All the effects so far discussed have related to the role of adsorbed organic materials at open water surfaces. There is, however, an important class of related effects which are still surface chemical in origin and which determine, to a large extent, the transport of organic material from the deeper layers of the oceans up to the surface: bubble coalescence effects and bubble velocity effects.

The measurements of Monahan (1969, 1971) demonstrated that whitecap coverage of fresh-water lakes is substantially less, for a given wind condition, than it is in the ocean situation. The experiments of Scott (1975) have shown that this effect is primarily due to the influence of the salt in the water on the coalescence behaviour of the clouds of bubbles formed by breaking waves. The details of the mechanism involved are still not clear, and although it is certain that a surface chemical mechanism is involved it seems that this is different from the surface stabilization effects so far considered in this paper.

The adsorption behaviour of inorganic salts is described as *negative* adsorption, i.e. the concentration of salt molecules in the interfacial layer is slightly less than in the bulk solution. It is this effect that causes the slight increase of surface tension in sea-water compared with pure water. Because of the higher bulk concentrations involved, any dynamic effects on the surface tension, analogous to those mentioned above, will occur only over very short time-scales. However, there are other mechanisms which may be of importance for the coalescence process.

The coalescence of two bubbles requires the rapid thinning and eventual rupture of the thin water film formed between them as the two approach closely. As described above, surface tension gradients set up by the shear flow of water out of the thin film will tend to retard the thinning, but other factors will also be important. One of these factors is that the dissolved salt molecules experience orientation at the surface, and form a layered structure which is not yet well understood. As the two bubble surfaces approach each other, the electrostatic fields of these layers begin to interact, and repulsive forces come into play which can have a major influence on the stability of the film.

As indicated above, such effects are only imperfectly understood. Comparison of the behaviour of a wide range of similar inorganic salts yields a confused picture, and several factors seem to be of importance. However, it seems likely that the mechanisms involved are related to those which result in the electrical charges found on droplets ejected into the atmosphere by the breaking of surface bubbles.

The net effect of the salt, therefore, is to produce a bubble population that is significantly different from that in fresh water. For a given wave-breaking event the bubbles will tend to be smaller, and therefore more numerous for a given volume of entrained air, and the available surface area for this volume of air will be greater. Two further important effects are 1) that the bubbles, being smaller, will rise more slowly, and 2) that the bubbles will be carried deeper by the breaking wave. These three effects together – greater surface area, decreased rate of rise to the surface, and greater depth of penetration – will lead to the more efficient scavenging of surface active organic materials to the surface.

The well-known effect of bubble rise retardation by an adsorbed organic layer will further aid the scavenging process. The retardation mechanism is a consequence of the elasticity of the adsorbed film that accumulates on a bubble during its rise. The shear flow of the water past the bubble will act to compress the surface film towards the lower side of the bubble, and the surface tension gradient so set up will make the bubble hydrodynamically solid in its surface behaviour in this lower region. As the accumulation process goes on organic molecules will be adsorbed on the upper area and will be swept down to augment the stagnant film below. The bubble will move more and more slowly as a consequence of the increased hydrodynamic drag this causes, and the scavenging will be, again, more efficient.

The inorganic salt concentration in sea water thus plays an important role in the transport of organic material to the ocean surface.

REFERENCES

Benjamin, T. B. (1964) Effects of surface contamination on wave formation in falling liquid films, *Archiwum Mechaniki Stosowanej* **16**, 615-626.

Davies, J. T. (1966) The effect of surface films in damping eddies at a free surface of a turbulent liquid, *Proceedings of the Royal Society*, **A290**, 515-526.

Goodrich, F. C. (1962) On the damping of water waves by monomolecular films, *Journal of Physical Chemistry*, **66**, 1858-1863.

Gottifredi, J. C. & Jameson, G. J. (1968) The suppression of wind-generated waves by a surface film, *Journal of Fluid Mechanics*, 32, 609-618.

Lucassen-Reynders, E. H. & Lucassen, J. (1969) Properties of capillary waves, *Advances in Colloid and Interface Science*, 2, 347-395.

Monahan, E. C. (1969) Fresh water whitecaps, *Journal of Atmospheric Sciences*, **26**, 1026-1029.

Monahan, E. C. (1971) Oceanic whitecaps, *Journal of Physical Oceanography*, **1**, 139-144.

Scott, J. C. (1972) The influence of surface-active contamination on the initiation of wind waves, *Journal of Fluid Mechanics*, **56**, 591-606.

Scott, J. C. (1978) The historical development of theories of wave-calming using oil, *History of Technology*, **3**, 163-186.

Scott, J. C. (1975) The role of salt in whitecap persistence, *Deep-Sea Research*, **22**, 653-657.

Scott, J. C. & Stephens, R. W. B. (1972) Use of moire fringes in investigating surface wave propagation in monolayers of soluble polymers, *Journal of the Acoustical Society of America*, **52**, 871-878.

Scott, J. C. (1979) 'Oil on Troubled Waters: A Bibliography on the Effects of Surface-Active Films on Surface-Wave Motions', Multi-Science Publishing Co. Ltd., London.

Toba, Y, & Kunishi, H. (1970) Breaking of wind waves and the sea surface wind stress, *Journal of the Oceangraphical Society of Japan*, **26**, 71-80.

Whitaker, S. (1964) Effect of surface active agents on the stability of falling liquid films, *Industrial and Engineering Chemistry Fundamentals*, 3, 132-142.

Wu, J. (1971) Evaporation reduction by monolayers: another mechanism, *Science*, **174**, 283-285.

A MODEL OF MARINE AEROSOL GENERATION VIA WHITECAPS AND WAVE DISRUPTION

E. C. MONAHAN, D. E. SPIEL & K. L. DAVIDSON
University College, Galway,
BDM Corporation, Monterey, CA 93940,
Dept. of Meteorology, Naval Postgraduate School, Monterey, CA 93940

1. Introduction

We have, over the past several years, as one element in the development of a time-dependent model of the aerosol population of the marine atmospheric boundary layer, attempted to define, in terms of aerosol droplet radius (r) and 10m-elevation wind speed (U), a model of open-ocean sea-surface aerosol generation. This source function is represented by the expression dF(r, U)/dr, which states the rate of production of marine aerosol droplets, per unit area of the sea surface, per increment of droplet radius. In the initial modeling efforts only the indirect aerosol production mechanisms associated with the bursting of whitecap bubbles (see Figure 1) were considered. The model for sea surface aerosol generation by the indirect mechanisms, first introduced in our Canberra SSAG-1 (Monahan, et al, 1979) and Manchester SSAG-2 (Monahan, 1980) papers, is given by Equation 1, where W is the

(1) $$dF_0/dr = W \tau^{-1} \, dE/dr$$

instantaneous fraction of the sea surface covered by whitecaps, τ is the time constant characterizing the exponential whitecap decay (measured in seconds), and dE/dr is the differential whitecap aerosol productivity, i.e. the number of droplets per increment droplet radius produced during the decay of a unit area of whitecap (expressed in $m^{-2}\ \mu m^{-1}$). The necessary expression for W(U) was obtained from shipboard photographic observations of whitecaps (Monahan, 1971; Toba and Chaen, 1973), while values for τ and dE/dr were derived from measurements made using the University College, Galway, whitecap simulation tank.

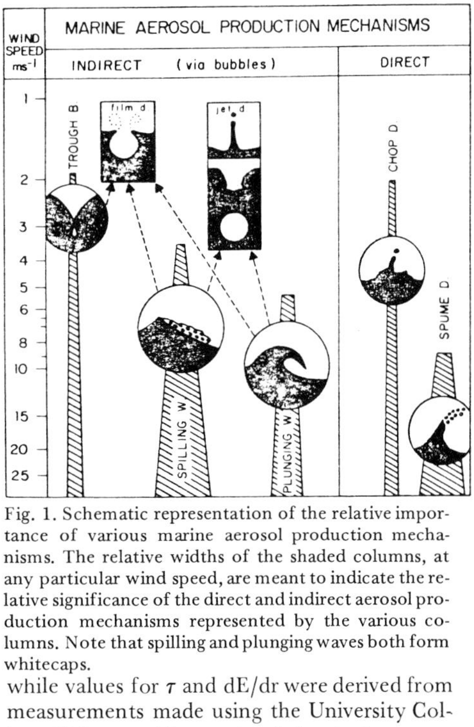

Fig. 1. Schematic representation of the relative importance of various marine aerosol production mechanisms. The relative widths of the shaded columns, at any particular wind speed, are meant to indicate the relative significance of the direct and indirect aerosol production mechanisms represented by the various columns. Note that spilling and plunging waves both form whitecaps.

The specific expression for W(U) used in the latest model of dF_0/dr, SSAG-3 (Monahan, et al, 1982), and given in Equation 2, was extracted

$$W(U) = 3.84 \times 10^{-6}\ U^{3.41} \quad (2)$$

E. C. Monahan and G. Mac Niocaill (eds.), Oceanic Whitecaps, 167–174.

from Monahan and O'Muircheartaigh (1980). In evaluating Equation 2, U, the 10m-elevation wind speed, must be expressed in ms^{-1}. Taking 3.53 s for τ (Monahan et al., 1982), a value that differs by less than 10% from the value obtained from an earlier simulation tank study (Monahan and Zietlow, 1969), and substituting Equation 2 in Equation 1, yields Equation 3. Note that it

$$dF_0/dr = 1.088 \times 10^{-6}\, U^{3.41}\, dE/dr \tag{3}$$

has been assumed that there is one value for τ appropriate for all whitecaps, no matter how large they are, and regardless of how they were formed. While the initial surface area of the whitecaps in the two sets of tank experiments referred to above differed greatly, these laboratory whitecaps all had initial areas that were small compared to typical oceanic whitecaps. The matter of whitecaps formed by spilling versus plunging breaking waves was addressed in Monahan, et al (1982), but clearly the aerosol productivity of whitecaps formed in each of these two ways needs ultimately to be determined separately.

The expression for dE/dr in the SSAG-3 model is given in Equation 4, where V is the volume of

$$dE/dr = V A_0^{-1}\, \Delta\,(dN/dr) \tag{4}$$

air within the hood enclosing the whitecap simulation tank (and beneath the lids of the side channels, in m^3), A_0 is the initial area of the water surface covered by a whitecap resulting from a splashing wave event (expressed in m^2), and $\Delta\,(dN/dr)$ is the change in concentration of aerosol droplets, per increment droplet radius, that occurs in the air within the hood as the result of the decay of a single whitecap (in $m^{-3}\ \mu m^{-1}$). It should be clear that the product of V times $\Delta\,(dN/dr)$ represents the total droplet production, per increment droplet radius, that results from the decay of one whitecap in the tank. In this formulation it has been assumed that the change in concentration of droplets that results from a whitecap decay is uniform throughout the air volume within the hood and beneath the lids.

Using Equation 5, in which the droplet radius (r) is expressed in μm, as the analytical expression

$$\Delta(dN/dr) = 2.38 \times 10^5\, r^{-3} \times 10^{1.19\, e^{-B^2}} \tag{5}$$

$$B = (0.380 - \log r)/0.650$$

describing the r-dependence of the $\Delta(dN/dr)$ measurements presented in Monahan, et al (1982), along with the V and A_0 values that pertained to the same set of simulation tank experiments, i.e. 1.85 m^3 and 0.349 m^2 respectively, makes it possible to evaluate Equation 4, and hence Equation 3. The resulting explicit form of this model, SSAG-3, for the generation of marine aerosols via the indirect, whitecap, mechanisms is given in Equation 6.

$$dF_0/dr = 1.373\, U^{3.41}\, r^{-3} \times 10^{1.19\, e^{-B^2}} \tag{6}$$

$$B = (0.380 - \log r)/0.650$$

2. *General Marine Aerosol Production Model*

The general model of marine aerosol generation which we herein present consists of two terms (Equation 7). The additional term in this

$$dF/dr = dF_0/dr + dF_1/dr \tag{7}$$

model, dF_1/dr, represents the contribution of spume drops (see Figure 1), resulting from the mechanical disruption of wave crests by the wind, to the total droplet production. Study of the low-elevation marine aerosol spectra recently derived from the measurements taken during the 1978 JASIN experiment (Monahan, et al, 1983) has brought us to the conclusion that a mechanism, distinct from those associated with bursting whitecap bubbles (Blanchard, 1963), is a significant contributor of large spray droplets at wind speeds in excess of 9 m s^{-1}. The energetics of mechanical disruption favour the production of just such large droplets as those that gave rise to the upswing, or 'tail', found at the large droplet ($r > 10\ \mu m$) end of the JASIN aerosol spectra associated with winds greater than 9 m s^{-1}.

The starting point in the development of the present SSAG-4 production model is the re-evaluation of dF_0/dr based on the aerosol data collected in the hood of the whitecap simulation tank during the UCG/NPS Experiment III, which was conducted in June and July 1981. A summary of these aerosol measurements, adjusted to a relative humidity of 80%, is to be found in Spiel (1983). An indication of how the instantaneous $\Delta(dN/dr)$ values are extracted from the actual measurements can be gleaned from Figure 2.

On Figure 3 are plotted the curves of $\Delta(dN/dr)$ versus r as determined for the two elevations at which sampling was carried out during the 1981 experiment, 0.015 m and 0.07 m. For comparison, an adjusted version of the $\Delta(dN/dr)$ curve derived from the measurements taken at an elevation of 0.063 m during the 1980 experiment is also plotted on Figure 3, as is a likewise adjusted form of Equation 5. This adjustment consists of multiplying the Δ (dN/dr) values obtained from the 1980 measurements by 1.191, to account for the fact that the enclosed volume of air, V, in the 1981 experiment was only 0.839 the enclosed volume of air in the 1980 experiment. This reduction in the value of V resulted from the use of a smaller hood, and the introduction of two PMS aerosol spectrometer probes within that hood, in the more recent experiment. A consideration

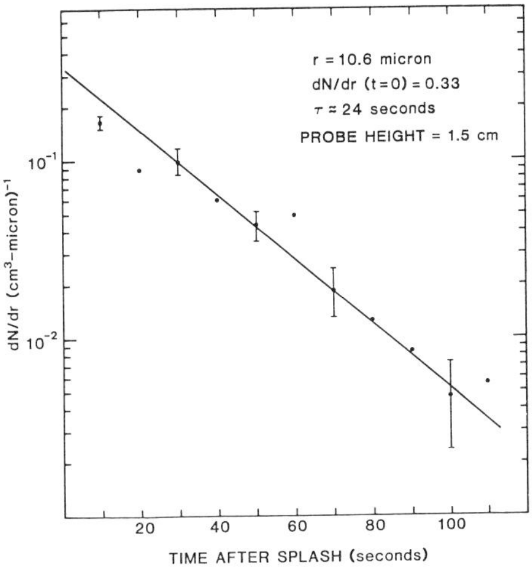

Fig. 2. $\Delta(dN/dr)$ as measured at various times after a breaking wave event. Note that $\Delta(dN/dr)$ at t = 0 s can be inferred by extrapolation from this semi-logarithmic representation.

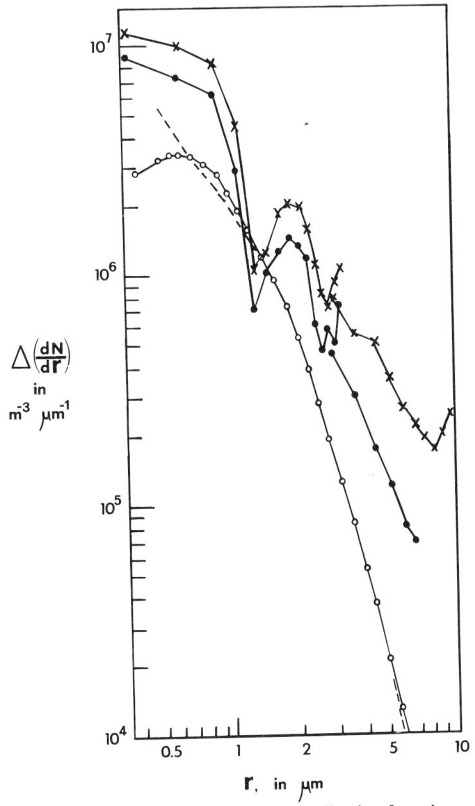

Fig. 3. Change in the aerosol size distribution function, $\Delta(dN/dr)$, in m^{-3} μ^{-1}, that resulted from a standard breaking wave event in the UCG whitecap simulation tank. X's, based on measurements made at an elevation of 0.015m in 1981 (Spiel, 1983). Filled circles, based on 0.07 m- elevation measurements of 1981 (Spiel, 1983). Open circles, based on 0.063 m-elevation measurements made in 1980 (event 387), adjusted to account for V_{80}/V_{81} ratio (see text). Dashed line, Δ(dN/dr) expression of Equation 5, likewise multiplied by V_{80}/V_{81}. The discontinuous nature of the first two curves reflects a shift of sampling range on the PMS spectrometers.

of Figure 3 will confirm the impression that the assumption, embodied in model SSAG-3, that the aerosol droplets produced by a decaying whitecap are mixed uniformly throughout the air enclosed above the tank by the hood and channel lids is not valid. Rather, it is now apparent that right above the water surface in the tank there is a high concentration of whitecap-produced droplets, and that this concentration falls off rapidly with height. This latter contention is based on the relative amplitudes of the 0.015 m- and 0.07 m-elevation curves. We now need to take into account in our dF_0/dr expression this previously overlooked thin layer of high droplet concentration. This can be done by assuming that the increase in aerosol concentration resulting from the decay of a whitecap in the simulation tank can be accounted for by two terms, the uniform $\Delta(dN/dr)$ term already included in the SSAG-3 model and a height-dependent term to describe the contribution of the low, thin, cloud discovered in the 1981 experiment. This leads to the replacement of Equation 4 by Equation 8.

$$dE/dr = A_0^{-1}\ V\,\Delta(dN/dr)_{0\cdot 63\ m} + \int_0^\infty \Delta(dN(z)/dr)dz \qquad (8)$$

Note that it is assumed in Equation 8 that the lateral extent of the low cloud corresponds to the initial horizontal dimensions of the whitecap which produced the aerosols that constitute this cloud. If it is assumed further that the concentration of aerosol droplets in the low, thin, cloud falls off exponentially with elevation above the water surface, then Δ (dN/dr) can be expressed by Equation 9. Substituting Equation 9 in the integral of Equation 8, and

$$\Delta(dN(z)/dr) = \Delta\,(dN(O)/dr)\,e^{-z/a} \qquad (9)$$

carrying out the integration, results in Equation 10. Using the 1981 results included in Figure 3, and additional data from Spiel (1983), the second term of Equation 10 has been evaluated in Monahan (1982). The relative significance of the two terms of Equation 10 is apparent from Figure 4. The ratio of the second term to the first term of Equation 10 can be approximated by the expression given in Equation

$$dE/dr = A_0^{-1}\ V\,\Delta(dN/dr)_{0\cdot 63\ m} + \Delta(dN(O)/dr) \qquad (10)$$

$$S = 0.057\, r^{1\cdot 05} \qquad (11)$$

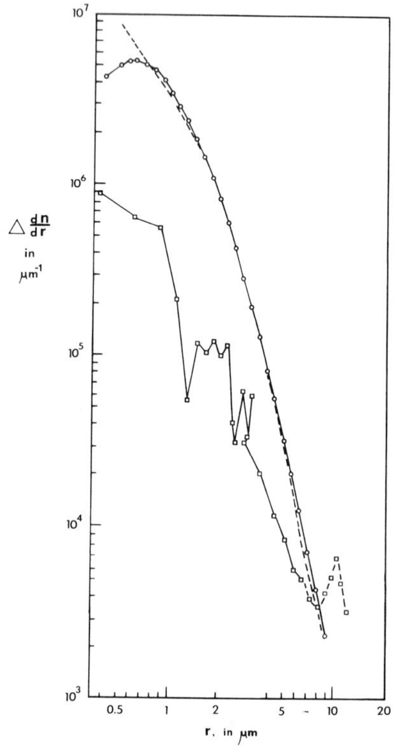

Fig. 4. The increase in aerosol numbers in the hood of the whitecap simulation tank as the result of a standard breaking wave event. Open circles, based on the UCG/NPS Experiment II results (1980), and assumption of uniform change in aerosol concentration throughout the trapped air volume. Dashed, Experiment II results when cast in form of Equation 5. Open squares, contribution of low, thin, cloud of aerosol droplets detected in Experiment III (1981).

11, as can be seen from Figure 5. Substituting Equation 11 in Equation 10, and the resulting expression in Equation 3, yields Equation 12,

$$dF_0/dr = 1.373 U^{3.41} r^{-3} (1 + 0.057 r^{1.05}) \times 10^{1.19 e^{-B^2}} \quad (12)$$

$$B = (0.380 - \log r)/0.650$$

the final form for the first term of the SSAG-4 model (Equation 7). In evaluating this term U must be expressed in m s^{-1} and r must be given in μm.

3. Estimation of Direct Production Term of SSAG-4.

Having concluded, as did Wang and Street (1978) previously, that the generation of spume drops by the mechanical disruption of wave crests becomes a significant aerosol production factor at high wind speeds, it remained for us to identify experimental results that could be used in defining dF_1/dr. Since the UCG whitecap simulation tank does not incorporate a wind tunnel as part of the hood enclosure, the measurements made with that facility could not be expected to shed light on the spume drop production mechanism. At this juncture it seemed appropriate to look carefully at the literature describing various laboratory tank studies, particularly at those papers which describe the results obtained with wind flumes, to see if any information on spume drop production could be extracted from these publications. The results of such a literature search are summarized in Monahan (1982).

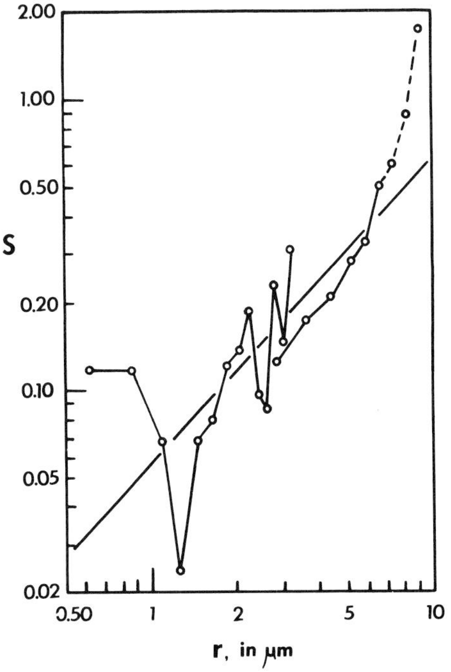

Fig. 5. S, the ratio of the contribution to dE/dr from the low cloud to the contribution from the uniform change in aerosol concentration. Expression for S(r) given in Equation 11.

One of the few studies reported in the literature in which spume drops apparently were being produced and where F_1, the production of droplets m^{-2} s^{-1} was measured as a function of wind speed, U, was that carried out by Wu (1973). The points from Figure 10 of Wu (1973) have been reproduced on Figure 6. The exponential increase of F_1 with U, as approximated by the straight line on the semi-logarithmic plot of Figure 6, is given in Equation 13.

$$F_1 = 2.87 \times 10^{-8} e^{2.08 U} \quad (13)$$

It was now necessary to obtain an expression for dF_1/dr from the results presented in Wu (1973). Having recourse to Wu's Figure 8, and some supplementary descriptions of the droplet spectra that appear in the text of that paper, we arrived at the dF_1/dr function shown on Figure 7, where the rapid drop-off in droplet concentration at radii below the peak in dF_1/dr, which we have taken to fall at a radius of 75μm, is here approximated by a simple cutoff. The U-dependence of M, the peak in the Wu spectrum, is given in Equation 14,

$$M = 1.53 \times 10^{-9} e^{2.08 U} \quad (14)$$

having been determined from the information presented on Figures 6 and 7 (Monahan, 1982). Combining Equation 14 with the Wu spectrum of Figure 7 yielded the initial expression for

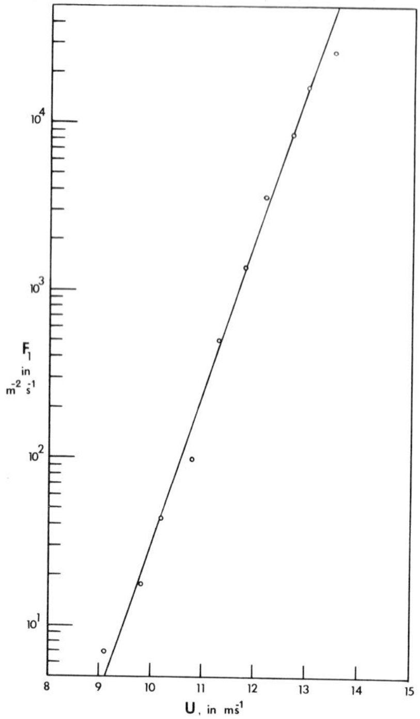

Fig 6. F_1, the number of drops produced per unit area of the mean water surface per unit time, in m^{-2} s^{-1}, versus U, the wind velocity, in m s^{-1}, based on Figure 10 of Wu (1973). The straight line fit, i.e. the exponential description since this is a semi-logarithmic plot, is given by Equation 13.

dF_1/dr presented in Equation 15. This model falls far short

$$dF_1/dr = 0, r < 75\ \mu m \tag{15}$$
$$= 4.83 \times 10^{-2}\ e^{2.08\,U}\ r^{-4} \quad 75\ \mu m \leqslant r < 100\ \mu m$$
$$= 4.83 \times 10^{6}\ e^{2.08\,U}\ r^{-8},\ 100\ \mu m \leqslant r$$

of answering our needs, as it predicts no 10 μm radius spume droplet production, as is required to explain the JASIN observations (Monahan, et al, 1983). But assuming that the cut-off, or low-r 'roll-over', at 75 μm radius in the Wu production spectrum is an artifact of measurement sensitivity (reflecting the detection threshold which was said to fall at 60 μm radius), then we are free to look elsewhere for a description of the dF_1/dr expression in the 10 μm - 75 μm range. Lai and Shemdin (1974) observed in their tank studies a drop size distribution of the form given in Equation 16,

$$dF_1/dr = C\,r^{-2} \tag{16}$$

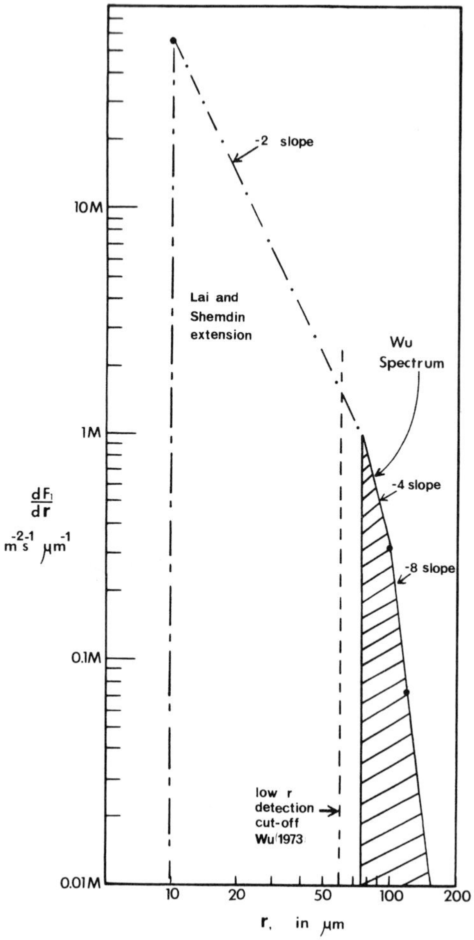

Fig. 7. Spume drop generation model, dF_1/dr. Solid line enclosing shaded region si the production function based on Wu (1973). Dash-dot line represents the extension of the production function towards smaller drop sizes based on work of Lai and Shemdin (1974). M is defined in text.

where most of their measurements were made of droplets with radii between 35 μm and 350 μm. We have adopted the Lai and Shemdin expression for dF_1/dr in the radius range of 10 μm to 75 μm, selecting the value for C so that the value of dF_1/dr at 75 μm determined from Equation 16 equals the value for dF_1/dr at 75 μm radius calculated using Equation 15. The resulting Lai and Shemdin low-r extension to the Wu drop production spectrum is included on Figure 7.

Equation 17 represents the present, and still tentative, statement of the dF_1/dr term for the SSAG-4 sea surface aerosol generation model.

$$
\begin{aligned}
(17)\qquad dF_1/dr &= 0, \; r < 10\ \mu m \\
&= 8.60 \times 10^{-6} e^{2.08U} r^{-2}, \; 10\ \mu m \leqslant r < 75\ \mu m \\
&= 4.83 \times 10^{-2} e^{2.08U} r^{-4}, \; 75\ \mu m \leqslant r < 100\ \mu m \\
&= 4.83 \times 10^{6} e^{2.08U} r^{-8}, \; 100\ \mu m \leqslant r
\end{aligned}
$$

Equation 18 is the result of an initial attempt to recast Equation 17 in analytical form. While the short-comings

$$
(18)\qquad \begin{aligned} dF_1/dr &= 6.45 \times 10^{-4} e^{2.08U} r^{-3} e^{-D^2}, \\ D &= 2.18\,(1.88 - \log r) \end{aligned}
$$

of Equation 18 as a description of the spume drop production term are such that its use is not recommended, it is felt that some such expression, i.e. a Junge-type distribution modified by a Gaussian envelope, will, in the light of future field observations, be found to be a suitable description of this direct production term.

4. *Conclusions*

The status of the two terms of the general aerosol production SSAG -4 model (Equation 7), as presented in Equations 12 and 17, are quite different. Equation 12 represents a refinement of the dF_0/dr term described in Equation 6. This further modification of the dF_0/dr term is in the sense foreseen in Monahan, et al (1982). Equation 12 has been incorporated in a computational model for the aerosol population of the marine atmospheric boundary layer, and the results from this initial application of our expression are encouraging (Burk, 1983).

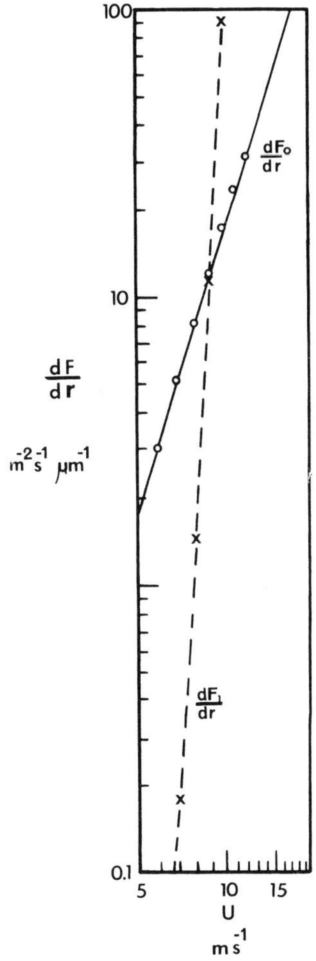

Fig. 8. Generation of droplets of 10 μm radius from the bursting of whitecap bubbles (dF_0/dr) and from the mechanical disruption of wave crests forming spume drops (dF_1/dr). Note that at winds of 9 m s^{-1} the contributions from the two droplet production mechanisms are comparable.

The preliminary expression for dF_1/dr given in Equation 17 is clearly in need of extensive revision to rid it of certain unrealistic features. But even in its present crude form it may have some merit. For example, when we calculate the relative contributions to marine aerosol production of the dF_0/dr and dF_1/dr terms, using Equations 12 and 17, we find that as the wind speed increases beyond 9 m s^{-1} the spume drops of 10 μm radius become more numerous than the bubble-produced drops of this same radius, as is illustrated on Figure 8. This fits well the picture we have drawn from the wind-dependent features of the JASIN 14 m-elevation aerosol spectra (Monahan, et al, 1983), and with the rapid increase in concentration at 0.13 m-elevation of drops with radii greater than 45 μm observed to occur at this wind speed (Monahan, 1968). Likewise, Gathman (1984) found that the wind dependence of the electric current density he calculated using both terms, i.e. Equations 12 and 17, of this SSAG-4 model was appropriate to explain the wind dependence of space charge density observed in the measurements taken over the ocean.

ACKNOWLEDGMENTS

Model SSAG-4 was first formulated by E. C. M. while supported by the Naval Environmental Prediction Research Facility (via ONR N00014-82-M-0071) and in residence at the U. of Maine at Orono. The U.C.G. whitecap simulation tank, and all other aspects of the whitecap research of E.C.M. at University College, Galway, are supported by the Office of Naval Research (N00014-78-G-0052 and N00014-82-G-0024). The participation of K. L. D. and D. E. S. in the Galway tank experiments, and the work at the Naval Postgraduate School in support of these experiments is also supported by O.N.R. (N00014-81-WR-10135). The reduction of the aerosol data was carried out by D.E.S. at BDM (N00014-79-C-0088). The authors acknowledge with thanks the assistance during UCG/NPS Experiment III of P. Bowyer, D. Doyle, F. Gaffney, M. Spillane, and J. Taper. A version of this paper was presented at the American Meteorological Society's Ninth Conference on Aerospace and Aeronautical Meteorology, held in Omaha, Nebraska, in June 1983.

REFERENCES

Blanchard, D. C., 1963: The electrification of the atmosphere by particles from bubbles in the sea, *Progress in Oceanography*, Vol. 1, Pergamon Press, 71-202.

Burk, S. D., 1983: A turbulence model of the generation, transport, and deposition of marine aerosols, *Proceed., 6th Symposium on Turb. and Diffusion*, A. M. S. (in press).

Gathman, S. G., 1986: Atmospheric electric space charge near the ocean surface (this volume).

Lai, R. J., and O. H. Shemdin, 1974: Laboratory study of the generation of spray over water, *J. Geophys. Res.*, 79, 3055-3063.

Monahan, E. C., 1968: Sea spray as a function of low elevation wind speed, *J. Geophys. Res.*, 73, 1127-1137.

—, 1971: Oceanic whitecaps, *J. Phys. Oceanogr.*, 1, 139-144.

—, 1980: Positive charge flux from the world ocean resulting from the bursting of whitecap bubbles, *Abstracts, VI Internat. Conf. on Atmos. Elect., Manchester*, 147-150.

—, 1982: Sea surface aerosol generation model No. 4, U. of Maine at Orono Tech. Rep., 54 pp.

—, K. L. Davidson, and D. E. Spiel, 1982: Whitecap aerosol productivity deduced from simulation tank measurements, *J. Geophys. Res.*, **87**, 8898-8904.

—, C. W. Fairall, K. L. Davidson, and P. Jones-Boyle, 1983: Observed inter-relationships amongst 10m-elevation winds, oceanic whitecaps, and marine aerosols, *Quart. J. Roy. Meteor. Soc.*, **109**, 379-392.

—, and I. O'Muircheartaigh, 1980: Optimal power-law description of oceanic whitecap coverage dependence on wind speed, *J. Phys. Oceanogr.*, **10**, 2094-2099.

—, B. D. O'Regan, and K. L. Davidson, 1979: Marine aerosol production from whitecaps, *I.U.G.G. XVII General Assembly, Canberra, Abstracts*, 423.

—, and C. R. Zietlow, 1969: Laboratory comparisons of fresh-water and salt-water whitecaps, *J. Geophys. Res.*, 74, 6961-6966.

Spiel, D. E., 1983: A study of aerosols generated in a whitecap simulation tank, BDM Tech. Rep. 006-83, Monterey, 35 pp.

Toba, Y., and M. Chaen, 1973: Quantitative expression of the breaking of wind waves on the sea surface, *Records, Oceanogr. Works Japan*, **12**, 1-11.

Wang, C. S., and R. L. Street, 1978: Measurements of spray at an air-water interface, *Dyn. Atmos. Oceans*, 2, 141-152.

Wu, J., 1973: Spray in the atmospheric surface layer: Laboratory study, *J. Geophys. Res.*, **78**, 511-519.

THE PRODUCTION AND DISPERSAL OF MARITIME AEROSOL

H. J. EXTON, J. LATHAM, P. M. PARK & M. H. SMITH
Department of Pure and Applied Physics
University of Manchester Institute of Science and Technology

and

R. R. ALLAN
Space Department
Royal Aircraft Establishment
Farnborough

Field experiments concerned with the characteristics and variability of maritime aerosol have been conducted, over several years, at a coastal site on the island of South Uist, in the Outer Hebrides. In the primary experiments aerosol size distributions (radii r 0.08 to 16μm) and basic meteorological parameters were measured from a 10m tower, for both maritime and continental air-streams. In all cases the volumetric aerosol loading V ($\mu m^3 cm^{-3}$) increased rapidly with windspeed U ($m\ s^{-1}$) according to

$$\ln(V) = a\,U + b,$$

where, for the larger aerosol in maritime air, a~0.2 and b~4. The size distributions could be sub-divided into two ranges separated at r~0.3μm, the lower corresponding to long-range aerosol and the higher to locally produced aerosol. In maritime air, this latter category is produced by bubble bursting.

In subsidiary field experiments evidence was found for: the production at the ocean surface of substantial quantities of aerosol in the radius band 20 to 60μm; a decrease of aerosol concentrations N and loading with altitude; an inverse relationship between N and mixing depth; an insensitivity of aerosol characteristics to the range of relative humidity experienced at South Uist - but not at an alternative site; and clear evidence from chemical and microscopic examination that the aerosol in maritime air-streams arises principally from the ocean.

Laboratory experiments revealed the sensitivity of bubble size-distributions to temperature and salinity. An assessment of the distribution of maritime aerosol loading over the North Atlantic indicates considerable spatial and temporal variability.

1. Introduction

Since 1979, scientists from RAE and UMIST have been collaborating in an extensive series of field investigations of atmospheric aerosol, of radius r exceeding ~0.1μm, at the coastal site of Ardivachar Point on the island of South Uist, in the Outer Hebrides. Its location is depicted in Figure 1. The size distributions and concentrations of the aerosol have been measured - over extended periods, at various times of year - as a function of wind-speed, air-mass trajectory, relative humidity, altitude, mixing-depth, temperature, tidal condition and other parameters. Some chemical analysis by scanning-electron microscopy of the aerosol was also performed. Most measurements were from the ground at altitudes of 2m and 10m, although a helicopter was employed on occasion for aerosol sampling at higher levels. These field studies have been supplemented by laboratory experiments concerned with air bubbles, aerosol production mechanisms, aerosol composition and other features of the overall problem. Theoretical work, and an assessment of the implications of the field studies to the North Atlantic as a whole, have also been undertaken.

The primary objective of this paper is to provide a cursory but wide coverage of these investigations , and thereby to elucidate the physical processes most important in determining the aerosol characteristics of the marine boundary layer. We envisage that more detailed and specific papers on this work will be submitted for publication in the near future.

E. C. Monahan and G. Mac Niocaill (eds.), Oceanic Whitecaps, 175–193.

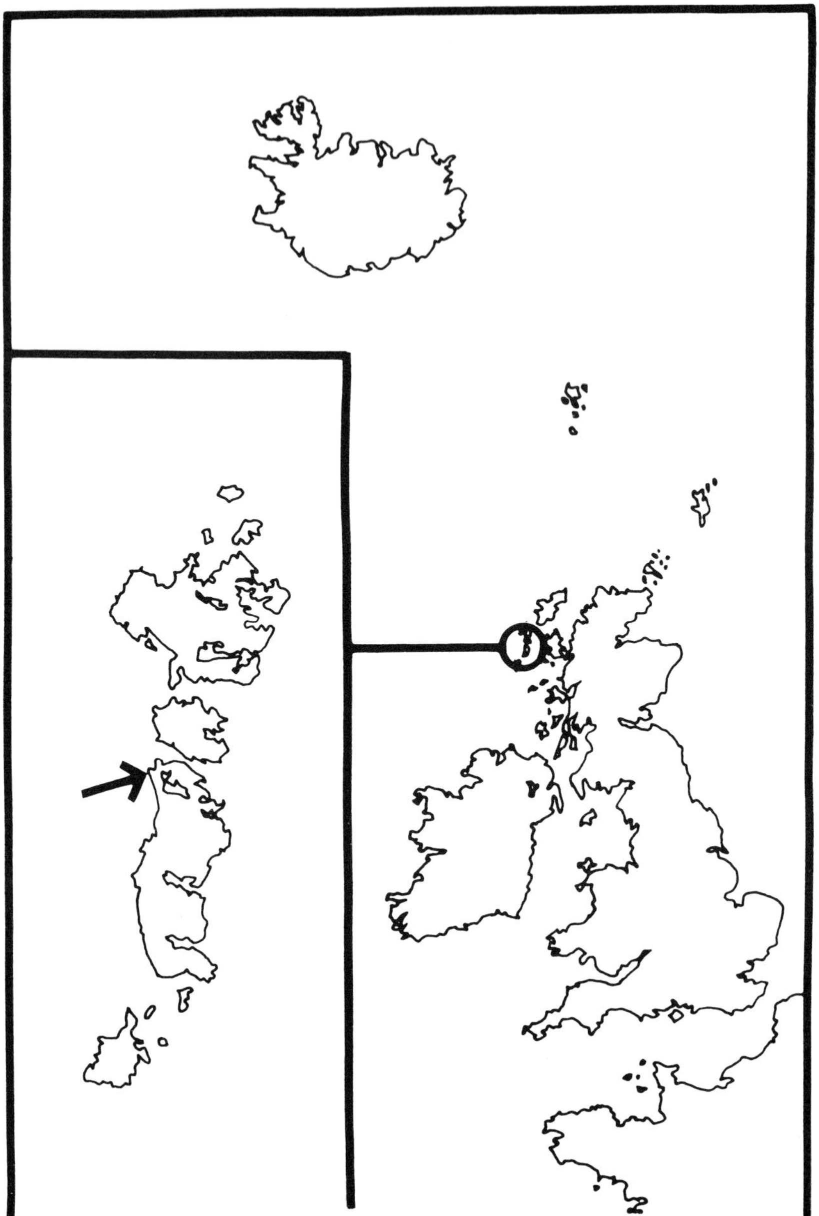

Fig. 1. Map of North-East Atlantic region. Inset: the Uist Islands, which are circled on the larger map. Ardivachar Point, the experimental site on south Uist, is indicated by an arrow.

2. Experimental procedures: Treatment of data

In this section we devote our attention to the basic apparatus and procedures employed in the primary experiments, in which the RAE group sampled the aerosol with two PMS probes (the ASAS and the CSAS). These two probes covered the radius range 0.08 to 16μm in six overlapping bands, each divided into 15 bins of equal width.

The probes were mounted at the top of a 10m tower, situated on a grassy bank at the head of a gently sloping beach. They were oriented automatically into the wind by means of a pivoted mount slaved to a weather-vane. In the first experiments they were operated on range 0 (combined radius band 0.3 to 16μm). Later they were automatically cycled through the various ranges to provide the maximum coverage, 0.08 to 16μm.

The probe data were recorded, using a PMS Data Acquisition System (DAS-64) and Hewlett-Packard mini-computer, on cassette tapes. Meteorological data were recorded concurrently; an anemometer was mounted on the 10m tower adjacent to the particle counters, whilst an EG & G humidity sensor was situated at 2m above ground-level near the base of the tower. Aerosol and meteorological data were recorded every five minutes for single range probe data and every 100 seconds in the autoranged experiments. During daytime hours, frequent manual meteorological measurements were recorded as a check on the automated system, together with notes on the beach and tide states, the synoptic situation and potential local sources of pollution.

Aerosol data for each day were classified as maritime or continental according to air trajectories drawn from surface weather maps. On about two-thirds of the total of 77 days of the primary experiment, conducted between April 1980 and March 1981, maritime air from

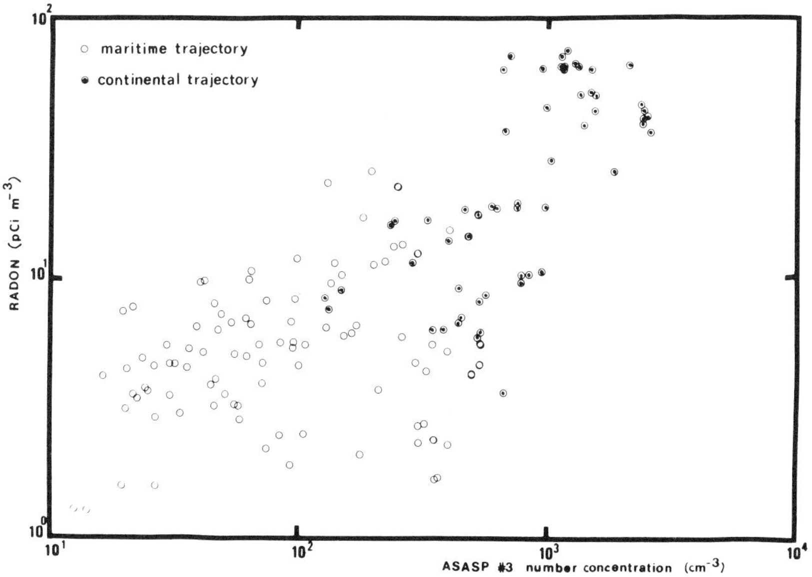

Fig. 2. Variation in radon concentration, R, with particle concentration, N, for ASASP Range 3 (0.08 to 0.15μm radius) for all air masses.

the region of the North Atlantic was encountered, and on the remainder, continental air masses traversing mainland UK or Ireland reached the site. On some occasions radon counting was used as a supplementary technique of characterising the air mass. Figure 2 shows a typically strong correlation between the higher values of radon count, R, and aerosol concentration, N, as measured with the ASAS probe. It is clear that our synoptic technique for air-mass characterisation, with the expectation that continental air will contain the higher values of N and R, is confirmed by these results.

Table 1 provides information on the meteorological characteristics of the five experimental periods (4 maritime, 1 continental) selected for closest examination herein. The four maritime data sets were recorded at different times of year, and covered wind speeds, U, in the range of approximately 0.5 to 20 ms^{-1}. During these periods relative humidity H varied between about 60% and 100%, with the majority of data collected with H close to the mean value of 85%. In order to smooth out short-time-scale variations, the probe and meteorological data were reduced to hourly-averaged values, from which spectral concentrations were plotted as number and volume distributions in the form dN/dlogR and dV/dlogR (Junge 1963). All number and volume concentrations used in correlations were hourly-averaged.

In plotting spectra, the response of the PMS probes was taken into account. As voltage settings on the first channels are close to the noise levels of these instruments, the probes have a tendency to overcount. Therefore channels 1 and 2 on the ASAS probe and channel 1 on the CSAS probe were disregarded in all ranges. Also, the Mie scattering transfer function in the region of 1μm gives rise to multi-valued responses (Pinnick and Auvermann 1979), so channels in the size band 0.5 to 3.0μm have been grouped. In the regions of overlap between probe ranges, the counts were combined to give a single value of particle concentration for each band.

3. *The primary field experiments: aerosol number and volume distributions*

The dN/dlogR spectra averaged separately

Table 1

Dates	Air Mass Category	No. hrs. data large (small)	Max + Min values U_{10} ($m\ s^{-1}$)	Max + Min values RH (p.c.)	Mean values U_{10} ($m\ s^{-1}$)	Mean values RH (p.c.)	T_{dry} (°C)	Sea Surf Temp. (°C)
15-28 Apr.'80	M	230 (0)	1.6-12.8	58-99	7.0 ± 2.8	82 ± 11	8.2 ± 1.4	9.0
19-25 Aug.'80	M	118 (0)	0.6-15.9	60-95	8.2 ± 4.1	79 ± 10	13.2 ± 0.9	14.0
2-15 Dec.'80	M	295 (292)	1.5-19.3	62-100	10.3 ± 3.9	85 ± 8	6.7 ± 2.2	11.0
6-9 Mar.'81	M	61 (61)	3.2-18.9	58-88	11.2 ± 3.8	78 ± 8	6.8 ± 1.3	8.0
24 Feb - 3 Mar.'81	C	160 (72)	3.3-18.1	49-98	11.2 ± 3.6	81 ± 8	4.3 ± 1.3	8.0
Combined maritime	M	704 (353)	0.6-19.3 (1.5-19.3)	49-100 (49-100)	9.4 ± 4.0 (10.1 ± 3.9)	82 ± 9 (84 ± 8)	7.5 ± 3.1 6.3 ± 2.2	–

Bracketed () quantities refer to the equivalent data sets for the small particles only

Table 1. Summary of meteorological conditions during the four maritime and one continental air mass periods under consideration. The number of hours data is quoted for both the large (0.25 to 16μm) and small (0.1 to 0.25μm) size ranges. U_{10} is the 10m wind speed. M indicates a maritime air mass trajectory and C a continental.

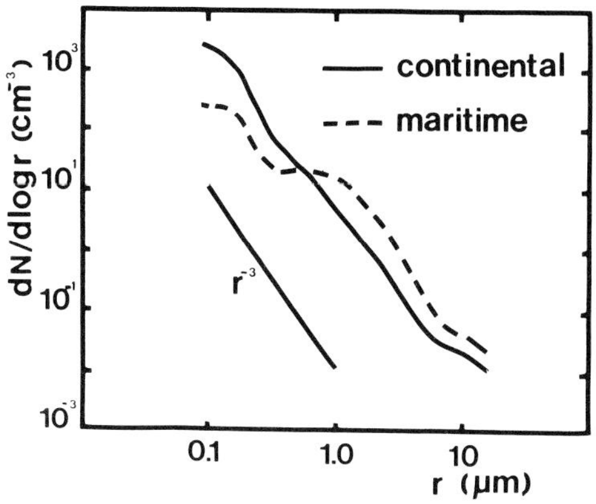

Fig. 3. Spectral number distributions averaged separately over the entire maritime and continental air mass sampling periods and plotted logarithmically in the form dN/dlogR against R. Averaged meteorological parameters for each data period are listed in Table 1.

over the entire periods of maritime and continental air mass sampling are shown in Figure 3. The data for the continental air approximate closely to the Junge r^{-3} power law, as illustrated in the figure. The average of the spectra for the four periods of sampling of maritime air - which agree well with each other - departs from the continental-type spectrum in two major features: the concentration of particles in the size band 0.1-0.25μm is lower by an order of magnitude; and the 0.25-16μm size band shows a substantially enhanced population. The reduction in the number of 0.1-0.25μm particles for the maritime cases is consistent with the idea that such particles are mainly continental in origin, and an extended maritime air trajectory immediately prior to sampling has reduced the concentrations by some process such as rainout, washout or coagulation. The excess number of 0.25-16μm particles under maritime conditions suggests that particles in this size range are produced at the sea surface.

The windspeed-related whitecapping phenomenon (Monahan 1971) is known to introduce large quantities of air into water, which results in the bursting of bubbles at the sea surface and the associated production of jet and film drops (Blanchard & Woodcock 1957). In order to investigate the relationship between aerosol concentration and windspeed, the data were divided into 5ms^{-1} wind-speed bands. Figures 4 and 5 present the resulting data in the form of volume distributions dV/dlogr for the maritime and continental data respectively.

The figures reveal that for both types of air-mass there is an increase of volumetric loading with increasing wind-speed, U, over the entire size-range examined. This is of similar magnitude in both cases. In the maritime case it is probable that this increase is attributable to a greater amount of bubble bursting at the ocean surface as U increases; Meszáros and Vissy (1974) found sub-micron NaCl particles in oceanic aerosol. Since the 'continental' air sampled flowed over water for part of its journey to the measurement site, it is likely that some significant contribution to the additional volumetric loading found with increasing U results from bubble-bursting. This argument is consistent with the observation, reflected in figures 4 and 5, that for both types of air mass dV/dlogr exhibits maxima at radii of about 0.2 and 2μm. We cannot eliminate the possibility that instrumental characteristics slightly modified these spectral shapes and contributed to this wind-speed effect, although they are unlikely to have been dominant.

The values of dV/dlogr at the radii corresponding to the two spectral peaks are indicative of the contributions of long-distance effects (for r~0.2μm) and localised ones (r~2μm) respectively. In the maritime case, the volumetric loading at r~2μm is about 100 times that at r~0.2μm, presumably because of an efficient large-particle production mechanism (bubble-bursting) occurring close to the site of measurement. In the continental case the volumetric loadings are similar at the two peak radii - the entire curve being much flatter than for maritime air - probably because the small-particle background loading is higher and larger particles tend to be deposited to ground during their journey over the island.

There is clear evidence (figure 4) for a rapidly increasing volumetric loading in maritime air as r increases from about 8μm to 16μm, the upper limit of detection by the CSAS probe. In order to determine the volumetric distribution at

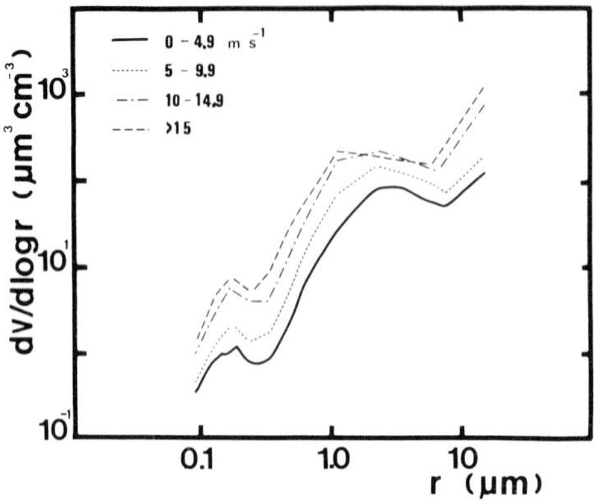

Fig. 4. Variation in volume spectra, dV/dlogR, with windspeed for maritime air masses (all maritime data summed).

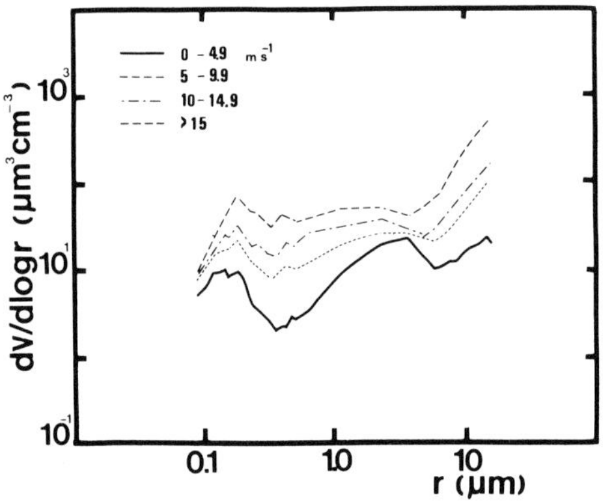

Fig. 5. Variation in volume spectrum with windspeed of a continentally influenced aerosol at South Uist (data for February 1981).

higher values of r, a PMS Optical Array probe was operated on some occasions, at the same site. Figure 6 presents some specimen results for a maritime airstream. They reveal a major contribution to the loading for particles of radii ranging from about 10 to 80μm. The peak at r $\sim$25μm fits well with the idea that jet-droplet

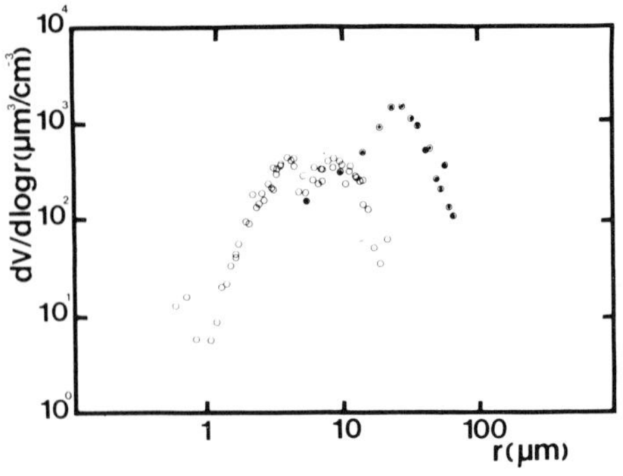

Fig. 6. Variation in volume spectra, dV/dlogR, with particle radius, r, for a maritime airstream. ASASP, FSSP (marked by o) and OAP (•) instruments operated at high water mark, 2 m above ground.

production from breaking waves is the dominant mechanism of production of these large aerosol. The weaker but definite evidence for large aerosol production in the case of 'continental' air flows (figure 5) is indicative of contributions from both soil erosion and jet-drop production.

Further support for the important role of bubble-bursting in determining the atmospheric loading for all sizes of aerosol in the maritime case is provided by figures 7, 8 and 9, which constitute a set illustrating the simultaneous variations, over a 3-day period, of aerosol number concentration N over all size ranges of the 3 probes operated at 2m by the UMIST group. A tidal signature is clearly evident - especially in the OA probe data - and there is seen to be a strong association between the values of N (for all r) and the distance between the water/beach boundary and the probes, which ranged from about 5m at high tide, when N was a maximum, to 270m at low tide.

4. The primary field experiments: effects of wind speed

The volumetric distributions displayed in figures 4 and 5 may conveniently be divided into the size-ranges 0.1 to 0.25μm ('small') and 0.25 to 16μm ('large'), which correspond to the two ranges identified by Junge. The particles in the

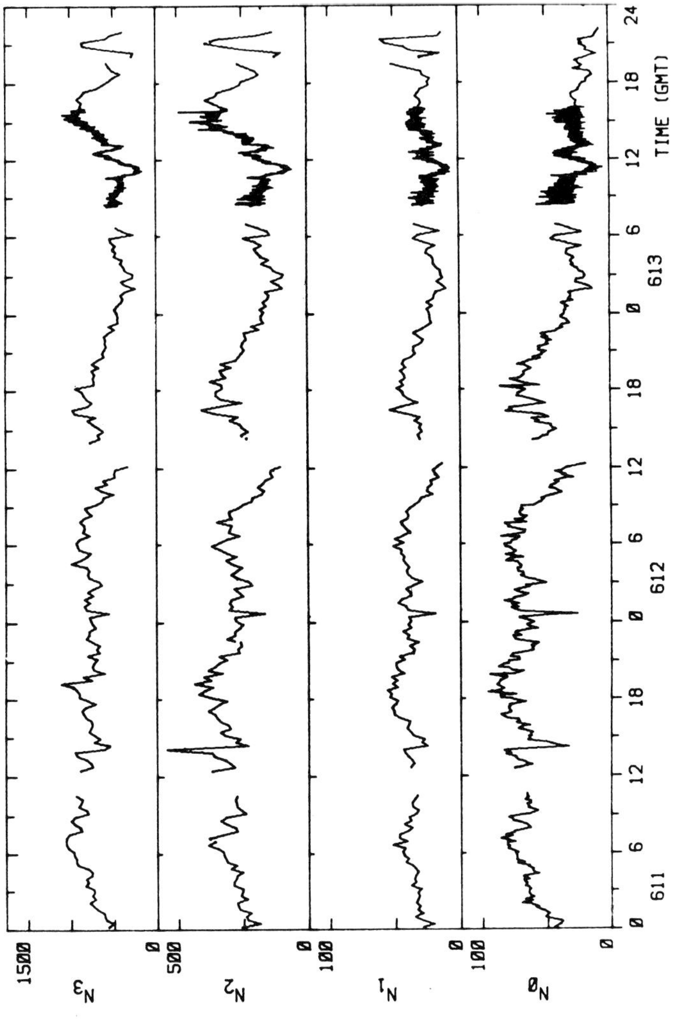

Fig. 7. Variation in aerosol number concentrations measured by the ASASP over a 3-day period for the radius bands N_0, 0.3μm to 1.5μm; N_1, 0.2 to 0.5μm; N_3, 0.10 to 0.3μm, and N_3, 0.08 to 0.15μm.

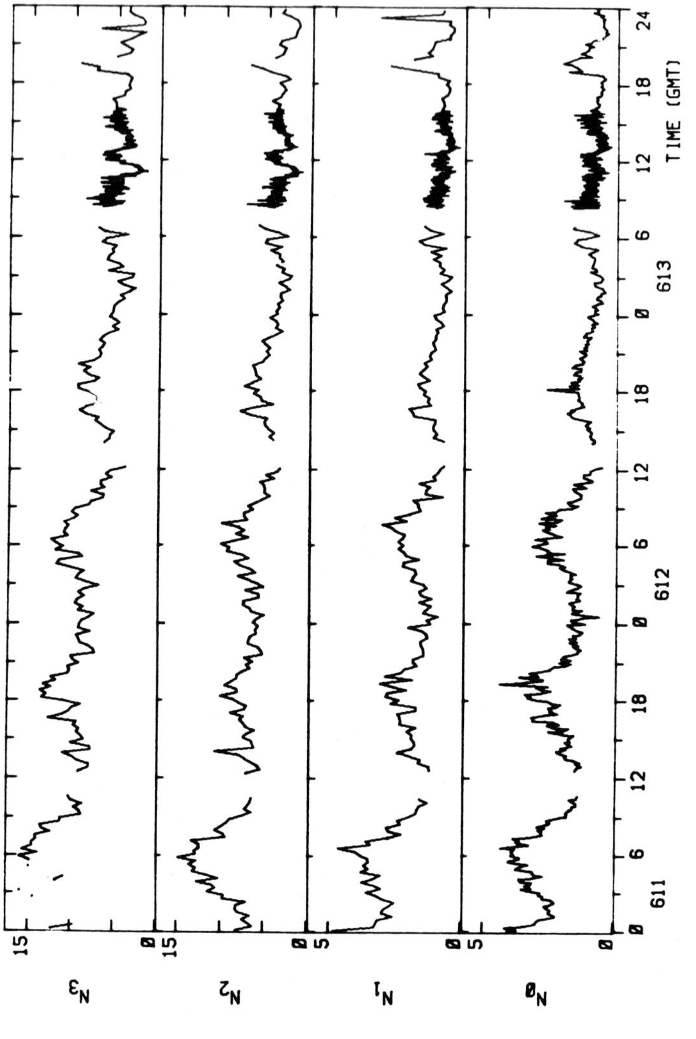

Fig. 8. Variations in aerosol particle concentration measured by the FSSP over a 3-day period for the radius bands N_0, 1.0 to 23.5μm; N_2, 1.0 to 16μm; N_2, 0.5 to 8μm and N_3, 0.25 to 4.0μm.

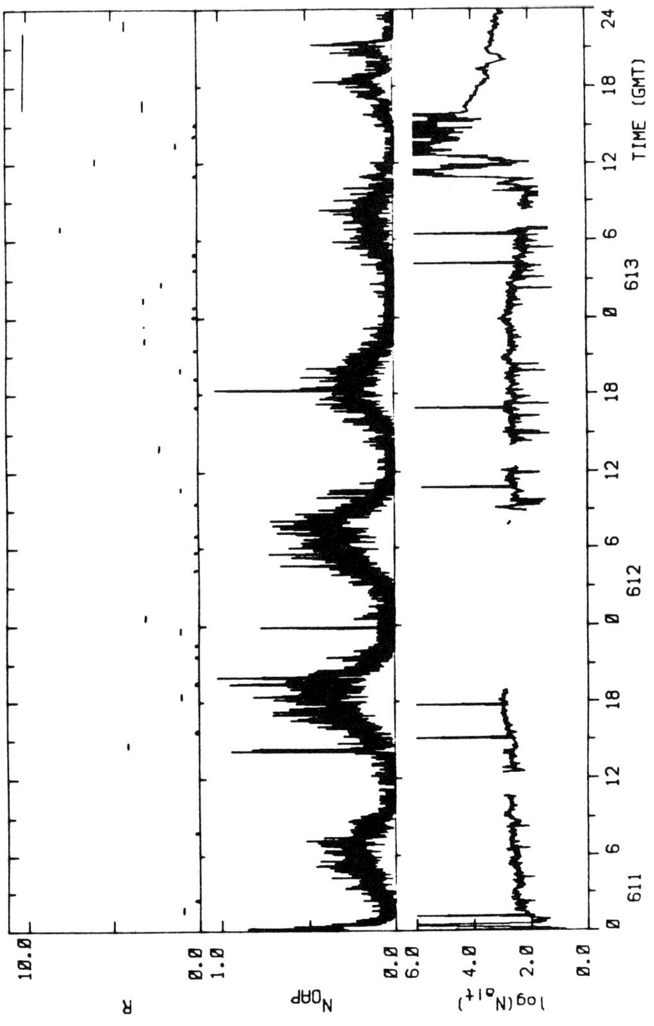

Fig. 9. Variation in aerosol concentrations measured by the CAP over a 3-day period for particle radii from 5 to 150μm. Bars indicate the intensity and duration of rain during each hour. The lower trace indicates the variation in Aitken nucleus concentration throughout the period.

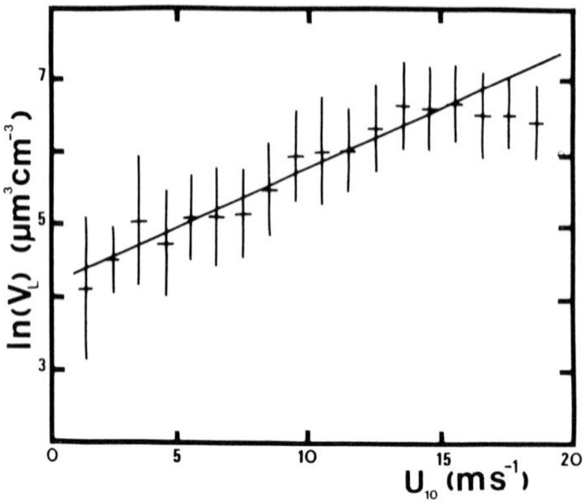

Fig. 10. Dependency of integrated 0.25 to 16μm particle volume (V_L) on wind-speed (U_{10}), plotted in $1ms^{-1}$ wind speed bands for all maritime data. Bars represent the standard deviation in each band, with single values indicated by -. The straight line is the linear regression fit to the data.

one range are small and capable of engaging in long-range transport, while those in the other range are of high mass and of identifiably local origin. In the case of a maritime air-stream this local source is the ocean surface.

Figure 10 presents the measured relationship between V_L, ($\mu m^3\ cm^{-3}$), the volumetric fraction for large particles, and windspeed, U_{10}, measured at the 10m level. This representation was obtained by summing all the data for the four maritime studies listed in Table 1. V_L is seen to increase rapidly with increasing U_{10} up to wind-speeds of around $13m\ s^{-1}$, above which V_L remains essentially constant. This flattening of the curve was found for each individual maritime case, and suggests a limit to the production of airborne aerosol of this size-range from breaking wave processes. This feature could be a consequence of increased turbulent dispersal of the aerosol as U_{10} increases.

Table 2 presents relationships, determined from the data, between the number concentration N_L, the volumetric fraction V_L, and the wind-speed U_{10} for each maritime case study, the continental case and the combined maritime data. These are of the form

$$\ln(V_L, N_L) = a\,U_{10} + b \qquad (1)$$

as suggested by Toba (1961). We see that N_L is somewhat more sensitive to U_{10} (the value of a is higher), than is V_L. This observation reflects the fact that as the wind-speed increases the number concentration increases most at the small-radius end of the size band. The values of a for the individual maritime case studies exhibit significant variation, which analysis shows was due largely to differences in measurement procedure employed in the 4 cases. A further contribution to this variability arises from the differences in the ranges and mean values of salient meteorological parameters, especially wind-speed. The continental air-mass studied also shows a strong dependence of both N_L and V_L upon U_{10} (with a larger in the former case, as was the case for the maritime air

Table 2

Data Set	Air Type	No. hrs. data	Volume concentration $\ln(V_L) = aU_{10} + b$	r	V_L at $U_{10}=0$	Number concentration $\ln(N_L) = aU_{10} + b$	r	N_L at $U_{10}=0$
Apr	M	230	$0.20U_{10} + 4.12$	.66	61.7	$0.25U_{10} + 1.33$	.79	3.8
Aug	M	118	$0.22U_{10} + 3.67$	.83	39.1	$0.29U_{10} + 0.36$	.94	1.4
Dec	M	295	$0.16U_{10} + 4.06$	.67	58.1	$0.19U_{10} + 1.33$	.76	3.8
Mar	M	61	$0.20U_{10} + 3.91$	.74	49.9	$0.23U_{10} + 0.86$	.85	2.4
Combined maritime	M	704	$0.17U_{10} + 4.14$	.69	62.9	$0.20U_{10} + 1.34$	.78	3.8
Feb	C	160	$0.14U_{10} + 3.01$	.56	20.2	$0.19U_{10} + 1.60$	.72	4.9

Table 2. Results of weighted linear regression fits for large particles (0.25 to 16μm radius) in the form $\ln(y) = aU_{10} + b$ where y = volume, V_L, ($\mu m^3\ cm^{-3}$) or number, N_L, (cm^{-3}) concentration. U_{10} is the 10 m windspeed ($m\ s^{-1}$) and a, b are constants. r is the correlation coefficient, and air type is defined as maritime (M) or continental (C).

masses), but the values of a which characterise these relationships are substantially lower than for maritime air. Table 2 shows that when the maritime data are combined N_L (cm^{-3}) and V_L ($\mu m^3\ cm^{-3}$) are related to U_{10} (ms^{-1}) by the expressions:

$$\ln(N_L) \sim 0.20U_{10} + 1.3, \quad \ln(V_L) \sim 0.17U_{10} + 4.1. \tag{2}$$

A similar analysis applied to the data for the small particles yields the curve shown in figure 11 relating the volumetric fraction, V_S, for small particles (0.1 to 0.25μm) in maritime air (all four studies combined), to the windspeed U_{10}. The relationships between V_S, N_S (number concentration of small particles) and U_{10} are presented in Table 3. The volumetric fraction is seen from the figure to increase with U_{10} at a slightly lower rate than for the large particles, and to level out at $U_{10} \sim 13m\ s^{-1}$. This observation is strongly suggestive of a significant contribution to the 'continental' aerosol of particles produced at the ocean surface. The table shows that the values of a are less for the small particle relationships than for the corresponding large-particle ones; for both N and V, maritime and continental. Also, for the small aerosol, a is greater for V_s than for N_s; and for maritime air than for continental air. The continental air is seen to have a larger background ($U_{10} = 0$) concentration of aerosol N_s than maritime air. Table 3 shows that when the maritime data are combined N_s and V_s are related to U_{10} by the expression:

$$\ln(N_s) \sim 0.12\,U_{10} + 3.5, \quad \ln(V_s) \sim 0.14\,U_{10} - 1.1 \tag{3}$$

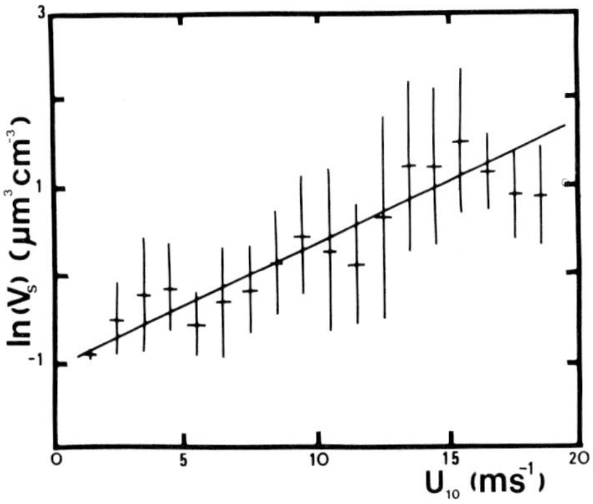

Fig. 11. Dependency of 0.1 to 0.25μm particle volume (V_S) on wind-speed (U_{10}) for the combined maritime data. Bars indicate the standard deviation in V_S in each $1ms^{-1}$ wind speed bin.

The similarities between the features revealed by figures 10 and 11 are remarkable; small particles display a similar increase in volume (mass) loading with windspeed to that shown by the large particles; but the rate of increase is lower for the small particles. The constancy of the volumetric fraction once U_{10} rises above $13m\ s^{-1}$ is found for both large and small particles. Also, the less prominent but clear features at about $4m\ s^{-1}$ and $9m\ s^{-1}$, where large-particle mass-loading appears to reach plateaus, also appear on the small-particle plots. Since V_L is measured mainly by the CSAS probe, and V_s exclusively by the ASAS probe, the features are unlikely to represent a purely instrumental effect. We conclude that a close relationship exists between small and large particle mass-loading. Figures 12 and 13 illustrate such a relationship, for maritime air, for the December

Table 3

Data Set	Air Type	No. hrs. data	Volume concentration $\ln(V_s) = aU_{10} + b$	r	V at $U_{10}=0$	Number concentration $\ln(N_s) = aU_{10} + b$	r	N_s at $U_{10}=0$
Dec	M	292	$0.15U_{10} - 1.09$	.60	0.34	$0.13U_{10} + 3.5$	.56	32.6
Mar	M	61	$0.14U_{10} - 1.37$	.57	0.25	$0.11U_{10} + 3.34$	.50	28.3
Combined maritime	M	353	$0.14U_{10} - 1.10$	.58	0.33	$0.12U_{10} + 3.50$	.53	33.0
Feb	C	72	$0.13U_{10} + 1.57$	.71	4.83	$0.9U_{10} + 6.44$	.59	628.7

Table 3. Linear regression fits for small particles (0.1 to 0.25μm) in the form $\ln(y) = aU_{10} + b$ where y is the volume, V_s, ($\mu m^{-3} cm^{-3}$) or number, N_s, (cm^{-3}) of small particles. Notation as in Table 1.

1980 and March 1981 data periods respectively. The second data set reveals an approximately linear relationship between V_s and V_L. The December set is better represented by two straight-line relationships, one similar to that of the March data, and the other associated with an enhanced degree of small-particle loading. V_s was greater than $5\mu m^3\ cm^{-3}$ from about 040 hours GMT on 9 December to about 0700 GMT on 10 December. These times coincide respectively with the passing of warm and cold fronts associated with a complex system of low pressure regions, and reflects the strong sensitivity of the small-particle loading to synoptic variability. This period was characterised by consistently high wind-speeds (>14m s^{-1}), which had persisted at this level since 1300 GMT on 8 December.

High values of relative humidity (H always >90%, for 7 hours above 95%) existed during the period of enhanced small-particle loading, but analysis of the data revealed that the changes in particle concentration were associated with fluctuating windspeed and not with changes in H. No correlation between U_{10} and H was found for this or any other period of data collection.

It is therefore concluded that the small-particle loading in the maritime environment is closely related to the large-particle loading, and to windspeed. As it is evident that large particles originate via the bubble-bursting phenomenon, it follows that the small-particle loading contains a certain proportion of sea-salt droplets, which presumably originate in the shattering of the bubble film (Day, 1964).

The linear correlation between small and large particles may be used to provide a crude indication of the fraction of the small-particle loading associated with continental influences. Taking the maritime data of Figure 13 for March 1981, the relationship between V_s and V_L can be expressed by:

$$V_s \sim V_L/500 + 0.4 \quad (4)$$

If the large particles are composed entirely of sea-salt droplets, then the value of V_s at zero V_L indicates the background loading of continental aerosol in the maritime environment. Assuming that the density of the continental material is $2g\ cm^{-3}$, then the continental mass-loading is $<1\mu g\ m^{-3}$; the estimated value for air of 'continental' origin is about $10\mu g\ m^{-3}$.

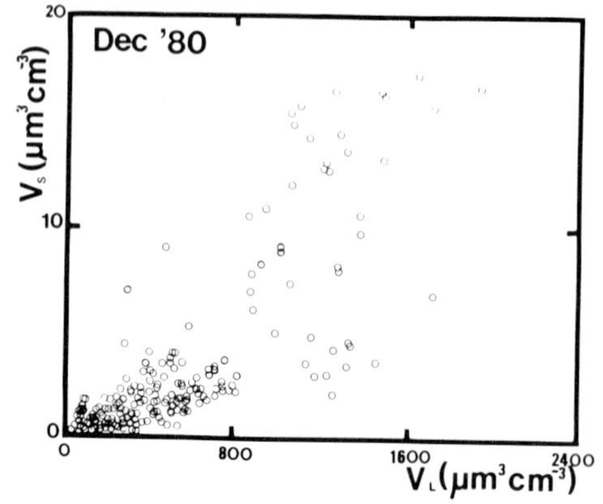

Fig. 12. Scatter plots of large (V_L) versus small (V_S) particle volume concentrations, for the maritime data of December 1980.

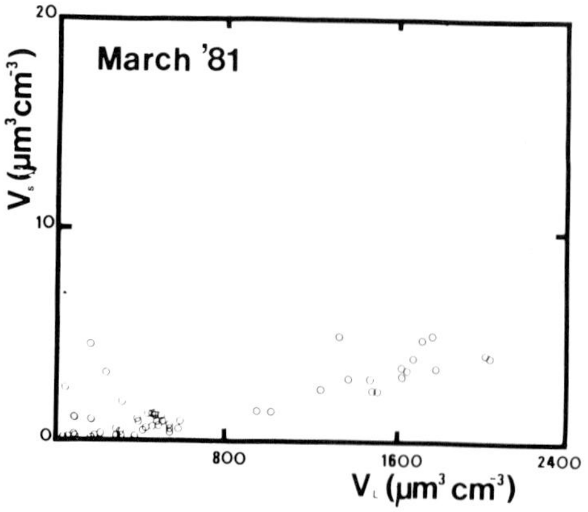

Fig. 13. As in Fig. 12, but for March 1981.

5. Subsidiary experiments: altitude, mixing depth, humidity, temperature, chemical constitution

In this section a brief outline is presented of field and laboratory studies subsidiary to the main experiments. These are concerned with various factors which influence aerosol production and concentration. More detailed accounts of this work will be described elsewhere.

A Wessex helicopter carrying a PMS FSS probe was employed during the period 20 to 28 August 1980 to study the vertical and horizontal variations of aerosol concentration in a maritime air-stream. Measurements were made over the ocean at altitudes of 10 and 30m during flights out to sea and back again. In this study the 'shore' region, which included the surf-zone, was defined as the first six nautical miles out to sea, and the 'sea' region as the zone 6 to 12 nautical miles. from shore.

Table 4 shows that for both shore and sea regions the aerosol concentration in each of the four size-ranges of the FSS probe is lower at the higher altitude by a factor which increases with aerosol size. This indicates that the ocean surface is an important source of these aerosol particles and that the greater masses/fall speeds of the larger particles reduce the efficiency with which they can accumulate at higher levels. The ratio N_{30}/N_{10} is lower for the smallest particles (range 3) over the sea than over the shore, suggesting that aerosol produced in the surf-zone tend to be smaller than those generated at the surface of deeper water.

Figure 14 shows that for the radius range 1-16μm the aerosol concentrations at 10m, measured at 'sea' by the helicopter FSS probe and with the CSAS probe mounted on the tower on the beach, were very similar, over a wide range of conditions. This was found to be so for other size ranges, departures from the 1:1 line being major only when the air-mass was continental.

The concentrations of maritime aerosol measured at low altitudes will clearly be dependent

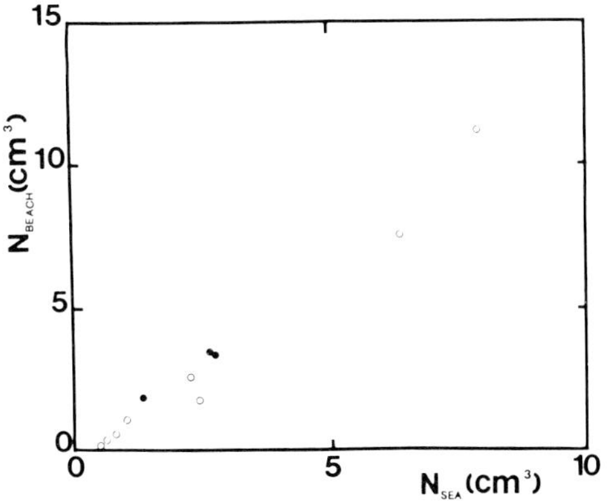

Fig. 14. Comparison of the number concentrations of aerosol particles in the radius range 1 to 16μm at 10m above the surface measured on the beach 10m tower and out to sea by means of a helicopter (•: $U > 8 ms^{-1}$).

upon the ease with which turbulence can remove the aerosol from its source at the ocean surface and distribute it throughout the boundary layer. Considerable effort has been devoted to investigating the relationship between aerosol concentration N and the mixing-depth, L, which was estimated from the Stornoway and ocean weather ship meteorological soundings. (Support for this method of estimating L was provided by studies performed at Great Dun Fell, in Cumbria, where the agreement between its predictions and more direct measurements - involving an instrumented glider and an acoustic sounder - was good.) Figure 15 presents a characteristic N/L relationship determined at South Uist for maritime air. As L increases the aerosol are dispersed over a greater depth of the atmosphere, and thus N is reduced.

Table 4

Mean ratio of N_{30}/N_{10} where N is number concentration (cm^{-3}) over each probe range.

Probe range	Radius range (μm)	'Shore'	'Sea'
#0	1-24	.75	.74
#1	1-6	.80	.79
#2	0.5-8	.91	.87
#3	0.25-4	.94	.88

Table 4.Mean ratios of particle concentrations, N, measured at 10m and 30m using the helicopter-mounted FSSP.

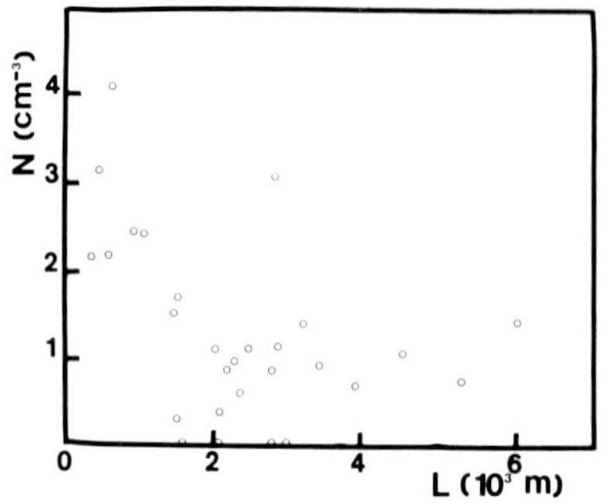

Fig. 15. Comparison of aerosol number concentration N (1 to 16μm) with mixing height L (derived from sea surface temperatures and meteorological soundings) for maritime air.

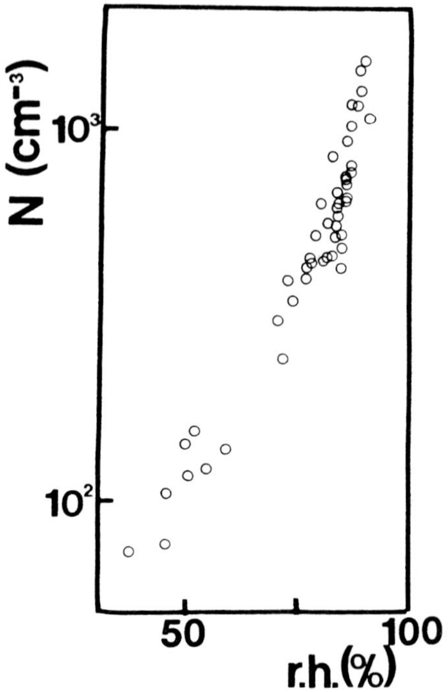

Fig. 16. Dependency of particle number concentration N (0.2 to 0.5μm) upon relative humidity for continental air sampled at Donna Nook on the east coast of England.

As shown in Table 1 the range of variation of relative humidity H in our Hebridean experiments was not great, and any influence that H may have had on the aerosol spectra was undoubtedly swamped by more powerful influences, notably the windspeed. No humidity effect could be identified. That H can, in some circumstances, influence the aerosol measurements in a profound way is illustrated in figure 16, where specimen data from a related project conducted by UMIST scientists at Donna Nook on the Lincolnshire coast are plotted. During the course of these experiments H varied diurnally from about 40 to 90% with no significant variation in any other relevant parameter. The accompanying increase in concentration, N, of particles within the radius-range 0.2μm to 0.5μm, was more than an order of magnitude, presumably because, as H increased, an increasing number of smaller aerosol (present in higher concentrations) grew into the range of detection.

A series of laboratory experiments has been performed to investigate the basic physical processes associated with the production of aerosol at the ocean surface. We mention here only studies of the influence of temperature, T, and salinity on the sizes and concentration of bubbles. Bubbles were produced in a deep tank filled with water by passing air through a filter-stick. The concentrations and sizes of the bubbles were determined photographically ovr the approximate temperature range 0° to 30°C, for both distilled water and sea water obtained from South Uist. Bubble production was much more efficient in sea-water than in distilled water. In sea-water, copious quantities of relatively small bubbles were produced, whereas in distilled water at the same temperature the concentrations were lower by about two orders

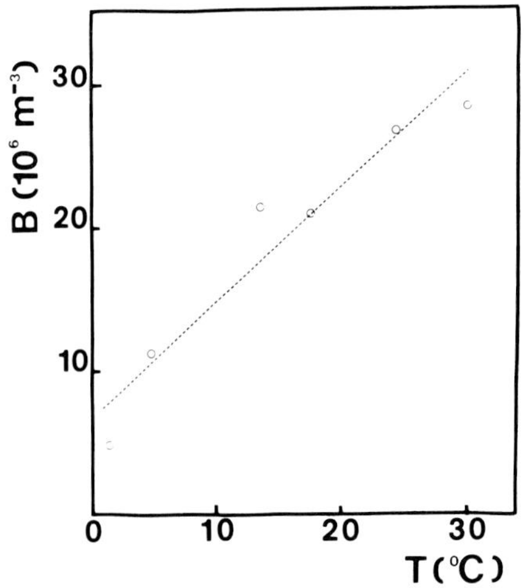

Fig. 17. Variation of bubble density, B, with sea water temperature, T, determined from laboratory experiments.

of magnitude, with a corresponding increase in bubble-size. The inference that bubble-coalescence is seriously impeded in sea-water (perhaps because of electrical effects associated with ionization) is supported by separate experiments in which air-bubbles were forced together under water. In distilled water they coalesced immediately, but it was impossible to promote their union in saline water. Figure 17 shows that as the temperature of the sea water in the tank was raised there was a rapid increase in the measured concentration of bubbles; a similar effect (with lower concentrations) was found for distilled water. The concentration of bubbles in sea-water at 31°C is seen to be much greater than that at 1°C. This temperature effect may be partially responsible for the discrepancy between the bubble concentration measurements of Johnson and Cooke (1979), who obtained a value of 10^6 m^{-3} at 2°C off the coast of Nova Scotia, and Blanchard and Woodcock (1957), who obtained a concentration of 10^8 m^{-3} at 30°C in Hawaiian waters.

A three-stage Junge-Jänicke impactor was used to obtain samples of aerosol at the South Uist site, which were subsequently analysed in order to determine their chemical constitution. Examination with an electron scanning microscope, supported by some crystal-growth experiments in the laboratory, revealed that the maritime aerosol consisted largely of NaCl cubic crystals and K_2SO_4 needle crystals. EDAX chemical analysis of the aerosol collected revealed significant concentrations of elemental sodium, chlorine, potassium and calcium - all of which are general constituents of sea-water.

6. Discussion

The field-experiments described in the foregoing sections have revealed that wind-speed U exercises a dominant influence upon the concentrations and atmospheric loading of maritime aerosol. The wind-speed effects found in our studies have been compared with those reported in four other investigations (Woodcock, 1953; Tsunogai, 1972; Lovett, 1978; and Kulkarni et al, 1982), in which impactor techniques were utilized for the estimation of NaCl loading. The recent important and comprehensive studies by Monahan et al (1983), in which PMS probes were used to measure aerosol size distributions, will be compared with our own results at a later date.

Information on these four earlier studies is presented in Table 5. In order to compare their data with our own the wet-volume of the Uist aerosol was converted to a dry mass by assuming that our aerosol was composed of an NaCl solution and that the relative humidity in our experiments was 85% (the mean value). The data for all four studies were expressed in the form of equation (1), and the associated values of a and b are presented in the table. The value of the exponent a determined from our studies is seen to fit well with those obtained by Woodcock and by Lovett. The latter agreement is particularly satisfying, as measurements were made at a similar height above sea level, at a similar latitude, and covered a similar windspeed range; yet with a fundamental difference in experimental method. The stronger dependency of salt concentration on windspeed reported by Tsunogai and by Kulkarni et al may be a consequence, in part, of the sampling conditions as discussed by Lovett. The measurements of Kulkarni et al were recorded 1.8km inland from

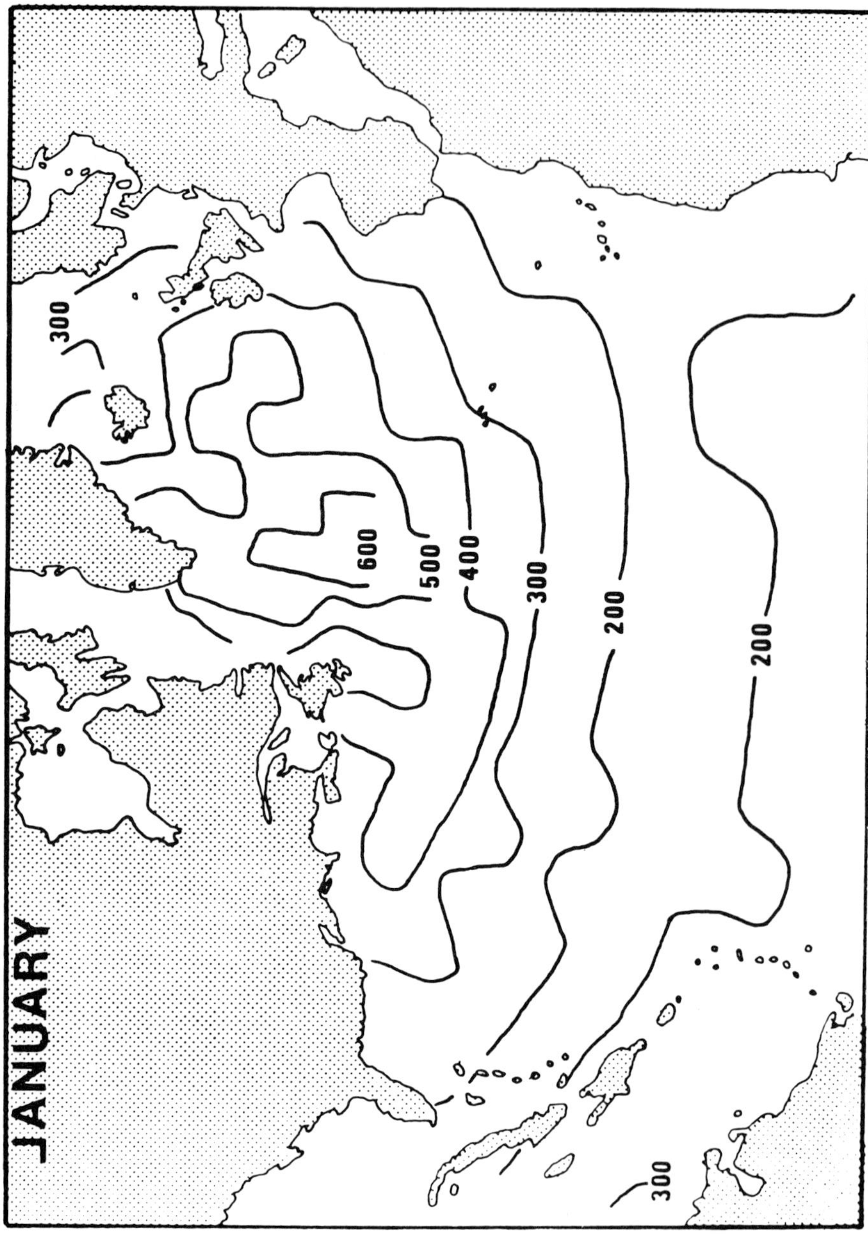

Fig. 18. Distribution of calculated mean surface volumetric aerosol particle loadings ($\mu g\ m^{-3}$) for the North Atlantic in January, based on 20 years of monthly average North Atlantic windspeed values, and calculated using the relationship $V_L = 0.17\ U_{10} + 4.1$.

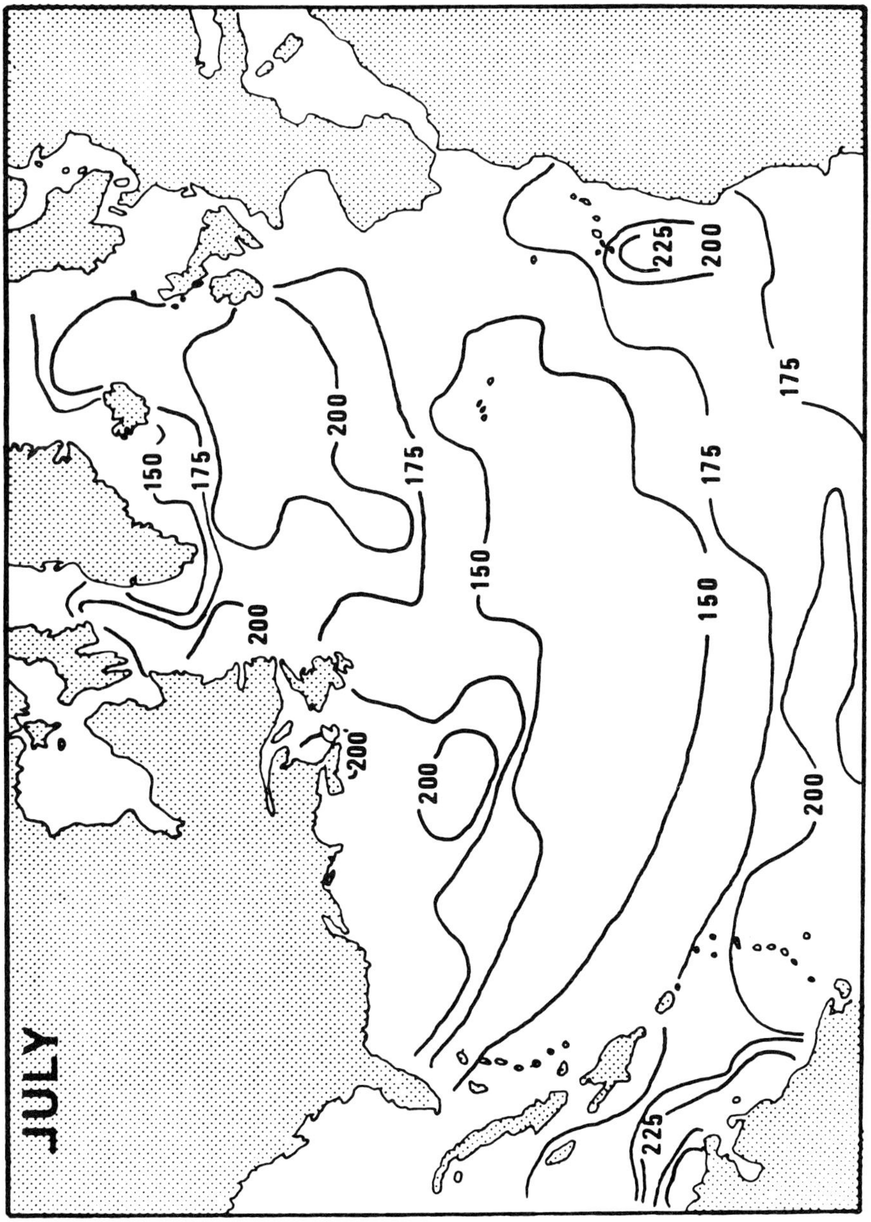

Fig. 19. As in Fig. 18, but for July.

Table 5

Authors	Latitude	Site and approximate sampling height	Max U ($m\ s^{-1}$)	a ($s\ m^{-1}$)	b ($\mu g\ m^{-3}$)
Woodcock 1953	20°N	Cloud base over Pacific Ocean (500m)	35	0.16	2.57
Tsunogai 1972 et al	40°N-50°S	Ship on Pacific Ocean	18	0.62	0.33
Lovett 1978	50°N-60°N	Weather ships in Atlantic Ocean (5-15M)	20	0.16	4.26
Kulkarni et al 1982	19°N	W. Indian coast 1.8km inland (1.2m)	8	0.27	5.35
Present Experiment	57°N	Hebridean beach facing Atlantic (15m)	20	0.17	14.30

Table 5. Published relationships between airborne salt concentration, θ ($\mu g\ m^{-3}$), and windspeed, U($m\ s^{-1}$), assuming an exponential relationship of the form $\ln\theta = aU + \ln(b)$ where a and b are constants.

the sea and, as the authors state, deposition between the source and site would have reduced the salt concentration almost by a factor of 2 at high wind speeds. This fact would indicate that the salt concentration over the sea depends even more strongly on wind speed than their results indicate. The discrepancy may be attributable to the limited wind-speed range and monsoon conditions.

Our estimated background (U = 0) salt loading is about $14\mu g\ m^{-3}$, much higher than those (column of b values, Table 5) measured in the earlier work. Analysis suggests that this discrepancy (especially in relation to the work of Lovett) is largely a consequence of differences in experimental technique.

Equation (2), together with meteorological information (in 5° grid squares) supplied by the Meteorological Office, has been used to estimate the distribution of volumetric loading of aerosol at an altitude of 10m over the North Atlantic. In our preliminary analysis, data for the period 1960 to 1980 were summed and monthly average distributions produced. The contours for January and July are presented in figures 18 and 19 respectively. In view of the particular nature of our measurement site, on South Uist, the quantitative values of volumetric loading must be subject to a considerable degree of uncertainty. The relative variations are likely to be more accurate. There is seen to be considerable variability, in both space and time.

ACKNOWLEDGEMENTS

The work described in this paper forms part of a series of investigations supported by the Procurement Executive, Ministry of Defence, under Agreements 2044/099 XR/SP and 2044/0117 ASWE.

The authors would like to express their gratitude to Mr P. C. Ashdown of the Royal Aircraft Establishment, Farnborough, for his invaluable assistance in maintaining and operating the equipment and the Manchester Weather Centre and the Benbecula station for supportive information.

REFERENCES

Blanchard, D. C. and Woodcock, A. H. 1957. Bubble fo.mation and modification in the sea and its meteorological significance. *Tellus* 9, 145-158.

Day, J. A. 1964. Production of droplets and salt nuclei by the bursting of air bubble films. *Quart. J. Roy. Met. Soc.* **90**, 72-78.

Johnson, B. D. and Cooke, R. C. 1979. Bubble populations and spectra in coastal waters: A photographic approach. *J. Geophys. Res.* **84**(c7) 3761-3766.

Junge, C. E. 1963. *Air Chemistry and Radioactivity.* Academic Press, NY.

Kulkarni, M. R., Adiga, B. B., Kapoor, R. K., and Shirvaikar, V. V. 1982. Sea salt in coastal air and its deposition on porcelain insulators. *J. Appl. Met.* **21**(3), 350-355.

Lovett, R. F. 1978. Quantitative measurement of airborne sea-salt in the North Atlantic. *Tellus* **30**, 358-363.

Meszáros, A. and Vissy, K. 1974. Concentration, size distribution and chemical nature of atmospheric aerosol particles in remote oceanic areas. *J. Aerosol Sci.* 5(1), 101-109.

Monahan, E. C. 1971. Oceanic whitecaps. *J. Phys. Oceanog.* **1**, 139-144.

Monahan, E. C., Fairall, C. W., Davidson, K. L. and Boyle, P. J. 1983. Observed inter-relations between 10m winds, ocean whitecaps and marine aerosols. *Quart. J. Roy. Met. Soc.* **109**, 379-392.

Pinnick, R. G. and Auvermann, H. J. 1979. Response characteristics of Knollenberg light scattering aerosol counters. *J. Aerosol Sci.* **10**(1), 55-74.

Toba, Y. 1961. Drop production by bursting of air bubble films on the sea surface (III). Study by use of a wind flume. *Memoirs Coll. Sci., Univ. Kyoto Ser. A*, 29, 313-344.

Tsunogai, S., Saito, O., Yamada, R. and Nakaya, S. 1972. Chemical composition of oceanic aerosol. *J. Geophys. Res.* 77(27), 5283-5291.

Woodcock, A. H. 1953. Salt nuclei in marine air as a function of altitude and wind force. *J. Met.* **10**, 362-371.

DYNAMICS AND MODELING OF AEROSOLS IN THE MARINE ATMOSPHERIC BOUNDARY LAYER

C. W. FAIRALL
Department of Meteorology
The Pennsylvania State University
University Park, PA 16802
and
K. L. DAVIDSON
Department of Meteorology
Naval Postgraduate School
Monterey, CA 93940

This paper is a review of some of the properties of marine aerosols and the relation of these properties to sources, sinks and turbulent transport characteristics. The ocean is a source of sea salt solution particles produced as sea water droplets by bursting air bubbles. Whitecaps are the primary mechanism for the creation of these air bubbles near the ocean surface. The ocean is a sink for aerosol particles brought into the marine boundary layer by a combination of advection and entrainment. All types of aerosol particles are removed from the atmosphere and dumped into the ocean by dry and wet deposition. Dry deposition can be estimated from simple (but essentially unverified) models based on diffusion sublayer and surface layer turbulence scaling theory (e.g. Slinn and Slinn, 1980). Wet deposition rates as a function of particle size and chemistry can only be crudely estimated at present. The dynamics of the marine aerosol size spectrum can be modeled by a conventional mixed-layer formalism where the boundary layer aerosol is considered to be a mixture of whitecap produced sea-salt particles and entrained continental background particles. Such a model requires specification of the ocean surface droplet source strength, S_i, as a function of wind speed. The height dependence of aerosol concentration in the surface layer and the mixed layer is discussed.

I. Introduction

The concentration and flux of aerosol particles over the world's oceans are of interest to investigators studying air pollution, climate and atmospheric optics. The dynamics of aerosols in the marine boundary layer are of particular importance because the ocean is a source of atmospheric salt particles and a sink for most other airborne particles. Given adequate knowledge of marine source/sink characteristics and atmospheric transport processes, one could, in principle, describe the evolution and chemistry of aerosol size spectra in the marine boundary layer as a function of position and height. This paper is a brief examination of these factors including oceanic particle production by whitecaps, the ocean surface removal efficiency (characterized by the dry deposition velocity, V_d), and the transport properties of the atmosphere.

The nature of the various atmospheric transfer processes permits us to identify certain height regimes where the analysis can be simplified by scaling arguments. For example, near the surface (within 10 meters of the ocean) the particle flux is independent of height. Recent review articles on the air-sea particulate transfer processes (Slinn et al., 1978; Coantic, 1980) have dealt with the relative importance of turbulent and diffusive transport mechanisms in this so-called constant flux layer. The diffusion dominated sublayer has been recently examined by Hasse (1980), Liu et al. (1979) and Wesely et al. (1981). Using a standard micrometeorological formalism, the surface source and sink properties can be described in a surface layer scaling context.

The dry removal of pollutant particles over the ocean was modeled by Slinn and Slinn (1980) for the case of zero concentration at the sea surface (no surface particle source) using simple parameterizations of gravitational settling, turbulent eddy diffusivity, molecular diffusivity and inertial impaction. Fairall and Larsen (1983) extended Slinn and Slinn's work to the surface source case (surface concentration

E. C. Monahan and G. Mac Niocaill (eds.), Oceanic Whitecaps, 195–208.

nonzero) and characterized the particle production rate of the ocean as it appears at a typical measurement height of 10 meters.

The dynamics of the aerosol spectra in the marine boundary layer are most simply described by mixed layer modeling where rate equations are developed that include the surface flux, the height of the boundary layer, the entrainment process and the assumed characteristics of the aerosols above the boundary layer (see Gathman, 1982 and Burk, 1983 for alternative approaches). The dynamic equations considered are appropriate for reasonably clean marine air over the open ocean where advection of non-marine aerosols in the boundary layer can be neglected. Thus, the aerosol particles present in the boundary layer must enter from the sea surface (sea salt particles) or by entrainment and gravitational fallout from the non-turbulent troposphere immediately above the marine boundary layer. The particles above the boundary layer are considered to be of non-local origin (primarily from land sources) and are termed 'continental' background aerosols. The deposition velocity obtained by Slinn and Slinn (1980) is assumed to be appropriate to describe the rate at which these aerosols (once in the boundary layer) are removed from the boundary layer by contact with the sea surface.

Although the assumption of two separate and identifiable aerosol components may be somewhat artificial (perhaps some aerosols present above the boundary layer originated from the ocean several days previously) it permits us to establish convenient boundary conditions. The background component is assumed to have zero surface production, zero surface concentration and some unspecified concentration above the boundary layer. The locally generated aerosols are assumed to have a surface strength, S_i, and zero concentration *above* the boundary layer. This surface source strength is related to the whitecaps and air bubbles near the sea surface (Monahan et al., 1982; Fairall et al., 1983).

The basis for modeling the evolution of aerosol properties is the continuity equation. In the Boussinesq approximation, some aerosol concentration variable, X, at the particle radius, a, can be written (Slinn et al., 1978)

$$\partial X/\partial t = -\vec{U}\cdot\nabla X - \nabla\cdot\vec{F} + S \tag{1}$$

where $\vec{U}$ represents the mean wind, $\vec{F}$ the particle flux and S a source or sink. We chose an aerosol concentration variable that is independent of the ambient relative humidity (e.g. dry mass concentration or concentration at fixed relative humidity). In the case of horizontal homogeneity

$$F = \overline{wx} - V_g X - D\partial X/\partial z \tag{2}$$

where small letters (w, x) denote turbulent fluctuations, W the vertical velocity, V_g the gravitational settling velocity (plus any other slip velocity), D the particle diffusion coefficient and z the height above the surface. Note that $\overline{xv_g}$ has been assumed to be negligible.

In its simplest one-dimensional form Eq (1) becomes

$$\partial X/\partial t + \vec{U}\cdot\nabla_H X + W\partial X/\partial z = -\frac{\partial}{\partial z}(\overline{wx} - V_g X - D\partial X/\partial z) \tag{3}$$

A passive scalar contaminant, C (e.g. water vapor, CO_2, O_3), can be described by a similar continuity equation but with $V_g = 0$.

$$\partial C/\partial t + \vec{U}\cdot\nabla_H C + W\partial C/\partial z = -\frac{\partial}{\partial z}(\overline{wc} - D_c\partial C/\partial z) \tag{4}$$

Very small aerosol particles (V_g negligible) can be considered passive scaler contaminants but, due to a much smaller diffusion coefficient, these particles encounter much greater diffusion sublayer resistance than typical gases.

II. Scaling Regimes

The relative dominance of the terms in Eq (2) as a function of height delineates three scaling regimes in the boundary layer and one above. These regions (see Fig. 1) are designated the troposphere above the mixed layer (p), the mixed layer (f), the turbulent surface layer (c), and the diffusion sublayer (d). The mixed layer, which constitutes about 90% of the boundary layer, is characterized by linear height dependence of the fluxes. The turbulent surface layer is close enough to the surface for the height dependence of the fluxes to be negligible. The diffusion sublayer is the region so close to the surface that the molecular diffusion dominates turbulent transport.

A. Surface Layer Turbulence Scaling

For turbulence scaling it is illuminating to examine the simplified, one-dimensional budget equation for the variance of particle concentration fluctuations, $\overline{x^2}$,

$$\tfrac{1}{2}\partial\overline{x^2}/\partial t - \tfrac{1}{2}V_g\partial\overline{x^2}/\partial z + \overline{wx}\partial X/\partial z - \tfrac{1}{2}\overline{x^2}\,\partial V_g/\partial z + \tfrac{1}{2}\partial\overline{wx^2}/\partial z = -\epsilon_x \qquad (5)$$

where ϵ_x is the variance dissipation rate. Dropping the time derivative term and multiplying by $\kappa z/(u_*x_*^2)$ the Monin-Obukhov surface layer similarity form can be obtained

$$(\kappa z/x_*)\partial X/\partial z - \tfrac{1}{2}\kappa z\frac{\partial}{\partial z}\left(\frac{\overline{wx^2}}{u_*x_*^2}\right) - \kappa z\epsilon_x/(u_*x_*^2) = \tfrac{1}{2}(\kappa z/u_*)(\partial V_g/\partial z)\overline{x^2}/x_*^2 + \tfrac{1}{2}\frac{\kappa z V_g}{u_*}\frac{\partial}{\partial z}(\overline{x^2}/x_*^2) \qquad (6)$$

where u_* is the friction velocity, κ is von Karman's constant, and the subscript, r, refers to a reference height in the surface layer ($z_r \sim 10$ m) and

$$\overline{wx}_r = -u_*x_* \qquad (7)$$

For particles sufficiently small such that $V_g/u_* \ll 1$, Eq. (6) therefore reduces to a form analogous to that for temperature (Wyngaard and Coté, 1971) or any passive scalar contaminant

$$\phi_x(\xi) - \phi_d(\xi) - \phi_{\epsilon_x}(\xi) = 0 \qquad (8)$$

where ϕ_x is the dimensionless gradient function (Businger et al., 1971), ϕ_d a dimensionless flux divergence term, ϕ_ϵ the dimensionless dissipation term, and $\xi = z/L$ where L is the Monin-

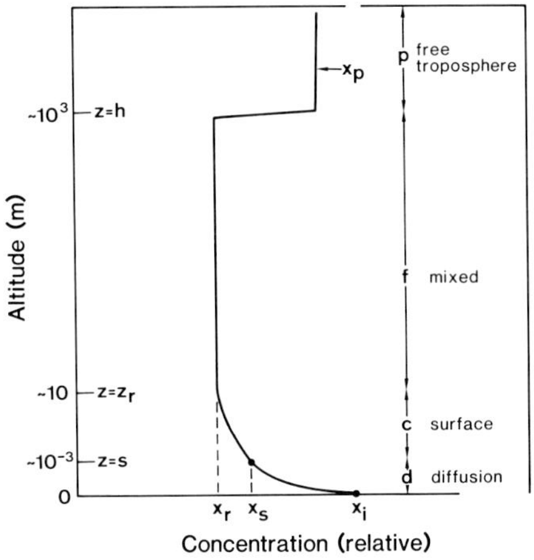

Figure 1. Schematic diagram of atmospheric scaling regimes (nonlinear scales).

Obukhov stability length

$$-L^{-1} = \kappa g\,\overline{w\theta}_{vr}/(Tu_*^3) \qquad (9)$$

where T is the temperature, g the acceleration of gravity and θ_v the virtual potential temperature. The implication of Eq (9) is that aerosol particles can be assumed to possess mean vertical profiles in the surface layer characterized by the known function, ϕ_x, provided $V_g/u_* \ll 1$.

B. Mixed Layer Scaling

A similar development for mixed layer scaling (Lenschow et al., 1980) involves multiplication of Eq (5) by $h/(w_*x_*^2)$ where w_* is the convective scaling velocity, h the boundary layer height and $wx_r = -w_*x_*$. By the same logic one concludes that the mean aerosol concentration at a given particle size will obey mixed layer scaling (Wyngaard and Brost, 1983) if $V_g/w_* \ll 1$. Since w_* is usually greater than 1 m/s, $V_g/w_* \ll 1$ for particles smaller than 30μm radius.

C. Diffusion Sublayer Scaling

Sublayer transport is a complex and controversial problem. One must consider both diffusional transport and inertial impaction. The traditional approach for diffusional transport has been to characterize the resistance of the sublayer, R_d, based on smooth flow as proportional to $S_c^{2/3}/u_*$ where S_c is the Schmidt number (ν/D) and ν is the kinematic viscosity of air (Slinn et al., 1978; Street et al., 1978). Slinn and Slinn (1980) modified this for rough flow to be consistent with surface renewal theory (Liu et al., 1979) so that $R_d \sim S_c^{1/2}/u_*$. Wesely et al. (1981) suggested $R_d \sim \ln(\kappa u_* z_0/D)/u_*$ where z_0 is the roughness length. The key parameter for inertial impaction is the Stokes number, S_t ($S_t = \tau u_*^2/\nu$ where $\tau = V_g/g$), which is essentially the ratio of the stopping distance of the particle to the sublayer thickness. If S_t is greater than unity, then the particle has a high probability of penetrating the sublayer by its residual turbulent velocity.

III. Sources and Sinks

A. Dry Deposition

Consider the dynamics of an aerosol species that has no local sources but is brought into the area by advective and turbulent processes and is characterized by some concentration, X_r at a reference height, z_r, near the top of the constant flux layer. If these particles 'stick' to the surface on contact, then they are removed from the atmosphere resulting in a downward particle flux

$$F = - V_d X_r \tag{10}$$

characterized by the dry deposition velocity, V_d. Slinn and Slinn (1980) derived an approximation for V_d (see Wesely et al., 1981 or Lewellen and Sheng, 1980 for alternative views)

$$V_d = \frac{k_c k_d}{k_c' + k_d' + V_{gd}} \tag{11}$$

where $k_x = k_x' + V_{gx}$ for x = c, d which designates sublayer (d) or surface layer (c) values (see Fig. 2).

The bulk transfer coefficients are given as

$$k_c' = (1-\kappa)^{-1} c_d^{1/2} u_* \tag{12a}$$

$$k_d' = \kappa^{-1} c_d^{1/2} u_* S_c^{-1/2} \tag{12b}$$

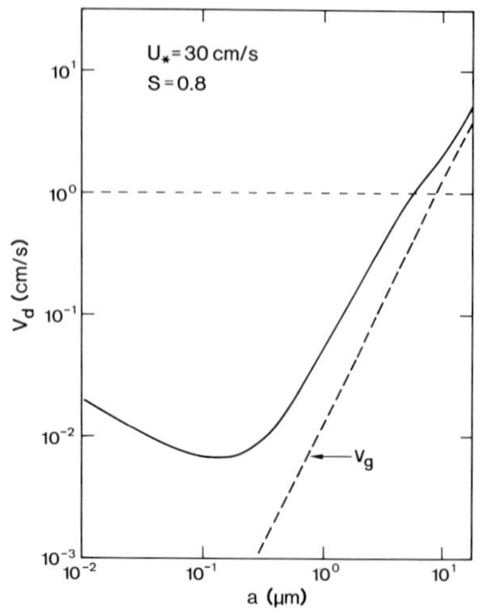

Figure 2. Dry deposition velocity, V_d (solid line) and V_g (dashed line), versus radius from Slinn and Slinn (1980). The dotted line represents the value $c_d^{1/2} u_*$ for this case.

where c_d is the momentum drag coefficient at z_r.

The gravitational fall velocities are different in each scaling regime because of the different relative humidities. Let us define V_g (dry) as the gravitational settling velocity of a dry particle,

$$V_g(\text{dry}) = 2\rho_0 g a_0^2/(9\eta) \tag{13}$$

where ρ_0 is the density of the dry particle, a_0 the dry radius and η the viscosity of air. In the boundary layer the particles are assumed to be in equilibrium with the local relative humidity (H is the water vapor saturation ratio), therefore,

$$V_g(H) = 2\rho_H g a^2/(9\eta) \tag{14}$$

and following Fitzgerald (1975)

$$a = \alpha a_0^{\beta} = G a_0 \tag{15a}$$

$$G = \alpha a_0^{\beta - 1} \tag{15b}$$

where a is the particle radius at H, α and β are parameters that vary with H and ρ_H is the particle density at H

$$\rho_H = (\rho_0 - \rho_w)/G^3 + \rho_w \tag{16}$$

for liquid water density ρ_w.

Typically H ~ 0.8 in the boundary layer ($G_r \cong 2$). In the sublayer Slinn and Slinn (1980) suggest $G_\delta = G(0.99)$ which is reasonable for particles with evaporation response times small compared to their sublayer mean residence time. A study by Fairall (1984) suggests that this assumption begins to break down for particles larger than about 5 μm radius. Since the consequences of this lack of equilibrium are unknown, the theory should certainly be considered incomplete for these particles. In this paper we have assumed $G_\delta = 4$ which is roughly equivalent to assuming the sublayer particles are sea water droplets. Only particles above the boundary layer are considered to be dry (G = 1, $\rho_0 = 2.2$). Therefore we specify

$$V_{gc} = (\rho_H/\rho_0) G_r^2 \, V_g(\text{dry}) \tag{17a}$$

$$V_{gd} = (\rho_{.99}/\rho_0) 16 \, V_g(\text{dry}) + \kappa^{-1} c_d^{1/2} u_* 10^{-3/S_t} \tag{17b}$$

The second term in Eq. (17b) is an additional slip velocity due to particle inertia which leads to a net transport of particles when they impact the sea surface.

B. Whitecap Droplet Production

The production of aerosol particles by whitecaps is well documented. In the present model formalism, we have just described this process in terms of a surface source, S, which is distributed vertically in a very thin layer of thickness, z_+, immediately above the sea surface. We can define an interfacial source strength, S_i, by

$$S_i = \int_0^{z+} S \, dz \tag{18}$$

where s_i has units of flux (e.g. particles per unit area per unit time). Fairall and Larsen (1983) showed that the total flux at the reference height can be written

$$F_r = S_r - V_d X_r = \overline{wx}_r - V_{gc} X_r \tag{19}$$

where

$$S_r = S_i / (1 + k_d'/k_c' + V_{gd}/k_c') \tag{20}$$

Thus, particles at the reference height are removed by various mechanisms characterized by V_d and are supplied (at that height) by an effective source strength, S_r, which is the actual source strength, S_i, diminished by a factor which is equal to one for small particles but rapidly increases for particles with $V_g/u_* > 1$ (an example is given in Fig. 3).

The determination of S_i (or S_r) as a function of wind speed is a major subject of research in whitecaps and air-sea interaction. Both laboratory (Monahan et al., 1982; Cipriano and Blanchard, 1981; Wu, 1982) and field measurements (Toba, 1965; Chaen, 1973; Blanchard and Woodcock, 1980; Sievering et al., 1982; Fairall et al., 1983) have been performed but

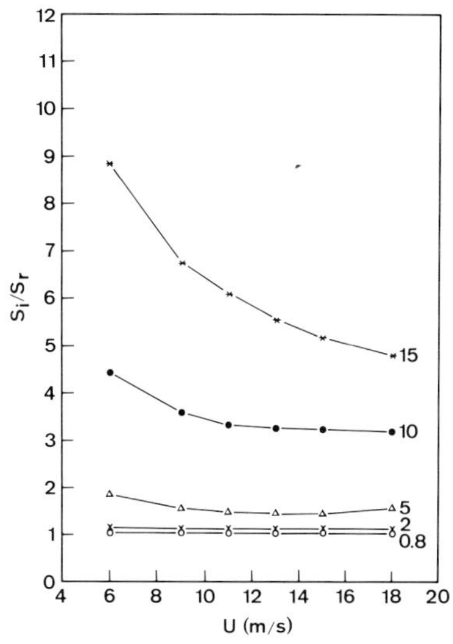

Figure 3. Ratio of effective source strength at $z_r = 10$m to surface source strength versus wind speed (at z_r) as a function of particle radius (at z_r) in μm.

TABLE I
Sea salt particle production rate, S_i, as a function of wind speed and particle radius (expressed at 80% relative humidity). The values in the table are the aerosol spectral volume flux ($\mu m^2/cm^2/s$) assuming the particles are at 80% relative humidity.

radius (μm)	0.8	2	5	10	15
wind speed (m/s)					
6	1.2	2.3	5.1	2.2	1.3
9	4.2	6.5	9.0	6.0	7.0
11	7.5	16.0	22.0	35.0	72.0
13	8.0	20.0	35.0	95.0	310.0
15	10.0	21.0	39.0	145.0	330.0
18	12.0	21.0	48.0	200.0	390.0

S_i still remains a major unknown. An estimate of S_i as a function of wind speed and particle size from Fairall and Larsen (1983) is given in Table I. For more discussion of this subject see Ling et al. (1980); Monahan et al. (1983), Cipriano et al. (1983) or Fairall and Larsen (1983).

C. Wet Deposition

Cloud condensation, coagulation and precipitation processes constitute aerosol particle removal mechanisms (sinks). See Chapter 12 of Pruppacher and Klett (1978) for more detail. The removal rate can be characterized by the wet deposition velocity, V_w, in terms of an effective removal flux

$$S_w = -V_w X \tag{21}$$

which is a weak function of the precipitating cloud thickness (among other things). Unlike the dry process (which removes particles only at the sea surface), the rain process removes particles from the upper regions of the boundary layer or the free troposphere.

The wet deposition velocity can be parameterized (Slinn et al., 1978) in terms of the rain rate mass flux, $\rho_w W_o$, and an average particle washout ratio, r,

$$V_w = r\, W_0 \tag{22}$$

which usually includes both in-cloud and below-cloud scavenging processes. The washout ratio is a strong function of particle size but a typical value for CCN particles is 0.3×10^6. In terms of the standard rain rate, R_r(mm/hr), a rough value of wet deposition velocity (in cm/s) is

$$V_w = 10\, R_r \tag{23}$$

which can greatly exceed the dry deposition velocity during periods of rain. Present knowledge of the washout ratios for marine aerosols as a function of particle size and of the relative magnitude of in-cloud versus subcloud contributions is woefully inadequate.

Subcloud wet removal processes are more logically scaled by the scavenging coefficient (Pruppacher and Klett, 1978), Λ_w, which is related to the subcloud wet deposition velocity, $V_{w\,sub}$, of a layer of thickness ΔZ_{sub} by

$$\Lambda_w = V_{w\,sub}/\Delta Z_{sub} \tag{24}$$

In-cloud processes are more logically scaled by a wet deposition velocity that is a function of the rain rate but not of the cloud thickness. Measured washout ratios contain contributions from both mechanisms. Thus, V_w is the sum of both mechanisms

$$V_w = V_{w\,cloud} + \Lambda_w\, \Delta Z_{sub} \tag{25}$$

IV. Marine Boundary Layer Structure and Dynamics

A. Mixed Layer Equations

The vertical structure of the atmospheric boundary layer is determined by the characteristics of the turbulent transport processes. The gradients are primarily confined to the upper and lower boundaries. The turbulent mixing process, which is most efficient in the interior of the boundary layer, greatly reduces the gradients through roughly 90% of the boundary layer in a region designated the mixed layer. The meteorological situation is depicted in Fig. 4, while the implications for a scalar contaminant are shown in Fig. 1. The marine mixed layer is cool and moist relative to

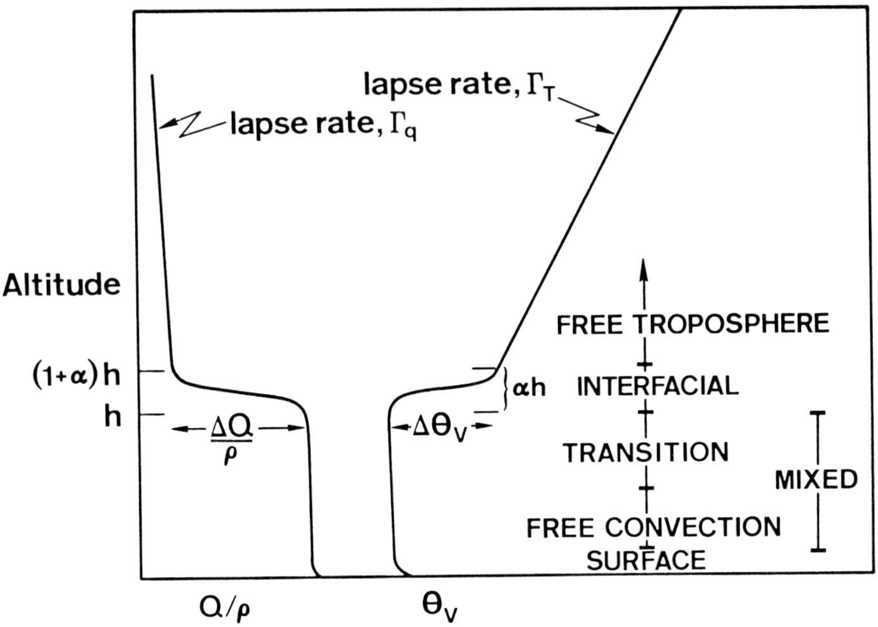

Figure 4. Marine boundary layer well-mixed structure in a cloud-free two-layer idealization. The height of the mixed layer is h, θ_V is the virtual potential temperature, Q the water vapor density and ρ the density of air.

the non-turbulent, stable air immediately above. The transition region between these two layers is referred to as the capping inversion or the entrainment layer.

There are a number of methods in use to model boundary layer processes. The mixed layer model is one of the simplest because it ignores the details of the vertical transport processes by assuming that the turbulence is strong enough to maintain a well-mixed boundary layer. This implies that the fluxes in the boundary layer have a linear dependence on height (Wyngaard and Brost, 1983) and that we need only specify the value of the flux at the bottom and top of the boundary layer. Thus, the time rate of change of aerosol particle concentration in an entraining mixed layer can be obtained by integrating Eq (3) from z_r through the capping inversion at z = h (Fairall et al., 1984)

$$h\partial X_r/\partial t = F_r - F_p - V_w X_r + W_e(X_p - X_r) \quad (26)$$
$$= \overline{wx}_r - V_{gc}X_r - V_w X_r - F_p + W_e(X_p - X_r)$$

where subscript p designates the value above the boundary layer and W_e is the entrainment velocity

$$W_e = \partial h/\partial t - W \quad (27)$$

Note that for notational simplicity we have dropped the horizontal advection terms. Using the surface layer definition of F_r from Eq (19)

$$h\partial X_r/\partial t = S_r - V_d X_r + V_{gp} X_p + W_e(X_p - X_r) - V_w X_r \quad (28)$$

B. Marine Regime

In principle, Eq (28) provides a general description of one-dimensional mixed layer aerosol dynamics. Horizontal advection effects are likely to be small in the open ocean boundary layer but they may dominate in the free troposphere. As stated in the introduction, the model is now applicable to the specific case of the open ocean with surface generated sea salt aerosols (X_s) and non-locally generated 'background' aerosols (X_b) as input from above the boundary layer. The total particle concentration, X, is the sum of X_s and X_b. The boundary conditions of the model are established by the following specifications and definitions:

Height	X_s	X_b	G
$z>h$	0	X_{bp}	1
$z=z_r$	X_{sr}	X_{br}	G_r
$z=0$	X_{si}	0	4

Simplified rate equations for each aerosol component can be obtained from Eq (26)

$$h\partial X_{sr}/\partial t = S_r - (V_d + W_e + V_w)X_{sr} \tag{29a}$$

$$h\partial X_{br}/\partial t = (V_g(dry) + W_e)X_{bp} - (V_d + W_e + V_w)X_{br} \tag{29b}$$

The free tropospheric equation (assuming we are considering a layer that is not within a precipitating cloud) is

$$\partial X_{bp}/\partial t = -\vec{U}\cdot\nabla_H X_{bp} - (W + V_{gp})\partial X_{bp}/\partial z - \Lambda_w X_{bp} \tag{30}$$

C. Boundary Layer Equilibrium

A state of dynamic equilibrium will exist for particles of some radius if $\partial X_r/\partial t = 0$. Such a state is unlikely during precipitation events, so we will consider only the dry situation. Near a state of equilibrium characterized by concentration, X_{re}, the rate equation becomes (the concentration in the mixed layer being assumed to be equal to the concentration at the reference height)

$$\partial X_r/\partial t = (X_{re} - X_r)/\tau_e \tag{31}$$

where (Fairall et al., 1983)

$$\tau_e = h/(W_e + V_d) \tag{32}$$

The boundary layer will tend to be near its equilibrium state if τ_e is much smaller than the time scales associated with typical temporal variations in the forcing terms (on the order of a few hours). For particles with $a < 1\mu m$ ($\tau_e \sim 1$ day) the condition is rarely met while for particles with $a > 15\mu m$ ($\tau_e \sim 5$ hours) the condition is usually met.

The boundary layer equilibrium state for the background component is defined in terms of the equilibrium concentration

$$X_{bre} = (V_g(dry) + W_e)X_{bp}/(V_d + W_e) \tag{33}$$

Equilibrium for surface produced particles implies a balance between production and removal mechanisms, thus

$$X_{sre} = S_r/(V_d + W_e) \tag{34}$$

Models that relate the distribution of marine aerosols to meteorological factors (e.g. Wells et al., 1977, Goroch et al., 1982) such as wind speed and boundary layer thickness are climatological equivalents to the equilibrium distributions.

V. Profiles

A. Historical Perspective

The height dependence of aerosol concentration is of considerable interest (see Blanchard and Woodcock, 1980, for a review). Toba (1965) considered the surface layer profile where the particle flux was assumed independent of height under horizontally homogeneous conditions

$$\partial X/\partial t = \frac{\partial}{\partial z}(\overline{wx} - V_g X) = 0 \tag{35}$$

which can be integrated to yield

$$\overline{wx} - V_g X = F_0 = \text{constant} \tag{36}$$

Toba assumed a balance of production and removal ($F_0 = 0$) with a neutral stability eddy diffusion coefficient, K,

$$\overline{wx} = -K\partial X/\partial z \tag{37}$$

where

$$K = \kappa z u_* \tag{38}$$

This equation can be solved to yield

$$\ln(X_r/X_0) = -\frac{1}{\kappa u_*}\int_{z_0}^{z_r}(V_g/z)dz \tag{39}$$

where X_0 and z_0 are integration constants.

The work of Toba was extended by Goroch et al. (1980) to include the effects of buoyancy by modifying Eq (38)

$$K = \kappa z u_*/\phi_H(z/L) \tag{40}$$

where ϕ_H is the gradient function ϕ_x in Eq (8).

In the special case of inert particles or negligible humidity gradients the solution is

$$\ln(X_r/X_{rn}) = -\frac{V_g}{\kappa u_*}(\ln(z/z_0) - \psi(z/L)) \tag{41}$$

where $\psi(z/L)$ is the profile stability function. The ratio of X_r to its neutral value is

$$\ln(X_r/X_0) = \frac{V_g}{\kappa u_*}\psi(Z/L)) \tag{42}$$

Davidson and Schutz (1983) used an entrainment equilibrium condition and, assuming $F_0 = W_eX_r \cong W_eX$, they derived

$$\ln(X_r/X_{rn}) = \frac{(V_g + W_e)}{\kappa u_*}\psi(z/L) \tag{43}$$

Since W_e is likely to be much larger than V_g for particles smaller than about 3 μm radius, one expects Eq (43) to be more realistic for smaller particles. This was verified by Davidson and Schutz (1983) for 1 μm radius particles.

It is clear that we are dealing with two separate assumptions with these solutions: the constant flux approximation and the equilibrium approximation. The essence of the constant flux assumption is that some of the parameters in Eq (36) have much more height dependence than the flux for some height region reasonably close to the surface and therefore a reasonable solution can be found by assuming F_0 is independent of height. The region for which this assumption is valid is called the constant flux or surface layer. The equilibrium condition is a reasonable assumption when the term $h\partial X_r/\partial t$ in Eq (28) is the small residual difference of two or more terms on the right hand side that are individually much larger than $h\partial X_r/\partial t$. Fig. 5 is an example of the relative

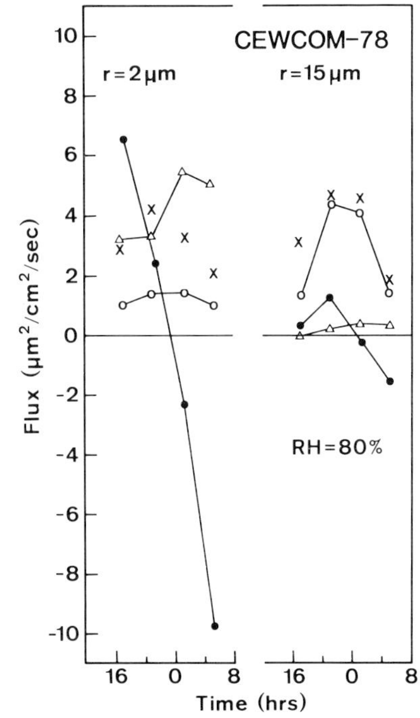

Figure 5. Time series of the terms of Eq (25) for 2 μm and 15 μm radius. The solid circle is $h\partial X_r/\partial t$, the open circle is $V_{gc}X_r$, the triangle is W_eX_r and the X is wx_r. Note that F_p, V_w and X_p are zero for this case.

magnitude of the terms in Eq (28) (from Fairall et al., 1983) where the time derivative term is small for 15μm radius particles but it is the largest term for 2μm radius particles.

B. Small Particle Profiles

The equilibrium assumption is not valid for smaller aerosol particles (the exact guidelines are somewhat difficult to nail down). If we do not assume a state of equilibrium, the profile can still be calculated for particles sufficiently small that their fall velocity is negligible compared to their turbulent transport,

$$F = \overline{wx} - V_g X \approx \overline{wx} = -u_* x_* \tag{44}$$

This implies a standard micrometeorological gradient relation

$$\frac{\partial X}{\partial z} = \frac{x_*}{\kappa z} \phi_H(z/L) \tag{45}$$

which can be integrated to yield

$$X = X_0 + \frac{x_*}{\kappa}(\ln(z/z_{ox}) - \psi(z/L)) \tag{46}$$

This can be written in the bulk aerodynamic form

$$x_* = c_p^{1/2}(X_r - X_0) \tag{47}$$

where the particle drag neutral coefficient is

$$c_p^{1/2} = \kappa / \ln(z_r/z_{ox}) \tag{48}$$

The small particle condition can be more precisely stated as $V_{gd}/k_c' \ll 1$ which implies that $S_r \rightarrow S_i$. A comparison of Eq (19) and Eq (47) yields

$$c_p^{1/2} = (k_d' + V_{gd} - V_{gc})/u_* \tag{49a}$$

$$X_0 = S_i/(k_d' + V_{gd} - V_{gc}) \tag{49b}$$

For very small particles we can neglect V_{gd} compared to k_d' and obtain

$$c_p^{1/2} = c_d^{1/2} S_c^{-1/2}/\kappa \tag{50}$$

Thus a 1 μm radius particle ($S_c^{-1/2} \sim 10^{-3}$) has a drag coefficient three orders of magnitude lower than a typical passive scaler. In terms of the neutral ratio (as in Eq (43)) we find

$$\frac{X_r}{X_{rn}} = 1 + \left(\frac{S_i}{\kappa u_* X_{rn}}\right)\psi(z_r/L) \tag{51}$$

The data of Fairall et al. (1983) can be used to estimate $S_i/(\kappa X_{rn}) \approx 1.5$ cm/s at r = 1μm. Fig. 6 shows a comparison of Eq (51) for $u_* = 15$ cm/s with the data given by Davidson and Schultz (1983).

C. Mixed Layer Profiles

The discussion in Section II implied that particles of less than 30 μm radius are expected to obey mixed layer scaling which is usually taken to mean the absence of a vertical gradient. An example of a vertical profile of aerosol total volume is given in Fig. 7 where the conserved quantity is nearly independent of height while the ambient volume increases with height in response to increasing relative humidity. Since the mixed layer formulation only requires $\partial/\partial t(\partial C/\partial z) = 0$, clearly a constant gradient is permissible. Wyngaard and Brost (1983) show that a mixed layer gradient would be on the order of

$$\frac{\partial C}{\partial z} = -\frac{1.5}{hw_*}[\overline{wc}_r + 2.5W_e(C_r - C_p)] \tag{52}$$

for a passive scalar contaminant. Note that the entrainment process tends to produce about 2.5. times the gradient of the surface flux. This is due to the difference in top-down and bot-

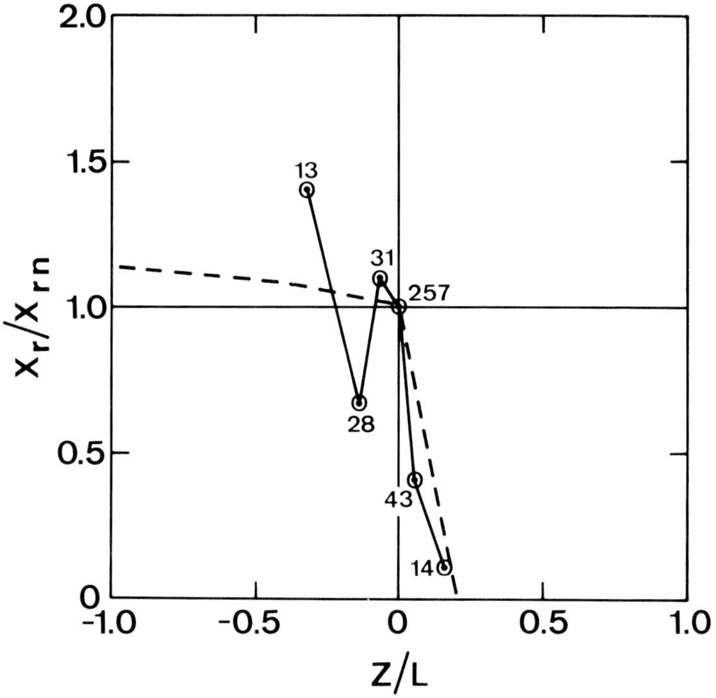

Figure 6. Effect of stability on aerosol concentration at 1 μm radius. The circles are data from Davidson and Schutz (1983) with the number of values indicated. The line is from Eq (50).

tom-up turbulent transport under convective conditions. For aerosols that obey convective scaling we can write Eq (52) as

$$\frac{\partial X}{\partial z} = -\frac{1.5}{hw_*}[S_r - V_d X_r + 2.5W_e(X_r - X_p)] \tag{53}$$

In terms of sea salt and background components

$$\frac{\partial X_s}{\partial z} = -\frac{1.5}{hw_*}(S_r - V_d X_{sr} + 2.5W_e X_{sr}) \tag{54a}$$

$$\frac{\partial X_b}{\partial z} = -\frac{1.5}{hw_*}(-V_d X_{br} + 2.5W_e(X_{br} - X_{bp})) \tag{54b}$$

VI. Conclusion

The general level of understanding of the dynamics and structure of the marine aerosol spectrum has progressed significantly in recent years. Much of this progress is due not to dramatic new discoveries but rather to the successful application of modern boundary layer physics and air pollution ideas to the aerosol problem. Through a consideration of scaling theory and source/sink processes, we have con-

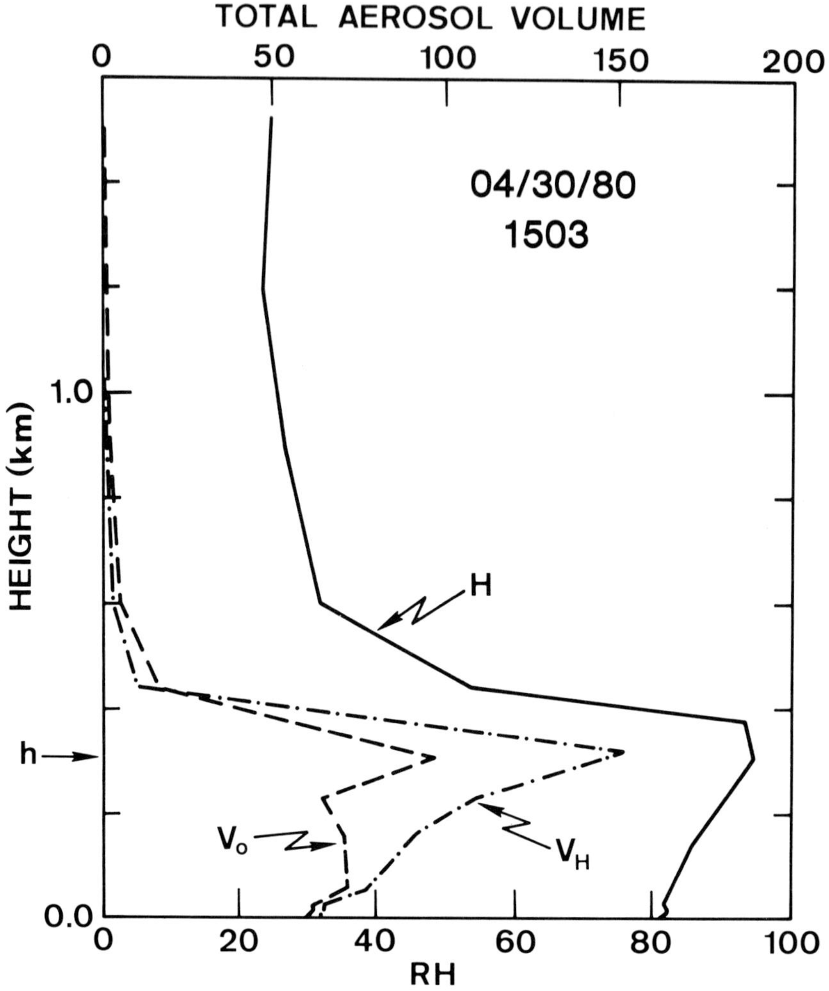

Figure 7. Height dependence of aerosol volume and relative humidity above the ocean. The curve labeled V is the ambient aerosol total volume ($\mu m^3/cm^3$), the curve label V_0 is the volume corrected to 80% reference humidity.

structed a mathematical formalism that can be used, admittedly with limitations, to model marine aerosols. Of course, the more general the theory the more difficult its application. Many complex processes have been reduced to a single parameter (e.g. wet deposition velocity) which must be specified as a function of particle size, chemistry and the meteorological situation. The accurate forecasting of marine aerosol spectra awaits the development of more reliable estimates or models of the simple variables we have discussed. However, many observable properties of marine aerosols can now be clearly accounted for in terms of real physical processes rather than simply being parameterized by least squares climatological averages.

ACKNOWLEDGEMENTS

This work was supported by the Office of Naval Research, the Naval Air systems command AIR-370) and the EO/MET program. The authors wish to acknowledge the contributions of Dr S. E. Larsen of RISO National Laboratory, Roskilde, Denmark.

REFERENCES

Blanchard, D. C. and A. H. Woodcock, 1980. The production, concentration and vertical distribution of the sea-salt aerosol. *Annals N. Y. Acad. Sci.*, **338**, 330-347.

Burk, S. D., 1983. A turbulence closure model of the generation, transport and deposition of marine aerosols. Proc. 5th AMS Conference on Atmospheric Turbulence and Diffusion, Boston, MA.

Businger, J. A., J. C. Wyngaard, Y. Izumi and E. F. Bradley, 1971. Flux-profile relationships in the atmospheric surface layer. *J. Atmos Sci.*, **28**, 181-189.

Chaen, M., 1973. Studies on production of sea-salt particles on the sea surface. *Mem. Fac. Fish. Kagoshima Univ.*, **22**, 149-195.

Cipriano, R. J. and D. C. Blanchard, 1981. Bubble and aerosol spectra produced by a laboratory 'breaking wave'. *J. Geophys. Res.*, **86**, 8085-8092.

Cipriano, R. J., D. C. Blanchard, A. W. Hogan and G. G. Lala, 1983. On the production of Aitken nuclei from breaking waves and their role in the atmosphere. *J. Atmos. Sci.*, **40**, 469-479.

Coantic, M., 1980. Mass transfer across the ocean-air interface: small scale hydrodynamics and aerodynamic mechanisms. *Phys. Chem. Hydro.*, **1**, 249-279.

Davidson, K. L. and L. Schutz, 1983. Observational results on the influence of surface layer stability and inversion entrainment on surface layer marine aerosol number density (1 micrometer). *Opt. Eng.*, **22**, 45-49.

Fairall, C. W., 1984. Interpretation of eddy-correlation measurements of particulate deposition and aerosol flux. *Atmos. Environ.* **18**, 1329-1337.

Fairall, C. W., K. L. Davidson and G. E. Schacher, 1983. An analysis of the surface production of sea-salt aerosols. *Tellus*, **35B**, 31-39.

Fairall, C. W. and S. E. Larsen, 1983. Dry deposition, surface production and dynamics of aerosols in the marine boundary layer. *Atmos. Environ.* **18**, 69-77.

Fairall, C. W., K. L. Davidson and G. E. Schacher, 1984. Application of a mixed layer model to aerosols in the marine boundary layer. *Tellus* **36B**, 203-211.

Fitzgerald, J. W., 1975. Approximation formulas for the equilibrium size of an aerosol particle as a function of its dry size and composition and the ambient relative humidity. *J. Appl. Met.*, **14**, 1044-1049.

Gathman, S. G., 1982. A time dependent oceanic aerosol model. Naval Research Laboratory Tech. Rept. 8536, Washington DC, pp 35.

Goroch, A. K., S. K. Burk and K. L. Davidson, 1980. Stability effects on aerosol size and height distributions. *Tellus*, **32**, 245-250.

Goroch, A. K., C. W. Fairall and K. L. Davidson, 1982. Modeling wind speed dependence of marine aerosols by a gamma function. *J. Appl. Met.*, **21**, 666-671.

Hasse, L., 1980. Gas exchange across the air-sea interface. *Tellus*, **32**, 470-481.

Lenschow, D. H., J. C. Wyngaard and W. T. Pennell, 1980. Mean-field and second-moment budgets in a baroclinic, convective boundary layer. *J. Atmos. Sci.*, **37**, 1327-1341.

Lewellen, W. S. and Y. P. Sheng, 1980. Modeling of dry deposition of SO_2 and sulfate aerosols. ARAP Tech. Rept. EA-1452, Princeton, NJ, pp 83.

Ling, S. C., T. W. Kao, M. Asce and A. I. Saad, 1980. Microdroplets and transport of moisture from the ocean. *J. Eng. Mech. Div.*, **6**, 1327-1339.

Liu, W. T., K. B. Katsaros and J. A. Businger, 1979. Bulk parameterization of air-sea exchanges of heat and water vapor including the molecular constraints at the interface. *J. Atmos. Sci.*, **36**, 1722-1735.

Monahan, E. C., K. L. Davidson and D. E. Spiel, 1982. Whitecap aerosol productivity deduced from simulation tank measurements. *J. Geophys. Res.*, **87**, 8898-8904.

Monahan, E. C., C. W. Fairall, K. L. Davidson and P. J. Boyle, 1983. Observed inter-relations between 10m winds, ocean whitecaps and marine aerosols. *Quart. J. Roy Met. Soc.*, **109**, 379-392.

Pruppacher, H. R. and J. D. Klett, 1978. *Microphysics of Clouds and Precipitation*, Reidel, Holland, pp 714.

Sievering, H., J. Eastman and J. A. Schmidt, 1982. Air-sea particle exchange at a near-shore site. *J. Geophys. Res.*, **87**, 11027-11037.

Slinn, W. G. N., L. Hasse, B. B. Hicks, A. W. Hogan, D. Lai, P. S. Liss, K. O. Munnich, G. A. Sehmel and O. Vittori, 1978. Some aspects of the transfer of atmospheric trace constituents past the air-sea interface. *Atmos. Environ.*, **12**, 2055-2087.

Slinn, S. A. and W. G. N. Slinn, 1980. Predictions for particle deposition on natural waters. *Atmos. Environ.*, **14**, 1013-1016.

Street, R. L., C. S. Wang, D. A. McIntosh, A. W. Miller, 1978. Fluxes through the boundary layers at an air-water interface: laboratory studies. *Turbulent Fluxes through the Sea Surface, Wave Dynamics and Prediction*, Ed. Favre and Hasselmann, Plenum, NY.

Toba, Y., 1965. On the giant sea-salt particles in the atmosphere, II. Theory of the vertical distribution in the 10 m layer over the ocean. *Tellus,* 17, 365-382.

Wells, W. C., G. Gal and M. W. Munn, 1977. Aerosol distributions in maritime air and predicted scattering coefficients in the infrared. *Appl. Opt.,* **16**, 654-659.

Wesely, M. L., D. R. Cook and R. M. Williams, 1981. Field measurements of small ozone fluxes to snow, wet bare soil and lake water. *Boundary-Layer Meteor.,* **20**, 459-471.

Wu, J., 1982. Sea spray: a further look. *J. Geophys. Res.,* **87**, 8905-8912.

Wyngaard, J. C. and O. R. Cote, 1971. The budgets of turbulent kinetic energy and temperature variance in the atmospheric surface layer. *J. Atmos. Sci.,* **28**, 190-201.

Wyngaard, J. C. and R. A. Brost, 1983. Top-down and bottom-up diffusion in the convective boundary layer. *J. Atmos. Sci.* (to appear).

WHITECAPS AND GLOBAL FLUXES

M. C. SPILLANE, E. C. MONAHAN, P. A. BOWYER,
D. M. DOYLE & P. J. STABENO
University College, Galway

1. Introduction

Over the past several years whitecap related research has been conducted at University College, Galway. While the details of several aspects of this research are covered in other workshop presentations, we preface this paper with an overview of our activities.

The research has both field and laboratory aspects, with further effort devoted to modelling and algorithm development. In recent years the whitecap photographic data sets of Monahan (1971), and of Toba and Chaen (1973), have been extended through participation in the JASIN and STREX experiments. Most recently both still and video images of oceanic whitecaps have been collected during MIZEX. Analysis of this imagery leads to estimates of the fractional coverage of the sea surface by whitecaps. Simultaneous measurements of wind speed, sea and air temperatures, and where possible aerosol particle spectra, form the data base used in the search for algorithms linking these parameters. The results from STREX are discussed by Doyle in his poster presentation.

Field data are also collected at a station on the Aran Islands, at the mouth of Galway Bay (Figure 1). Here a lattice tower and instrument shelter overlook the Atlantic in the face of the prevailing westerlies. The station is manned in short intensive periods of data collection. A large volume air sampler was operated at the site, on behalf of the University of Rhode Island, to monitor chemical composition as an indicator of air mass motions.

The laboratory effort centres on the whitecap simulation tank, shown schematically in Figure 2. Single whitecaps are formed, in a reproducible manner, in the deep central well through the collision of opposing walls of water, released from the end reservoirs. The tank is instrumented, with cameras and aerosol probes, which monitor various aspects of the whitecapping process. Bowyer describes, in his poster, data being collected in the tank to study the electrical phenomena associated with breaking waves. Earlier work resulted in estimates of aerosol productivity, and of the characteristic decay time of whitecaps. Video images collected during the MIZEX cruise should throw further light on the latter process under oceanic conditions. The simulation tank is used extensively in cooperative studies with the Naval Postgraduate School, Monterey, California.

The development of algorithms, linking whitecap coverage and aerosol spectra to environmental parameters, has involved both standard least squares procedures and more sophisticated statistical techniques that recognise the non-uniform distribution of the data. The latter aspect is discussed by O'Muircheartaigh and Monahan in their presentation. Suitable algorithms have been combined with available climatological wind and stability data to produce a world whitecap atlas. This has been used in turn to model gas exchange at the air-sea interface (Monahan and Spillane, 1983a), and in an estimate of the influence of whitecaps on ocean albedo. The latter process is modelled

E. C. Monahan and G. Mac Niocaill (eds.), Oceanic Whitecaps, 209–218.

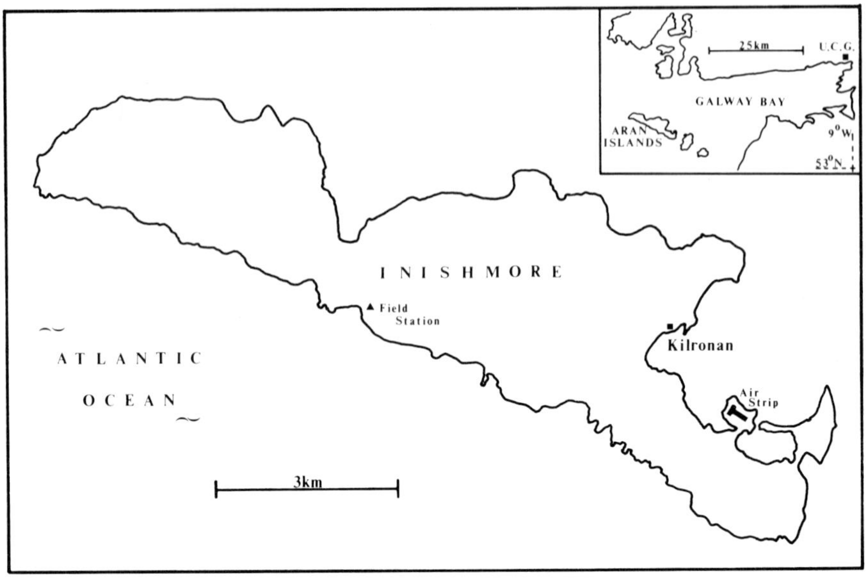

Fig. 1. Location of the field station on Inishmore, in the Aran Islands at the mouth of Galway Bay.

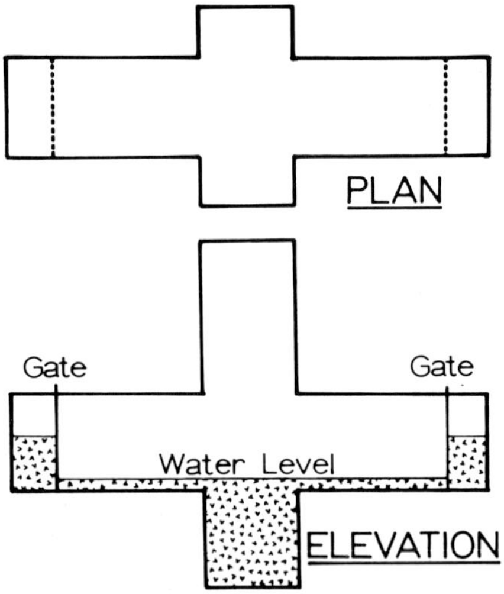

Fig. 2. Schematic illustration of the plan and elevation of the whitecap simulation tank.

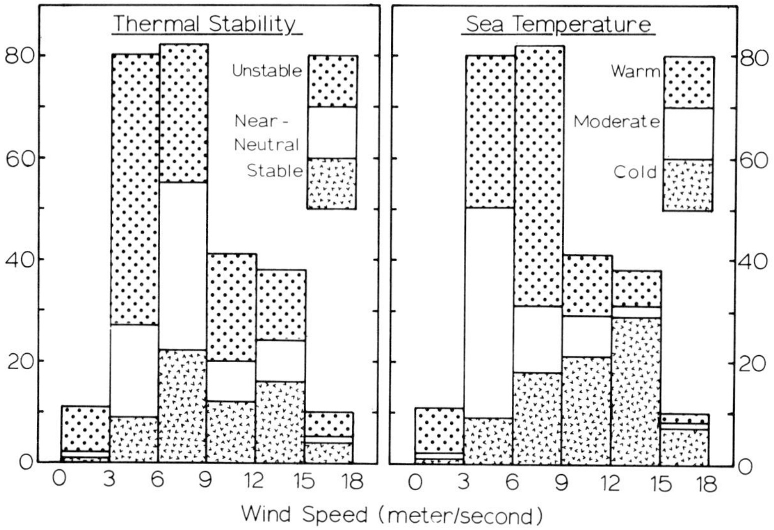

Fig. 3. Distribution of the whitecap observation to date (262 cases) with wind speed, water temperature and thermal stability. The classifications used are defined in the text.

by Stabeno and Monahan in the reserve paper for the workshop, published in these proceedings.

Another major effort has been the modelling of sea surface aerosol generation. Data collected in the simulation tank have been used to formulate a model of the aerosol fraction that originates in the bursting of bubbles at the sea surface (Monahan, Spiel and Davidson, 1983). This has been employed to estimate global fluxes of salt and sea-water into the marine atmosphere (Monahan and Spillane, 1983b), and to revise the estimates of electrical current due to bubble-produced charged droplets, following the work of Blanchard (1963). Details of these resplts are presented below, following an examination of the current status of the whitecap data base.

2. *Whitecap Observations*

As of June 1983 the combined file of whitecap photo-analyses consists of 262 cases, covering a wide range of oceanic and meteorological conditions. The distribution is illustrated in Figure 3. Here the wind speed U refers to the ten-metre level, T_W is the sea surface temperature and DT = $T_W - T_a$, where T_a is the air temperature) is a measure of the thermal stability over the water. The data have been classified as follows:

Sea surface temperature	$T_W < 12.5°C$	COLD
	$12.5 < T_W < 14.0°C$	MODERATE
	$T_W > 14.0°C$	WARM
Thermal stability	$DT < -0.4°C$	STABLE
	$-0.4 < DT < 0.6°C$	NEAR NEUTRAL
	$DT > 0.6°C$	UNSTABLE

The temperature classes are chosen at natural breaks in the data. The ‘cold’ category consists entirely of STREX results (with a lowest tempterature of 5.11°C). ‘Moderate’ conditions represent those encountered during JASIN, while the ‘warm’ class covers both the early Monahan

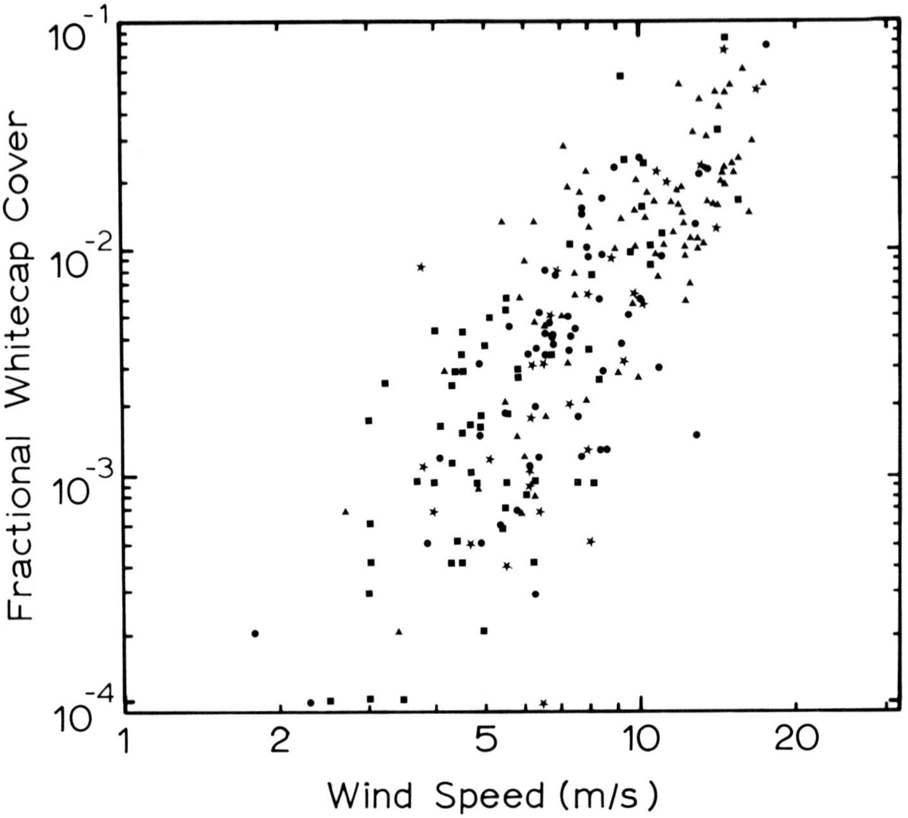

Fig. 4. Variations of fractional whitecap coverage with wind speed at the ten meter level. Excluded are the 31 cases of zero coverage estimates. Circles represent data from Monahan (1971), stars the data from Toba and Chaen (1973), while squares and triangles denote data from JASIN and STREX respectively.

data (largely obtained during BOMEX), and the East China Sea data of Toba and Chaen. The highest sea surface temperature included is 30.55°C. Some combinations of conditions are poorly represented. There are for example no cold, unstable cases at low wind speeds. Such deficiencies detract from the reliability of W (U, T_W, DT) algorithms computed from these data.

Figure 4 shows the distribution of fractional whitecap coverage with wind speed, the major influence. Also indicated is the ordinary least squares fit based on the minimization of the mean squared error

$$MSE = \Sigma(W_i - AU_i^B)^2 / N$$

with respect to the coefficient A and exponent B. The optimum power law

$$W = 2.692 \times 10^{-5}\ U^{2.625},\ MSE = 8.56 \times 10^{-5}$$

has a much lower exponent than that (3.41) derived from the earlier data base by Monahan and O'Muircheartaigh (1980). A piecewise power law dependence, subdivided by sea surface temperature

$$W = \begin{cases} 9.279 \times 10^{-5}\, U^{2.112}, & \text{COLD} \\ 4.755 \times 10^{-5}\, U^{2.525}, & \text{MODERATE} \\ 3.301 \times 10^{-6}\, U^{3.479}, & \text{WARM} \end{cases}$$

leads to an improved mean squared error (MSE = 7.76×10^{-5}), and is suggestive of a temperature dependence of the exponent. Attempts to quantify the latter point have been unsuccessful to date. Since the temperature classification matches largely the breakdown by experiment, it is possible that other influences are at work - for example wind duration, fetch or the presence of surface films.

3. *The Whitecap Atlas*

Since whitecaps play a role in several oceanic processes, a useful product of a suitable whitecap algorithm is a climatological atlas of world whitecap coverage. A useful source of surface observations of wind speed and stability, over the past century, is the TDF-11 file of ship observations, compiled by the National Climate Center at Asheville, North Carolina. A convenient summary of these observations (in excess of 30 million), by month and 2°x2° square, has been provided by Hellerman (of GFDL, Princeton), and is discussed by Hellerman and Rosenstein (1983). Of relevance to the atlas are the data products

$$U^3, U^4, U^5, U^3 DT, U^3 DT^2, U^4 DT$$

Also provided is the rate of wind work at the sea surface, computed using a wind speed and stability dependent drag coefficient C_D based on the data of Bunker (1976).

$$W_t = \rho C_D (U, DT) U^3 \tag{1}$$

The preliminary version of the whitecap atlas was derived from the global distribution of the rate of wind work function, using the algorithm

$$W = 6.667 \times 10^{-6} W_t \tag{2}$$

This form is based on the Monahan (1971) and Toba and Chaen (1973) data and when tested against the present data set results in a mean squared error that is 50% above optimum. This deficiency will be corrected in future versions of the atlas.

The current atlas is displayed in Figure 5. The projection used for these monthly charts assigns equal area to each 2°X2° element. The area covered, from 70° to 70°, is essentially free from pack ice on average. The overall maximum of fractional whitecap coverage occurs in the North Atlantic, with a lesser peak in the North Pacific. The relatively low values in Southern Ocean in the Austral winter are probably due to sampling problems. Ships of passage will, if possible, avoid areas of storm activity with a consequent underestimation of the climatological wind conditions. In Figure 6 the seasonal cycle of zonally averaged fractional whitecap cover is contoured. Total whitecap area for the area covered by the atlas is listed, by month, in Table 1.

A piecewise fit, in the form of Equation 2, for each category of thermal stability does not lead to an improvement in mean squared error. The form of the Hellerman data suggests an alternate algebraic form. A multiple regression for W, based on the six wind speed/stability moments above, leads to a significant improvement of 20% in the MSE. This suggests an avenue of approach for future versions of the atlas.

4. *Applications*

The whitecap atlas has been applied to several different problems to date. A simple model of gas exchange at the sea surface (Monahan and Spillane, 1983a), where whitecaps are treated as low impedance vents to the atmosphere from the ocean interior, has been tested against the radon evasion data from the GEOSECS and TTO (Transient Tracers in the Ocean) experiments. Due to the time scale of several days for the gas evasion process, the measured piston velocities are better correlated with the climatological whitecap coverage than with the short term (24-48 hour) local wind estimates (Peng et al., 1979), or whitecap coverage values based on these.

A series of Sea Surface Aerosol Generation (SSAG) models have been proposed, as des-

April
May
June
January
Febuary
March

Figure 5. Atlas of climatological whitecap coverage, by month, for the region 70° to 70°. This first version is based on the rate of wind work derived from the TDF-11 file of ship observations (Hellerman and Rosenstein, 1983), using the algorithm defined in the text. Contour values indicate the computed percentage whitecap coverage.

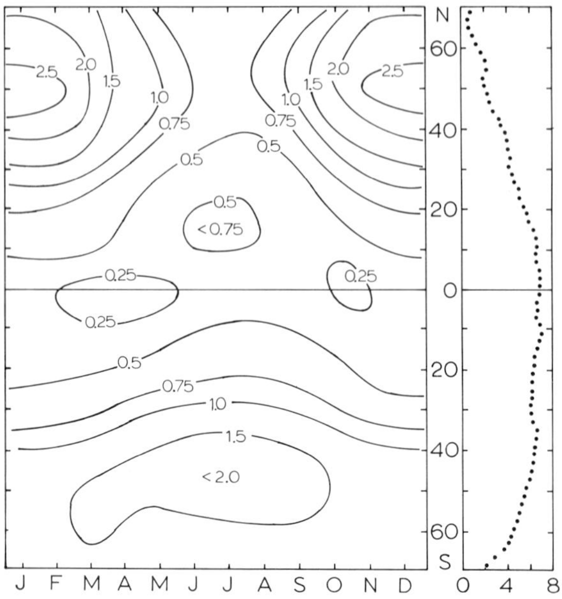

Fig. 6. Seasonal cycle of zonal averaged fractional whitecap coverage (expressed as percentage), derived from the climatological atlas. Also shown is the latitudinal variation in sea area (expressed in $m^2 \times 10^{-12}$, per 2° zonal band), used in the computation of total global whitecap cover.

Table 1. Climatological estimates of global whitecap coverage and fluxes of mass, particle surface area and electrical charge from the sea surface. Based on the whitecap atlas and the model of aerosol generation given by Equation 3 in the text. The estimates are derived for the region from 70° to 70°, and bubble-ejected aerosol particles whose 80% relative humidity radii are in the range 0.8 to 10 microns.

	Whitecap Area ($m^2 \times 10^{-12}$)	Salt Mass Flux ($kg/s \times 10^{-5}$)	Water Mass Flux ($kg/s \times 10^{-6}$)	Droplet Area Flux ($m^2/s \times 10^{-9}$)	Electrical Current (A) Case 1	Case 2
January	3.28	1.21	3.45	1.60	250	175
February	3.25	1.19	3.39	1.58	247	174
March	3.17	1.16	3.31	1.54	241	169
April	2.82	1.04	2.96	1.38	215	151
May	2.74	1.01	2.88	1.34	209	146
June	2.95	1.08	3.08	1.43	225	158
July	2.93	1.08	3.08	1.43	223	157
August	2.90	1.07	3.05	1.42	221	155
September	2.85	1.05	2.99	1.39	217	152
October	2.92	1.07	3.05	1.42	222	156
November	3.17	1.16	3.31	1.54	241	169
December	3.27	1.18	3.36	1.56	249	175
Annual Totals:		3.50×10^{12} kg	9.98×10^{13} kg	4.64×10^{16} m^2		

cribed by Monahan et al. (1983). Here direct mechanisms of aerosol formation are distinguished from the indirect mechanisms associated with the rupture of bubbles at the sea surface. The direct processes (by chop and spume) were estimated from a literature search and appear to become increasingly important at higher wind speeds. Indirect aerosol production is modelled through Equation 3

$$dF_0/dr = 3.585 \times 10^5 W r^{-3} S(r) 10^{B(r)} \tag{3}$$

where

$$S(r) = 1 + 0.033 r^{1.179}$$

and

$$B(r) = 1.19 \exp\{-((0.38 - \log r)/0.65)^2\}$$

which is based on data collected in the whitecap tank during cooperative UCG/NPS experiments. This productivity spectrum gives the number of droplets, per unit radius range, per second, ejected from a square meter of sea surface, when the fractional whitecap coverage is W. Droplet radii in microns are referred to a standard relative humidity of 80%. The inverse cubic factor results from the transformation of data from aerosol volume distribution (dV/dr) to number density. The Gaussian form of the exponent B(r) reflects the symmetric, bell-shaped dV/dr distribution in log-log space. The correction term S(r) is a modification of that included in SSAG-4, which represents the effect of a vertical gradient of particle concentration in the hood of the whitecap simulation tank. Equation 3 (SSAG-5) models aerosol generation, in the radius range 0.8 to 10μm, by the indirect process of bubble rupture at the sea surface.

Knowledge of the aerosol productivity spectrum allows the computation of flux estimates for salt, sea water and electrical charge. The algorithms of Fitzgerald (1975) coupled with density considerations lead to the relations

$$r_d = 0.50r; \qquad r_0 = 1.96\,r$$

for the radii of the dry salt particle (r_d) and the newly ejected droplet (r_0), in terms of its 80% relative humidity radius (r). Integration of the flux spectra over the radius range 0.8 to 10μm, coupled with the climatological global whitecap cover, derived from the atlas, lead to the global flux estimates tabulated in Table 1. The electrical current estimates are based on the charge-radius expressions of Blanchard (1963), with a bubble age of 8 seconds. Following Monahan (1980) two cases are evaluated. In cases 1 and 2, the divisions between jet and film droplet populations are set at 0.75 and 1.5 microns respectively.

5. Conclusion

The collection of whitecap imagery is likely to be accelerated by a recently acquired video system. Details of the breaking wave process can be examined, for example differences in the characteristics of whitecaps caused by spilling or plunging waves. The increased data flow will require a move to automated image processing. Current photo-analysis follows the manual protocol described by Monahan (1969). Selection of data collection expeditions should reflect the gaps in the existing data set, with perhaps improved wind recording to flesh out possible influences of growth or decay in sea state. Long term records from weather ships or island stations should be examined to correct deficiencies in the TDF-11 file of surface observations and improve future versions of the whitecap atlas. Direct mechanical disruption as a source of aerosols would appear, from SSAG-4, to be of increasing importance at higher wind speeds (Monahan et al., 1983). It should be noted that the flux estimates derived above relate only to the indirectly produced aerosol mechanisms, and to a limited range of radii. As such, these estimates must be considered only as lower bounds, subject to revision following future improved SSAG models and whitecap atlases.

ACKNOWLEDGEMENT

The study of whitecaps and their influence on the marine atmosphere at University College, Galway is being funded by the U.S. Office of Naval Research, Grant #N-00014-78-G-0052.

REFERENCES

Blanchard, D. C., 1963. The electrification of the atmosphere by particles from bubbles in the sea. *Progress in Oceanography*, Vol. 1, 71-202. Pergamon Press.

Bunker, A. F., 1976. Computations of surface energy flux and annual air-sea interaction cycles of the North Atlantic Ocean. *Mon. Weather Rev.* **104**, 1122-1140.

Fitzgerald, J. W., Approximation formulas for the equilibrium size of an aerosol particle as a function of its dry size and composition and the ambient relative humidity. *J. Appl. Meteor.* **14**, 1044-1049.

Hellerman, S., and M. Rosenstein, 1983. Normal monthly wind stress over the world ocean with error estimates. *J. Phys. Oceanogr.* (in press).

Monahan, E. C., 1969. Fresh water whitecaps. *J. Atmos. Sci.* **26**, 1026-1029.

Monahan, E. C., 1971. Oceanic whitecaps. *J. Phys. Oceanogr.* **1**, 139-144.

Monahan, E. C., 1980. Positive charge flux from the world ocean resulting from the bursting of whitecap bubbles. Abstracts, VI Int. Conf. on Atmos. Elect., Manchester.

Monahan, E. C. and I. O'Muircheartaigh, 1980. Optimal power-law description of oceanic whitecap coverage dependence on wind speed. *J. Phys. Oceanogr.* **10**, 2094-2099.

Monahan, E. C., D. E. Spiel and K. L. Davidson, 1983. Model of marine aerosol generation via whitecaps and wave disruption. Preprint of Extended Abstracts, IX Conf. on Aerospace and Aeronautical Meteorology, Omaha, Nebraska, 147-150.

Monahan, E. C. and M. C. Spillane, 1983a. The role of oceanic whitecaps in air-sea gas exchange. Proceedings, Int. Symp. on Gas Exchange at Water Surfaces, Cornell U., (In press).

Monahan, E. C. and M. C. Spillane, 1983b. The role of oceanic whitecaps in the exchange of mass across the air-sea interface. Abstracts, XVIII General Assembly, I.U.G.G., Hamburg.

Peng, T.-H., W. S. Broecker, G. G. Mathieu and Y.-H. Li, 1979. Radon evasion rates in the Atlantic and Pacific Oceans as determined during the Geosecs Program. *J. Geophys. Res.* **84**, 2471-2486.

Toba, Y. and M. Chaen, 1973. Quantitative expression of the breaking of wind waves on the sea surface. *Records, Oceanogr. Works Japan* **12**, 1-11.

COMPARISONS BETWEEN ELECTRICAL PROCESSES OCCURRING OVER LAND AND OVER WATER

B. VONNEGUT
Atmospheric Sciences Research Center
State University of New York at Albany
Albany, New York 12222

Introduction

There are many differences between atmospheric electrical processes that take place over the continents and over the oceans. Perhaps the most striking distinction, one documented by observations from ships at sea (Sanders and Freeman, 1982), is that atmospheric electrical activity as indicated by the frequency of lightning discharges is much less intense over the oceans than over the land. Observations from spacecraft confirm this (Sparrow and Ney, 1971; Turman, 1978; Orville, 1981; Radio Research Laboratories, Japan, 1981).

Although lightning from ordinary, large thunderclouds is less frequent over the oceans, other electrical phenomena appear to be more common over the water. Another less common variety of lightning, warm cloud lightning, which is produced by small clouds that do not extend above the freezing level, apparently occurs more often over the ocean than over land. There are three references to warm cloud lightning over the ocean (Foster, 1950; Petrowski, 1960; Moore, et al., 1960), but only one report of its occurring over the land (Michnowski, 1963). Similarly, despite the much higher population density over land, a high proportion of the sightings of St. Elmo's fire have been made at sea.

The relative importance of the various processes that initiate and maintain the electrification of clouds is not known. As a result, it is impossible at present to understand the reasons for the differences that are observed between the electrification process taking place in clouds over the land surfaces and in those over the ocean. It is the purpose of this paper to consider some of the contrasts that exist between the land-atmosphere interface and the ocean-atmosphere interface and to speculate on their possible role in affecting the electrification of clouds.

Characteristics of the Air-Sea Interface and Their Possible Influences on Electrical Processes

Convection

Undoubtedly an important reason for there being less lightning over the oceans than over land is that, because of the smaller temperature contrasts at the water-air interface and the lack of orographic perturbations, intense vertical convection is less common over water surfaces than over land. As Workman and Reynolds (1949) have shown, strong convective activity appears to be a necessary requirement for the electrification process. The rate at which clouds produce lightning increases markedly with their height and the intensity of their updrafts and downdrafts (Shackford, 1960). It has long been recognized that as the earth rotates, global thunderstorm frequency and the magnitude of the negative charge carried by the earth are at a maximum when the sun is shining on the continents and convective activity is most intense, and at a minimum when the sun is shining on the oceans and convection is less intense.

Even though the intensity of convection is probably of dominant importance, there are

E. C. Monahan and G. Mac Niocaill (eds.), Oceanic Whitecaps, 219–226.

other differences between processes taking place over land and water that may have important influences on atmospheric electrification.

Atmospheric Aerosols

Often there are large differences between the atmospheric aerosol populations over land and over the ocean. Because of various natural and anthropogenic sources, the number concentration of aerosol particles over the continents is usually an order of magnitude or more larger than that over regions of the ocean far removed from land.

Not only is the number concentration of aerosol particles in the lower atmosphere over the ocean much lower than over land, but there are frequently large differences in the character of these particles. As Woodcock (1952) and Blanchard and Woodcock (1957) have shown, the bursting of bubbles at the ocean surface leads to the production of large hygroscopic salt particles that are more effective cloud drop forming nuclei than those usually encountered in continental air masses. These differences between the aerosol population over the land and over the sea can be expected to affect the characteristics of the atmospheric electrical phenomena that are taking place.

Atmospheric aerosols play an important direct role in atmospheric electrical processes because they have a large influence on the electrical conductivity of the atmosphere. The small fast ions that are produced either by ionization of atmospheric gases by cosmic rays or radioactive decay rapidly become attached to aerosol particles that may be present in the atmosphere. As a result, in the high concentrations of aerosol particles that frequently occur over continents, the electrical conductivity of the air can be reduced as the result of ion attachment to aerosols, sometimes by as much as a factor of 10 (Moore, et al., 1962). Over the ocean remote from land areas the atmosphere frequently has such low concentrations of aerosol particles that they do not play a significant role as a sink for fast ions. As a consequence, even though the ionization rate over the ocean is usually somewhat smaller than that over land, the atmospheric electrical conductivity may be significantly higher than that usually measured over land areas. Furthermore, because of the absence of high gradients in aerosol concentrations, such as commonly occur over the land, the electrical conductivity of air over the ocean is probably far more uniform than that over land.

Atmospheric aerosols have an indirect effect on atmospheric electrical processes because of the large influence they exert on the microphysical characteristics of clouds. There are usually fewer but more effective drop forming nuclei in the air over the oceans than over land. As a result, the drop size distribution in maritime clouds is often significantly different from that found in clouds over the land. There is usually a lower concentration of cloud droplets and the droplets that are present are larger than those occurring in continental clouds. Because of the lower concentrations of cloud droplets in maritime clouds, it is to be expected that fast ion attachment to these particles would occur at a lower rate and that therefore the transport of charge by the movement of fast ions could occur more effectively in maritime than in continental clouds.

Because of their different particle size distribution, it is to be expected that in maritime clouds the cloud electrification processes taking place might be quite different from those that occur in continental clouds. It seems probable that the electrification that occurs, either as the result of induction or contact electrification, when a precipitation particle collides with a cloud particle, would be strongly dependent on the dimensions of the colliding cloud particles and probably would increase rapidly with their size. One might expect on this basis that maritime clouds would build up charge more rapidly than continental clouds. On the other hand, because of the lower number concentration of the cloud particles the rate of charge separation resulting from drop collisions might conceivably be less than that in continental clouds.

Because of their low number concentration, the specific surface area available to carry electric charge and the space charge densities formed by ion attachment will be significantly less in maritime clouds (Brown et al., 1971). If charged particles formed by ion attachment play an important role in the formation of charged regions, as is suggested by Grenet (1947) and Vonnegut (1955), then it is to be expected that the dimensions of the charged regions required to produce the intensity of electric fields required for lightning might be significantly larger in maritime than in continental clouds.

Another factor that may be of importance in producing cloud electrification is the concentration of ice-forming nuclei that are present. Since more ice-forming nuclei originate over land than water, it seems probable that formation of the ice phase, which may be of importance in cloud electrification, will take place at lower temperatures in clouds over the ocean than over land.

Atmospheric Electrical Conductivity

The electrical conduction process in fine weather just above the ocean differs from that over land. This is not only because the aerosol concentration is lower over the oceans, but because of differences in the air ionization process. Over land surfaces ionizing radiation from rocks and soils, radioactive gases and aerosols contributes significantly to the ionization of the lower atmosphere. Here the ionization rate caused by radioactivity is of the same order of magnitude as that caused by cosmic rays. At times radioactivity in the region a few tens of centimeters above a land surface is even more intense and is capable of causing a several-fold increase in conductivity. Over oceans away from radioactive sources originating from land masses, ion pair formation is due almost entirely to cosmic rays, and radioactivity plays a very minor role. With the introduction into the atmosphere by man of significant quantities of Kr85, an isotope having a half-life of ten years, ion pair production by radioactivity has recently begun to assume a significant role in the ionization process over the oceans (Boeck, 1976).

Space Charge

Two factors that influence atmospheric electrical conductivity, the sources of aerosols and radioactivity, are not nearly as localized or concentrated over the ocean as they are over land. Ionization from radioactivity in the water and from radioactive gases and aerosols in the atmosphere is much less intense and the number concentration of aerosol particles is lower. As a consequence, the gradients in conductivity are usually not nearly as steep as those that occur over land. As a result, the shallow layers of negative space charge often observed near the surface of land are absent, as are the layers of dense space charge of both polarities that form above and below stratiform regions of aerosol particles introduced by air pollution (Moore, et al., 1962).

Because of the smaller local effects of aerosols and radioactivity, the electrode effect is much more pronounced over the ocean than over land and the air close to the water surface usually carries a positive space charge. Blanchard (1963) has shown that the region of positive space charge near the surface of the ocean is further intensified by positive particles that are produced when bubbles break at the ocean surface. Small jets are formed that produce highly electrified, small, positive droplets that play an important part in the transfer of positive charge into the atmosphere. Studies carried out with an airplane by Markson (1975) have shown variations in the electric field over the ocean that are produced when the positive space charge formed near the water surface is carried upward by organized convection currents, confirming Blanchard's (1963) suggestion that these particles are an important contribution to the positive charge in the lower atmosphere.

It appears possible that when convection carries this space charge into clouds, it may play a role in the initiation of their electrification. Conceivably, this oceanic space charge may be important in the electrification process that leads to the formation of the lightning that is sometimes observed to form in warm oceanic clouds.

When thunderstorms occur and intense electric fields begin to develop over the ocean surface, the distribution of space charge is undoubtedly quite different from that existing under continental thunderclouds or that prevailing over the ocean in fair weather. Because the space charge under thunderstorms over the ocean is intimately related to the electric field at the ocean surface, it will be discussed in the following section.

Electric Field and Charge Transfer

The electric field at the interface between the earth and the atmosphere is greatly influenced by its geometric form. There are large differences between land surfaces and water surfaces. For the most part, land surfaces have elevated, electrically conductive structures such as vegetation and man-made artifacts that often terminate in sharp points. The induced electric charge on such structures greatly modifies the ambient electric field. By concentration it can in a very limited region increase the field intensity by as much as 5 orders of magnitude. The surface of the ocean is in marked contrast to that of the land. Whereas the interface over land changes slowly with alterations in the vegetation during the course of a year, the ocean interface is subject to very rapid changes depending on waves generated by winds in the atmosphere and currents in the water. When the water surface is calm, no local intensification or reduction of the electric field occurs, and it is uniformly the same over large areas. On the other hand, when the surface is agitated by the wind and waves are formed, the situation becomes more complicated. Measurements made under conditions of moderate wind during fair weather show that the electric field 0.5m above the water surface can be twice as great over the crest of the wave as over the trough (Vonnegut, 1974). Very near to the crest the field is doubtless more intense. Because waves are constantly in motion, the sea surface produces periodic variations in time and position of the intensity of the electric field that are detectable above the ocean surface (Vonnegut and Markson, 1975). Little or no information is available on the perturbations in the electric field that result when high winds cause breaking and colliding waves. It seems probable that under certain conditions these can produce transient, vertical plumes of water many meters high, which being electrically connected to the earth, cause large electric field concentrations at their extremities.

The electric field at the ocean-atmosphere interface will also be influenced by distortions of the surface other than waves. When particles such as raindrops, snow, hail, or large spray drops fall on the surface, they will create Worthington jets of water (1908) that will intensify any existing electric field. Similarly, bubbles and bursting bubbles will disturb the interface in such a manner as to intensify ambient electric fields (Blanchard, 1963). When marine organisms, vegetable or animal, project above the ocean surface, they will also cause field intensification. Floating ice or debris will exert a similar effect.

The few measurements that have been made beneath a thundercloud over a body of water show that the ambient electric field can reach intensities far larger than those generally observed over land. Toland and Vonnegut (1977) have reported field intensities as high as 130 kv m^{-1} over a lake. This value is approximately 10 times greater than that usually measured over land. Perhaps such intense electric fields are responsible for causing the impressive displays of St. Elmo's Fire that are sometimes observed on the rigging of ships.

Without doubt, this large difference arises from the fact that over land point discharge occurs from the extremities of vegetation and other elevated structures when the ambient field reaches about 2 kv m^{-1}. When this happens, the resultant space charge that is released into the atmosphere forms a screening layer that limits the average field to less than 10 kv m^{-1} (Standler and Winn, 1979). Because sharp points and elevated structures are usually absent from lake or ocean surfaces, far more intense electric fields are required to produce dielectric breakdown and corona. As a result, ambient electric fields over water can rise to values far higher than over land.

It is conceivable that the high electric field intensities, in excess of 100 kv m^{-1}, that have been measured over water are limited to this value by some process such as point discharge that causes the emission of charged particles into the atmosphere. This appears to be unlikely, however. Laboratory experiments show that even a water surface roughened by falling rain requires a field in excess of 180 kv m^{-1} to produce corona (Griffiths et al., 1973), and one roughened by bursting bubbles, 260 kv m^{-1} (Latham, 1975).

A satisfactory understanding of the behavior of water surfaces under the influence of the strong electric fields beneath thunderclouds will require further laboratory and field investigations. It appears quite probable that under intense storms when the electric field intensity becomes very high and the ocean surface is greatly agitated by winds and precipitation that it will release charged particles, either in the form of ions produced by corona and/or electrified water droplets produced by electrical induction (Blanchard, 1963) and by electrical atomization (Vonnegut and Neubauer, 1952; Griffiths, et al., 1973).

The transfer of charge from the ocean beneath thunderstorms is of importance to our understanding of the role that such storms play in maintaining the negative charge that resides on the surface of the earth. Blanchard (1963) has estimated that under thunderclouds inductive charging of jet drops could result in a global charging current as large as 150 amperes. Measurements of the charge transfer process made on thunderclouds over land indicate that perhaps the largest transport of charge to the earth is taking place as the result of corona (Wormell, 1927). In view of the great difference in the corona characteristics of the ocean, it will be of interest to determine whether maritime storms are as effective in transferring charge to the earth as those over land, and by what processes this transfer takes place. Measurements made over land show that falling precipitation plays only a minor role in carrying charge to the earth, transferring only of the order of 0.1 amperes of positive charge in contrast to the 1 ampere of negative charge estimated to be brought to the earth by point discharge. Some evidence indicates that the positive electrification of the rain is produced by the attachment of positive ions that have been introduced into the atmosphere as the result of point discharge. It is therefore possible that if ions or other charged particles are not being released from the surface of the ocean under a thunderstorm, the rain over the ocean may carry negative instead of positive charge and act to increase rather than to decrease the negative charge of the earth.

Quantitative measurements of the charge transfer taking place at the sea surface are essential to an understanding of the processes responsible for the electrification of maritime clouds. If the corona current acts to dissipate electrical energy of the cloud, as is assumed in those mechanisms of electrification based on falling precipitation, then clouds over the water that never produce a corona should produce more lightning than comparable clouds over land. On the other hand, if the space charge produced by corona is carried up by convection where it acts to intensify positive charge in the upper part of the cloud, as has been proposed in the convective theory (Vonnegut, 1955), then it is to be expected that those maritime clouds in which corona is absent would produce less lightning than those over land.

Lightning

Surprisingly, little information is available concerning the relationships between lightning and the air-sea interface. Observations of lightning strikes to a fresh water lake (Vonnegut, 1976) indicate that when the lightning discharge makes contact with the surface of the water, it throws up a large plume of spray similar to that produced by the fall of a large object. Presumably this is the result of the shock wave produced in the atmosphere near to the channel, and the vaporization of water near the surface produced by electrical heating. The effects produced when lightning strikes sea water may be different from those observed in the fresh water case, for sea-water's electrical conductivity is 6 orders of magnitude greater and the resistive heating would be much less.

In connection with the large difference between the conductivity of fresh and salt water, it is worth noting that electrical shock from lightning probably presents a far greater hazard to animals immersed in fresh, as opposed to salt, water. While the electrical conductivity of body tissues and fluids is generally of the same order of magnitude or even less than that of salt water, it is many times greater than that of fresh water. As a result, in salt water electric currents produced by lightning tend to flow around animals rather than through them. In fresh water the situation is just the opposite and the electric current preferentially passes through the organisms. Irving Langmuir told the author of having experienced a strong electrical shock when lightning struck nearby while he was swimming in Lake George.

Studies of lightning that have been carried out over land surfaces show that the initiation of the lightning discharge usually begins within the thundercloud and leads to the formation of a stepped leader that progresses toward the surface of the earth. Photographs show that when the stepped leader approaches within tens of meters of the earth's surface, luminous ionized streamers from vegetation and other elevated structures rise to meet the advancing stepped leader. When contact is made, a return stroke ensues. Presumably, a similar phenomenon takes place over the surface of the sea. In this case, it seems probable that the streamers would originate from the crests of waves.

Studies over land show that only very rarely is the lightning discharge initiated by objects on the surface beneath the cloud. This phenomenon is observed when the lightning event originates in the very intense electric fields that prevail above very tall structures, such as the Empire State Building, television antennas, and sharp mountain peaks. Experiments have shown that in a similar fashion lightning discharges can be triggered from the ground by firing a rocket that trails a wire up into a thundercloud.

Presumably it is only rarely that lightning is triggered from the ocean, for the water surface very seldom provides conducting structures that are sufficiently elevated to initiate lightning. It should be noted, however, that because the electric fields may be far more intense over water than over land, much smaller vertical structures might be capable of initiating lightning. It has been observed that the explosion of a depth charge in Chesapeake Bay triggered a lightning discharge when the electrically conducting plume of water that it produced rose to a height of approximately 70m (Young, 1961). It is conceivable that under conditions of very intense winds and very intense electric fields under a thundercloud it may be possible for lightning discharges to be initiated at the ocean surface.

Summary

Our information on the electrical processes taking place at the sea surface is obviously fragmentary. It is clear, nevertheless, that what is going on is quite different in many respects from what has been observed to take place over the land. When the sea surface is smooth and not disturbed either by precipitation or waves it appears probable that electric fields can develop that are even more intense than the maximum of 130 kv m^{-1} that has been measured.

When it is considered that electrical forces and the electrical energy density vary as the square of the electric field intensity, it is evident that phenomena may take place over the surface of the ocean that are unknown over land. For example, under the influence of the very intense electric field, electrical atomization may be occurring at the top of Worthington jets (1908). It appears conceivable, as Blanchard has suggested (1963), that the droplets produced at the ocean surface may be so highly charged that under strong electric fields they are levitated and carried up into the cloud by electrical forces.

It is possible that some of the water particles produced in strong electric fields under oceanic thunderclouds may carry electric charges close to the Rayleigh limit (1896). When this is the case, and the droplet decreases in size because of evaporation, it may well become unstable and break up and thus form a larger number of smaller droplets (Doyle, et al., 1964). If significant numbers of highly charged water droplets from the ocean surface are carried up into the cloud they may be of importance in accelerating the growth of precipitation by coalescence by processes similar to those envisioned in the rain gush effect (Moore, et al., 1964).

It appears that few, if any, electrical measurements have been made under thunderstorms when there are high winds, large waves, and heavy precipitation. Under such conditions, it is possible that ions produced by corona or charged water drops released from the ocean surface may limit the electric fields to intensities much lower than that indicated by the few measurements reported, which were made when the water surface was relatively undisturbed. Because of the large variations possible in the state of the ocean surface, there may be two varieties of maritime thunderstorms. In one the surface may be sufficiently agitated that charged particles are emitted from the ocean surface. In the other the surface may be so smooth and free of bubbles that no production of ions or charged particles can take place. The two types of storms might be expected to exhibit quite different electrical behavior.

Making measurements of the various processes transferring charge from an ocean surface beneath an active thunderstorm promises to be even more difficult and dangerous than making similar measurements over land. It will not be easy to estimate the relative magnitudes of the rate at which charge is being transferred by lightning, by falling precipitation, by corona, and by the detachment of charged water particles from the ocean surface. Probably the best way of making comparisons between the charge transfer processes of continental and oceanic storms will be to observe the electric current that flows in the clear air over the top of the storm. From considerations of continuity it follows that this current must be equal to the sum of all the currents that are flowing to the ocean surface beneath the storm.

Measurements from aircraft will add greatly to our knowledge of the electrical processes over the ocean. It will also be necessary, however, to investigate the details of the complicated electrical phenomena at the ocean-atmosphere interface by using instrumentation on the surface. Because electrical storms are much more rare over the sea than over land and because there are no orographic features to trigger their formation at a given location, it will be difficult to be at the right place at the right time to make these measurements. Despite the many problems, the results that will be obtained should be well worth the effort for the new insights that they will provide concerning this neglected area of atmosphere-ocean interactions. In addition, when we learn more about the differences between maritime and continental thunderstorms it is likely that we will have new and invaluable clues concerning the nature of the processes that are responsible for causing some clouds to become highly electrified.

REFERENCES

Blanchard, D. C., 1963: The electrification of the atmosphere by particles from bubbles in the sea. *Progr. Oceanogr.*, **1**, 73-202.

Blanchard, D. C., and A. H. Woodcock, 1957: Bubble formation and modification in the sea and its meteorological significance. *Tellus*, **9**, 145-158.

Boeck, W. L., 1976: Meteorological consequences of atmospheric Krypton-85. *Science*, **193**, 195-198.

Brown, K. A., P. Krehbiel, C. B. Moore, and G. N. Sargent, 1971: Electrical screening layers around charged clouds. *J. Geophys. Res.*, **76**, 2825-2835.

Doyle, A., D. R. Moffett, and B. Vonnegut, 1964: Behavior of evaporating electrically charged droplets. *J. Colloid Sci.*, **19**, 136-143.

Foster, H., 1950: An unusual observation of lightning. *Bull. Amer. Meteor. Soc.*, **31**, 140-141.

Grenet, G., 1947: Essai d'explication de la charge électrique des nuages d'orages. *Ann. de Géophys.*, **3**, 306.

Griffiths, R. F., C. T. Phelps, and B. Vonnegut, 1973: Charge transfer from a highly electrically stressed water surface during drop impact. *J. Atmos. Terrest. Phys.*, **35**, 1967-1968.

Latham, J., 1975: Possible mechanisms of electrical discharge involved in biogenesis. *Nature*, **256**, 34.

Markson, R., 1975: Atmospheric electrical detection of organized convection. *Science*, **188**, 1171-1177.

Michnowski, S., 1963: On the observation of lightning in warm clouds. *Indian J. Meteor. Geophys.*, Delhi, **14**, 320-322.

Moore, C. B., B. Vonnegut, B. A. Stein, and H. J. Survilas, 1960: Observations of electrification and lightning in warm clouds. *J. Geophys. Res.*, **65**, 1907-1910.

Moore, C. B., B. Vonnegut, R. G. Semonin, J. W. Bullock, and W. Bradley, 1962: Fair-weather atmospheric electrical potential gradient and space charge over central Illinois, summer 1960. *J. Geophys. Res.*, **67**, 1061-1071.

Moore, C. B., B. Vonnegut, E. A. Vrablik, and D. A. McCaig, 1964: Gushes of rain and hail after lightning. *J. Atmos. Sci.*, **21**, 646-665.

Orville, R. E., 1981: Global distribution of midnight lightning–September to November 1977. *Mon. Wea. Rev.*, **109**, 391–395.

Petrowski, E. L., 1960: An observation of lightning in warm clouds. *J. Meteor.*, **17**, 562–563.

Radio Research Laboratories, Japan, 1981: *World distribution of thunderstorm activity obtained from ionosphere sounding satellite-b observations June 1978 to May 1980*, 57pp (available from Radio Research Laboratories, Ministry of Posts and Telecommunications, 10-2 Nukui-Kitamachi 4-Chome, Koganei-shi, Tokyo 184, Japan).

Rayleigh, Lord, 1896: *The Theory of Sound*, Vol. 2, 2nd ed., Macmillan (New York), 374.

Sanders, F., and J. C. Freeman, 1982: Thunderstorms at Sea (Chapter 3). *Thunderstorms: A Social, Scientific, and Technological Documentary*, Vol. 2, *Thunderstorm Morphology and Dynamics*, E. Kessler, ed., U.S. Department of Commerce, NOAA Environmental Research Laboratories.

Schackford, C. R., 1960: Radar indications of a precipitation-lightning relationship in New England thunderstorms. *J. Meteor.*, **17**, 15–19.

Sparrow, J. G., and E. P. Ney, 1971: Lightning observations by satellite. *Nature*, **232**, 540–541.

Standler, R. B., and W. P. Winn, 1979: Effects of coronae on electric fields beneath thunderstorms. *Quart. J. R. Meteor. Soc.*, **105**, 285–302.

Toland, R. B., and B. Vonnegut, 1977: Measurement of maximum electric field intensities over water during thunderstorms. *J. Geophys. Res.*, **82**, 438–440.

Turman, B. N., 1978: Analysis of lightning data from the DMSP satellite. *J. Geophys. Res.*, **83**, 5019–5024.

Vonnegut, B., 1955: Possible mechanisms for the formation of thunderstorm electricity. *Proceed. Conf. Atmosph. Electr., Geophys. Res. Papers No. 42*, Geophysics Research Directorate, AFCRC, Bedford, MA, 169–181.

Vonnegut, B., 1976: Lightning strikes to fresh water. *Weather*, **31**, 69.

Vonnegut, B., and R. Markson, 1975: Electrical detection of waves at water-air surface. *Proceedings Civil Engineering in the Oceans*, Vol. 3, Amer. Soc. of Civil Engineers, New York, 1137–1149.

Vonnegut, B., and R. Neubauer, 1952: Production of monodisperse liquid particles by electrical atomization. *J. Colloid Sci.*, **7**, 616–622.

Woodcock, A. H., 1952: Atmospheric salt particles and raindrops. *J. Meteor.*, **9**, 200–212.

Workman, E. J., and S. E. Reynolds, 1949: Electrical activity as related to thunderstorm cell growth. *Bull. Amer. Meteor. Soc.*, **30**, 142.

Wormell, T. W., 1927: Currents carried by point-discharges beneath thunderclouds and showers. *Proc. Roy. Soc.*, **A 115**, 445–455.

Worthington, A. M., 1908: *A Study of Splashes*. Longmans, London, republished 1963 by Macmillan, New York.

Young, G. A., 1961: A lightning strike of an underwater explosion plume. U.S. Naval Ordnance Laboratory Report 61–43.

ATMOSPHERIC ELECTRIC SPACE CHARGE NEAR THE OCEAN SURFACE

STUART G. GATHMAN
Naval Research Laboratory,
Washington, DC 20375

The atmosphere near the air-sea interface is electrified by different processes. The space charge associated with this electrification resides on three classes of particles or droplets each of which have a different origin and history.

It has been shown that jet drops resulting from the bursting of small bubbles in sea water have a positive electric charge. These droplets contribute only part of the space charge found near the air-sea interface and only during times when white water phenomena exist upwind of the observation site.

Spray droplets separated from the water surface by the stress of wind action on waves may also be charged. The charge on these drops is produced by electrostatic induction and the polarity and magnitude will depend on the direction and magnitude of the geoelectric field at the point of their separation from the water's surface. As the electric field at sea is usually of a direction to produce negative inductive charge, we can say that except for rare instances (i.e. thunderstorms) spray droplets are negatively charged.

The sea surface acts as a negative electrode. In the region away from the electrode, ions of both polarities formed in the atmosphere by cosmic rays flow under the action of the electric field and carry the air-earth current. Near the electrode however, there is no compensating flow of negative ions leaving the sea surface and the total current must be carried by positive ions. In a simplistic description this produces a region of net positive space charge called the 'electrode effect' in the lowest tens of meters. The process is complicated by the existance of aerosol and turbulence in the marine boundary layer which must be included in the complete formulation. These factors determine the vertical extent of electrode effect space charge.

Over the ocean the electrode effect and the bubble electrification process both produce positive space charge at the air-sea interface. Therefore special experimental arrangements must be used to separate these mechanisms when looking at the electrical effects of whitecaps at the surface in terms of marine boundary layer parameters.

Experiments performed over a large body of fresh water, on the other hand, assist in the differentiation of these processes in that the fresh water jet drop mechanism produces a charge different from that of sea water, while the turbulent electrode effect and the inductive charging of the splash drops remains the same. Alternatively, surf zone measurements may be profitably employed to study white water electrification since the surf can produce space charge which may in special cases obtain values an order of magnitude or more greater than can be expected from the electrode effect process.

Introduction

The sea surface is not usually thought of as a source of electrification. It was not until Blanchard (1955) noted from laboratory measurements that the breaking of bubbles at the sea surface produced charged jet drops, which were mechanically ejected into the air away from the sea surface, that the role of the sea in the field of fair weather atmospheric electricity came into focus. Previously the sea was thought of as a gigantic quiescent electrode to which most of the electric current generated by world-wide thunderstorm activity returned to the earth. Blanchard's discovery renewed interest in the electrical effects operating at the sea surface and field experiments were designed in the effort to detect the relationship between the marine space charge level and whitecap activity. The bubble bursting mechanism as a source of charged jet droplets was expected to be a function of whitecap activity upwind of the

E. C. Monahan and G. Mac Niocaill (eds.), Oceanic Whitecaps, 227–243.

measurement site.

Blanchard (1958, 1966) set up experiments on the shore of the island of Hawaii where he measured both the atmospheric space charge and the potential gradient. He observed that both of these electrical parameters were variable, positive, and of higher value when the air reaching the measurement site came from over the surf zone a few meters upwind than when it came from other directions. He, of course, associated these results with the white water in the surf zone. These measurements underline the fact that there is indeed space charge associated with the marine boundary layer.

Between the times of Blanchard's publication of these results, Mühleisen (1962) reported similar observations on electrical measurements along the shore of the North Sea, but he attributed the measured large positive values to the electrode effect. Either one or both mechanisms could have been the cause of the shoreline observations and a more sophisticated set of experiments was needed in order to understand the role each mechanism played over the ocean.

In this paper I wish to discuss in more detail the physics of these two mechanisms, the experiments which were done to isolate the mechanisms in operation over open water, and to discuss some of the applications of these results to other aspects of marine science.

Whitewater Electrification - Open Ocean

The charged jet drop mechanism has been well investigated by Blanchard (1963). The purpose here is not to reiterate the details of his research but to touch on aspects of this work which are applicable to the space charge over the ocean and its relationship to whitecap activity. He found that the charge on the top jet drop produced by the bursting of an air bubble at the sea surface has a natural electric charge. The magnitude and polarity of this charge depends on the age of the bubble producing it, the chemistry of the water and the size of the droplet. A charge is also induced on the droplet by the earth's natural electric field as it leaves the earth's surface. Blanchard showed that in the case of the normal marine environment, the natural positive charge of the bubble produced jet droplet is more significant than the negative induced charge.

These laboratory measurements of the electric charge on individual jet droplets can be used to estimate the net current density carried by the bubble produced jet droplets from the sea when the drop spectrum from whitecaps is known. Certain simplifying assumptions are necessary however in making this step. Among these are:

(1) Drops once produced will stay airborne.

(2) Only top drops are involved in charge transfer.

As the aerosol size distribution leaving the sea surface is a function of wind speed, the net current density leaving the sea surface can be estimated in terms of the wind speed by weighting the size distribution by the charge for each size class. Figure 1 is a plot of the estimated droplet current leaving the sea surface as a function of wind speed. The net current nominally flowing to the earth from the upper atmosphere is shown by the horizontal dashed line and has a nominal value of 2×10^{-12} amperes per square meter.

The hatched area on the figure represents the electric current density predictions based on the droplet flux from the latest sea surface aerosol generation model, SSAG (Monahan et al. 1983). In this plot there is a differentiation between the droplets which are introduced indirectly into the atmosphere by the bursting of bubbles mechanism and those introduced directly by the break up of wave tops in the chop and spume droplet producing mechanisms. The natural positive charge is placed only on the jet droplets described by the indirect term of the model, whereas a negative induction charge is placed on droplets from both terms of the model.

Although Blanchard (1963) has investigated the induction charge produced on jet drops leaving the sea surface, there has not been much work to my knowledge on the inductive charging of directly produced droplets from the breakup of wave tops. He showed, however, that for jet drops the negative induced charge was only 1.1% and 4.3% of the natural positive charge for normal fair weather electric fields for 5 and 20 micron droplets respectively. The

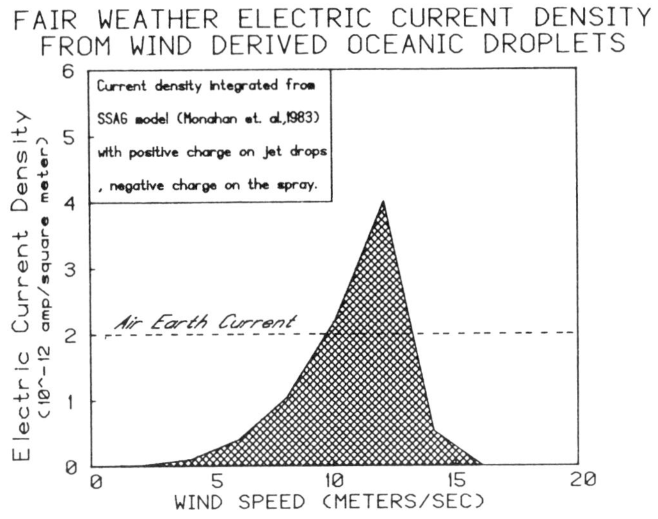

Fig. 1. Wind dependence of the electric current density from charged droplets leaving the sea surface under a fair weather atmospheric condition.

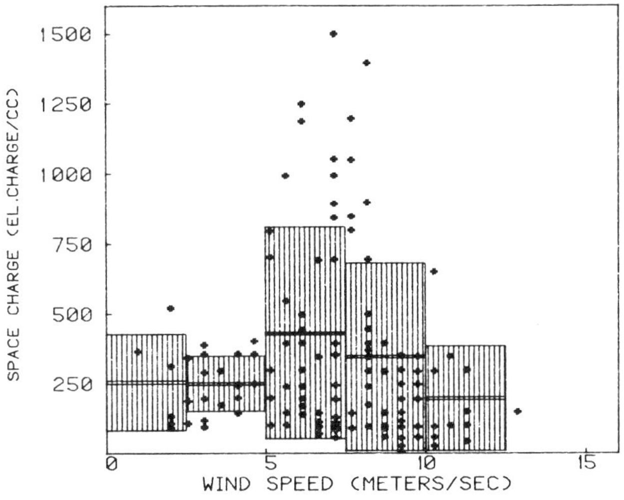

Fig. 2. The wind dependence of atmospheric space charge density over the open ocean from the cruise of the USNS *Eltanin* Nov.-Dec., 1967.

calculations of induced charge used in Figure 1 are based on the estimate that the magnitude of the induced negative charge would be 1/10% of the natural positive charge value that a similar jet drop would contain. The plot shows that the mean current from the ocean approaches the value of the normal air-earth conduction current at wind speeds of about 7 or 8 meters per second and turns sharply around when the large direct aerosol mechanism starts in full operation at about 12 meters per second. Indeed, one would expect that over the open ocean the space charge density in the marine boundary layer would be solely under the influence of the positive jet drop production mechanism for only a relatively narrow window of wind speeds.

Although the precise amount of induced charge resulting from the interaction of the geoelectric field, wave shape, droplet spectrum and wind speeds is not know at present, the fact that at high wind speeds the quantity of droplets produced by the direct, spume, droplet mechanism becomes so large, according to the SSAG Model that even if they are only slightly negative it is enough to cause this term to override the positive current of jet drops at some high wind speed value.

In November–December 1967, the research ship, USNS *Eltanin* was equipped to measure atmospheric electric space charge over the open ocean with a six foot cubic Faraday cage. The instrument and experimental procedure were described by Gathman and Trent (1968). The space charge density measurements from this cruise are plotted as a function of wind speed in Figure 2. In this plot, a large scatter in points is noted at a wind speed of about 5 meters per second and continues to about 10 meters per second. When these data were first taken, the decrease in space charge density at the higher wind speeds could not be reconciled with the ever increasing sea surface current with respect to wind speed which would result if only jet drops were involved in the electrification process.

In the figure the mean space charge over a band of wind speed values is shown by the plotted horizontal bands in the hatched rectangles. These mean calculations show an approximate doubling in value to about 400 elementary charges per cc for the band of wind speeds between 5 and 10 meters per second compared to the values at lower wind speeds. The mean value again returns to lower values at the higher wind speeds. About each of these mean levels is a hatched rectangle representing the standard deviation in space charge for that particular band of wind speed values.

Although the mean values of sea surface space charge density are greater within the bands covering the wind speed range of 5 to 10 m/sec than in the lower wind speed bands, the scatter of the data is also significantly increased. The question then must be asked:

> Are the mean values of space charge density between 5 and 10 meters per second significantly different from the normal sea surface space charge density found over water where there is obviously no white water phenomenon?

This is answered by the statistical evaluation of these data using the student-t parameter to estimate the probability for each wind band, that the differences between the mean space charge for this wind band and the mean space charge determined by combining all bands could be expected to be the result of random differences in the data selection. With this system of evaluation, the probability being small indicates a highly significant difference, whereas large probabilities indicate that the observed differences are probably due to random selection processes in the data. This analysis for different wind speeds is shown in Table I where P is the probability based on the student-t statistic that the observed mean space charge density was not significantly different from the usual value of 200 elementary charges per cc.

These statistics show that the peaks in the observed means which occur within the band of

Table I

Wind Speed	Space Charge	Std. Dev.	t	p
1.25	254.8	172.9	.18	.5
3.75	249.7	99.1	2.18	.05
6.25	430.1	378.9	3.98	.001
8.75	344.9	335.1	2.73	.01
12.25	196.6	187.2	.06	.1

5 to 10 meters per second wind speed are indeed significantly different from the non-whitecapping value. This observation is exactly what is expected from the prediction, shown in Figure 1, that the sea surface generation of positive electric current is a maximum in this wind speed range. It is suggested that this observation of positive space charge in the marine boundary layer over and above that found during calm periods is indeed the result of positively charged jet drops being ejected from the sea surface for wind speeds in the band of 5 to 10m/sec.

It is believed that the major amount of scatter in the measured values is due to variations in fetch and duration resulting in variations in sea state at each given wind speed.

Electrode Effect Electrification

Another source of electric space charge at the air-sea surface is the electrode effect. One of the important conceptual tools in atmospheric electricity is the powerful Maxwellian equations which can be used to relate different types of electrical variables. The action at a distance character of the electric forces can be utilized to determine the location of electric charges from non-in-situ measurements. Thus from Poisson's equation, we can determine space charge from the gradient in electric field. Near an approximately flat conducting surface (such as the earth's surface) we can determine the one dimensional space charge distribution by differentiating the measured electric field profile data. Within this framework the quantities of electric field and potential gradient differ from each other only by a negative sign and are often used interchangeably. The electrode effect is a seemingly elusive process at the earth's surface which has been postulated to exist on theoretical grounds beginning with the work of Schweidler (1908). Others, notably Behacker (1910) and Swann (1913), have made similar calculations differing from each other only in their assumptions on the recombination of ions. All of these early calculations predict that the electric field at 1 meter above the earth's surface is 30% less than that at the earth's surface. This change in electric field from Poisson's equation is an indication of a

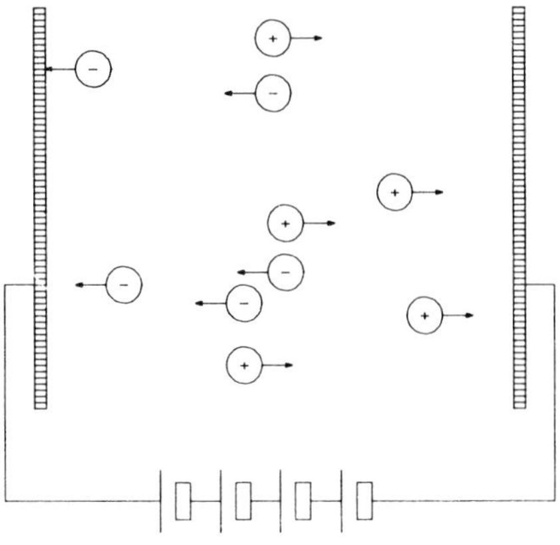

Fig. 3. Simplified diagram of ionic distribution in a parallel plate capacitor to illustrate the physical basis of the simple electrode effect. See text for a more detailed explanation.

considerable amount of space charge near the earth's surface. Conceptually this process is illustrated in Figure 3 which shows a gigantic capacitor in the center of which is the free atmosphere where equal numbers of atmospheric ions of both polarities are constantly being formed by external radiation. Any current that flows between the plates is carried by both kinds of atmospheric ions, each going in a different direction, the positive ions flowing toward the negative electrode and the negative ions flowing toward the positive electrode. Near the electrodes however, there tends to be a build up of a particular polarity as there is no counter acting flow to a volume of space near the electrode from the electrode material itself to compensate for the flow away from the electrode. While such a process is well known in vacuum tubes and other nonatmospheric electrical situations, the experimental verification of the theoretical electrode effect over land sites was never fully realized for many years. The reasons usually invoked to explain this apparent descrepancy between theory and observation were the nonuniform ion production in the atmosphere produced by ground radioactivity, the action of atmospheric turbulent diffusion, and atmospheric aerosol.

The micrometeorological situation over the

ocean is simpler than over land in that the problem of ground radioactivity is eliminated and the aerosol population is reduced, although the processes of atmospheric turbulence are still in action.

Balloon measurements of potential gradient by Mühleisen (1961) over land sites and over water sites (Lake Constance), showed that something approximating a classical electrode effect appeared to be in operation over the water sites but were not observable over the land sites. The problem remains to determine what part of the space charge over the ocean is from jet and spray drops and what part is due to the electrode effect.

Hoppel (1967, 1968) solved numerically the system of second order non-linear differential equations that governs atmospheric electric variables. This solution can be adapted to the case of the marine environment where the aerosol and eddy diffusion are accounted for. In the special case of a nonturbulent and aerosol free atmosphere, the solution matches the analytical solutions of the early investigators mentioned above. Adaptations of the ionization profile also can be used to show why the classical electrode effect cannot be observed over most land sites. The numerical solution has such an advantage in its flexibility and adaptability that it can even be used to invert the problem. Atmospheric electrical measurements were thus used to determine the eddy diffusion and aerosol content of the marine boundary layer (Hoppel and Gathman, 1971, 1972).

Figure 4 shows the solution for a nonturbulent electrode effect over water where electric field, positive and negative small ion density and positive and negative charged aerosol concentration are plotted as a function of altitude. In this nonturbulent case, most of the electrification takes place in the lowest 5 meters of the atmosphere. The space charge density for this case can be either obtained by the differentiation of the electric field variable or by calculating the difference between the positive and negative charged small ions and charged aerosol. The points 'A' and 'B' in the figure are actual small ion measurements made over Lake Superior at a height of approximately 1 meter on a very calm, stable day.

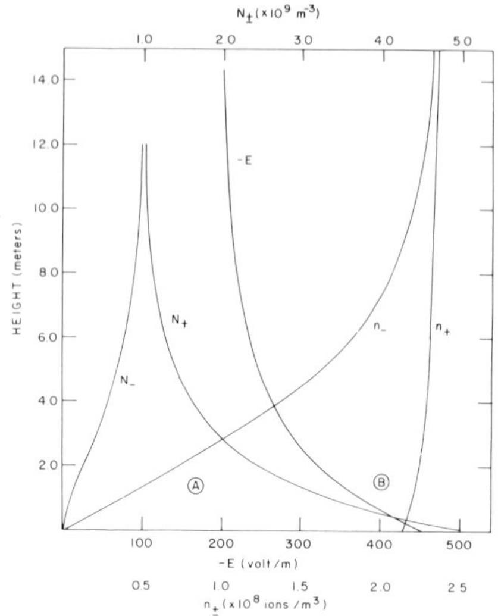

Fig. 4. Theoretical nonturbulent profiles of positive and negative small ion densities, positive and negative charged nucleus densities and electric field for conditions present at Eagle Harbor, Michigan, 1600 local time, 15 July 1968. Points 'A' and 'B' represent measured small ion densities at this time.

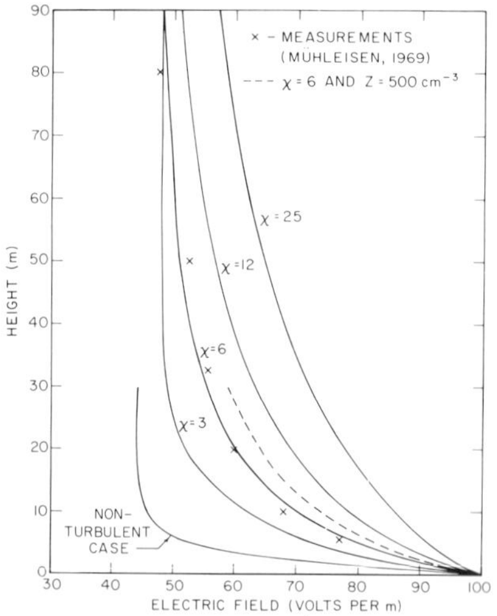

Fig. 5. Numerical solutions of the turbulent electrode effect with calculated electric field profiles for various values of the eddy diffusion coefficient, 'X'.

Figure 5 shows the solution of the electrode effect for a wide variety of cases that might apply to the marine environment. In these solutions the eddy diffusion coefficient is represented by:

$$K(z) = (X \cdot z + d) / (z + 100) \tag{1}$$

which reduces to the ionic diffusion value of 'd/100' at the surface and reaches a constant value of 'X' at relatively high altitudes. With this scheme, the intensity of turbulence can be represented by the parameter 'X'.

As the turbulence parameter, 'X' increases, the vertical extent over which the electrode effect space charge is spread also increases. The turbulence essentially 'mixes up' the available space charge. The solid lines in the plot assume that the background value of the large Aitken nuclei count, 'Z', is negligible for the calculation and the single dashed curve is for the case where Z is 500 per cubic centimeter with a turbulence parameter of X = 6. For comparison with this theory, an actual set of mean electric field profile measurements over the ocean (normalized to 100 v/m at the sea surface) is plotted as the symbol 'x' in the plot (Muhleisen, 1969).

Gathman, 1972b, describes a very sensitive field mill which was designed to be operated from a small tethered balloon. This device shown schematically in Figure 6 was in the form of a split cylinder rotating about its horizontal axis. Such a device is insensitive to self charge and gives a signal proportional to the external electric field. Figure 7 shows the noise level and sensitivity of this device when operating in a calibration field. In the marine environment, however, the turbulent processes in operation cause rapid variations in the observed values. These rapid fluctuations will average out in time to produce the well-known mean profiles predicted by the numerical calculations using 'K' theory. Figure 8 shows the mean plot of the electric field profile from the open ocean near Barbados, West Indies. The striking feature of this plot is the extent of the horizontal lines about each mean value which show the standard deviation of the electric field values at each level. The mean points fit one of the calculated electrode effect profiles for X = 6 and Z = 500/cc. It is interesting to note that the variations in the data decrease with altitude. This is a graphic example of the necessity of using mean values when working with 'K' theory.

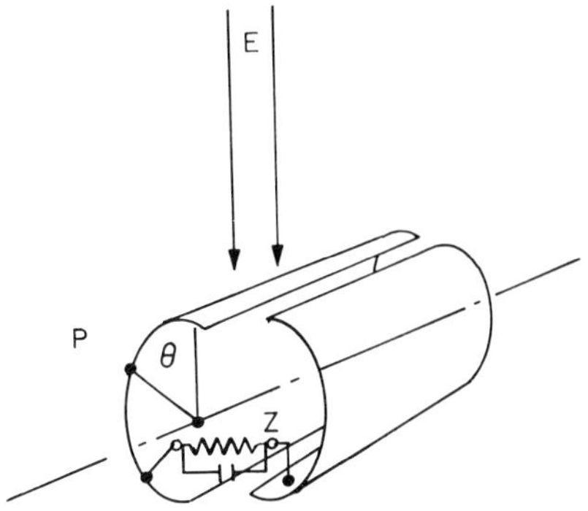

Fig. 6. Simplified diagram of the cylindrical field mill suspended in the free atmosphere. The split cylinder rotates about its axis and the ambient electric field, E, induces an alternating charge, P across the plates which results in an alternating current flowing through impedance 'Z'. The amplitude of this current is proportional to the magnitude of the electric field.

The same mean profile information may also be expressed in the value of the electric field and its slope at any particular altitude. In this method, the slope of the electric field profile is expressed as the mean space charge density at that altitude. From a knowledge of the aerosol count, mean electric field and mean space charge density at any altitude, the 'X' factor can be determined from nomograms such as Figure 9. With the assumption that white water generation processes are not in operation, these atmospheric electrical measurements can be used to measure the distortion of the mean electric field profile and the eddy diffusion parameter necessary to produce that distortion.

Because the parameter X/100 is an eddy diffusion term for electrical parameters, we can make the assumption that it is approximately proportional to the eddy viscosity obtained by the bulk aerodynamic method. This assumption was born out from the results of Hoppel and Gathman (1972) where, using the measurements collected during 6 days of experiments off shore of Barbados in 1971, the ratio of the

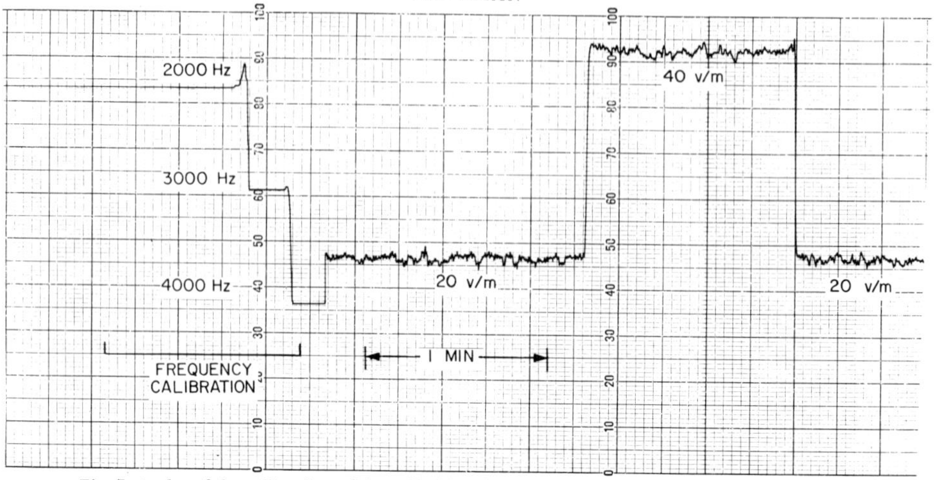

Fig. 7. A plot of the calibration of the cylindrical field mill showing the sensitivity and low noise level of the device when placed in an artificial calibration field.

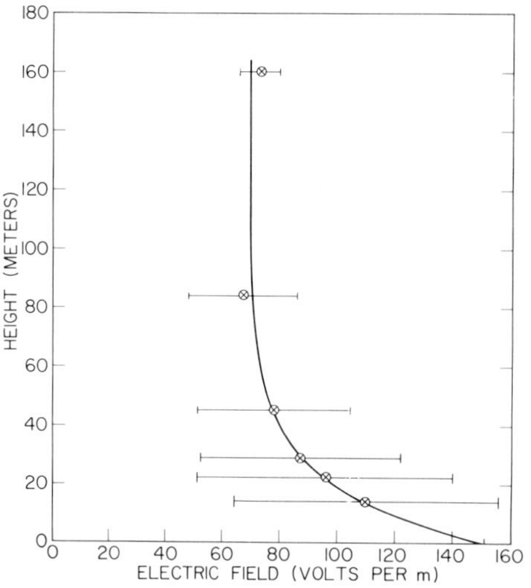

Fig. 8. A plot of the electric field profile obtained with the cylindrical field mill mounted from a tethered balloon over the Atlantic ocean near Barbados in March 1971. The circles represent the means of the electric field readings at each level and the horizontal lines represent the standard deviation in the measurements of the geoelectric field under the turbulent electrode effect. The curve is the numerical solution of the turbulent electrode effect equation for the case of 'X' = 6 and an Aitken nucleus count of 500 per cubic centimeter.

eddy diffusion coefficient measured by the atmospheric electrical distortion method to that from the bulk aerodynamical method was found to be 1.38 with a standard deviation of 0.2.

With the assumption of a linear relationship between X and wind speed, we can use the theoretical electrode effect solutions shown in Figure 5 to plot the space charge density at any particular altitude as a function of wind speed. Figure 10 shows a plot of space charge density produced by the electrode effect for several different altitudes. As expected, these data all show decreases in the space charge density with an increase of wind speed. This decrease is not shown in the data of Figure 2 where space charge density increases with wind speed to wind speeds of from 12 to 15 knots. We therefore must attribute the large amounts of space charge in the intermediate wind speed region to a process other than the electrode effect, such as the droplet electrification phenomenon.

Surf Electrification

The jet drop charging mechanism ought to be especially easy to observe in and around surf action. The air bubble production in areas where waves are feeling bottom and spilling over is a strong source of electrified jet drops. When space charge density observations were made at

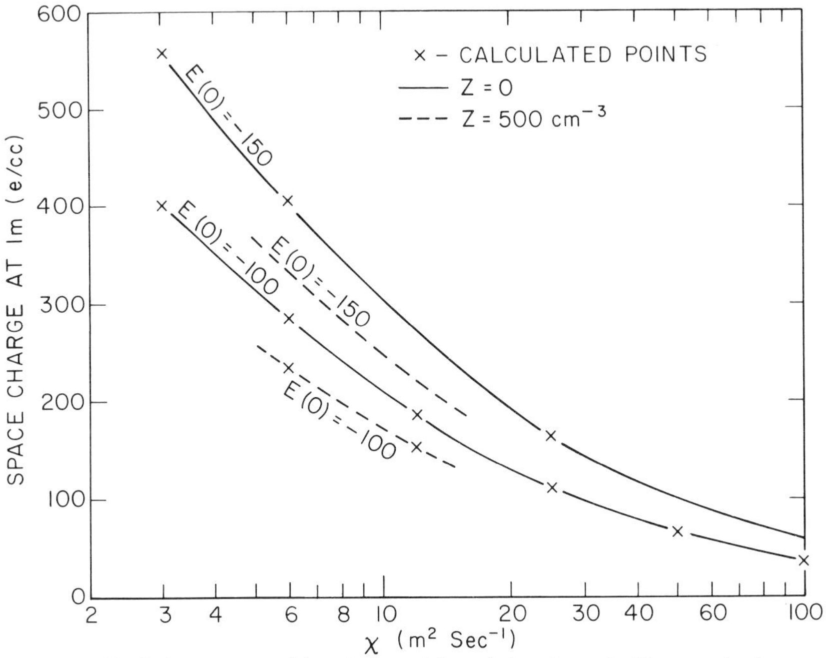

Fig. 9. A nomogram of the solutions of the turbulent electrode effect equation for an altitude of 1 meter showing the application whereby the eddy diffusion coefficient can be determined from measurements of atmospheric space charge density, atmospheric electric field and the Aitken nucleus count.

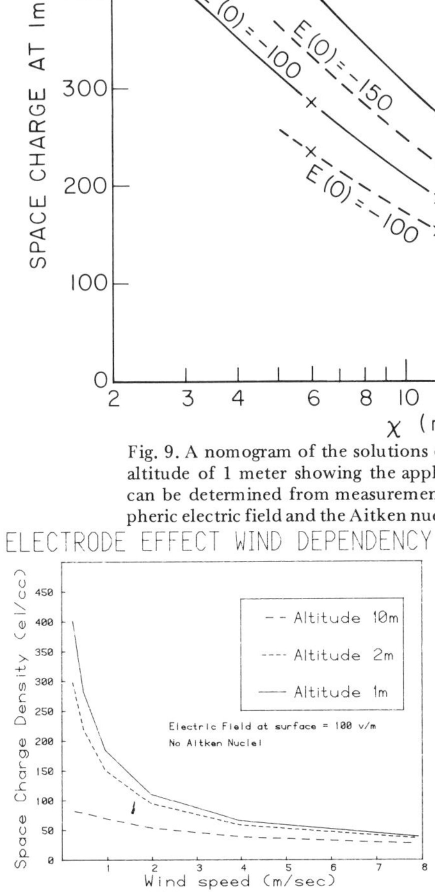

Fig. 10. The wind dependence of the turbulent electrode effect plotted for several altitudes.

shore sites by Blanchard and Mühleisen, large quantities of positive space charge were observed. The source of this charge was demonstrated by both investigators to be related to wind blowing to the observational site from the sea. Whether or not the observed space charge was originally produced by the electrode effect further out to sea or by the jet drop process in the nearby surf zone could not be ascertained by these experiments. What was needed was a demonstration of what the space charge density was both outside the surf zone and at the shore during periods of on-shore breezes.

Gathman and Hoppel (1970b) performed such an experiment in 1969 on the shores of Barbados. The site on the east coast of Barbados at Tent Bay was an ideal location for these experiments. Figure 11 shows the surf breaking on a coral reef approximately 300 meters off-shore. This provides a natural break-water for the local fishing fleet which anchors small wooden fishing boats shorewards of the reef. A

Fig. 11. The surf breaking on the coral reef at Tent Bay, Barbados.

Fig. 12. A time exposure of surf zone produced by waves breaking over the coral reef at Tent Bay, Barbados.

Fig. 13. The fishing boat equiped with a Faraday cage for measuring space charge, and a potential gradient apparatus for measuring the geoelectric field, in and around the surf zone at Tent Bay, Barbados.

narrow channel was blasted into the reef structure itself to allow safe passage of the boats between the open ocean fishing grounds and the relatively calm anchorage. Barbados, being the most eastward of the Lesser Antilles, is the first land-fall of the trade winds as they blow across the central Atlantic. The background aerosol is strictly marine in nature with the possible exception of an occasional outburst of Sahara dust which sometimes makes its way from the old world to the new at relatively high altitudes. Figure 12 is a time exposure of a portion of the surf zone at Barbados. The photo was taken with a neutral density filter while the shutter remained open for a period of 20 minutes. It is evident that the surf zone amplifies the white water area and becomes a strong localized source of jet droplets.

The experiments of Gathman and Hoppel were made with a small fishing boat (Figure 13), equipped with a 6-foot Faraday cage and a potential gradient device, which was able to traverse the surf zone from the open ocean into the relatively quiet waters inside of the reef by following the narrow channel between the open ocean and the anchorage area. During times when the wind was blowing from the open ocean towards the land, measurements of atmospheric potential gradient and atmospheric electric space charge were taken both inside and outside of the reef, The atmospheric electric potential measurements were made with very high impedance, guarded voltage followers. These devices (Gathman, 1972a) were used to measure the electrostatic potential in the center of the cage as well as the electrostatic potential at a distance outside of the cage. Figure 14 shows the actual data plots of field and space charge taken on 10 May 1969 at Tent Bay Barbados. The figure contains the data from both the space charge density and potential gradient instruments plotted on the same recorder in a time shared fashion. It shows the values and variations in both measurements from both inside and outside of the surf zone. The data show conclusively that the large space charge is being produced in the surf zone itself. A frequency of occurrence plot of the space charge density observed upwind and downwind of the surf zone is shown in Figure 15. The data taken

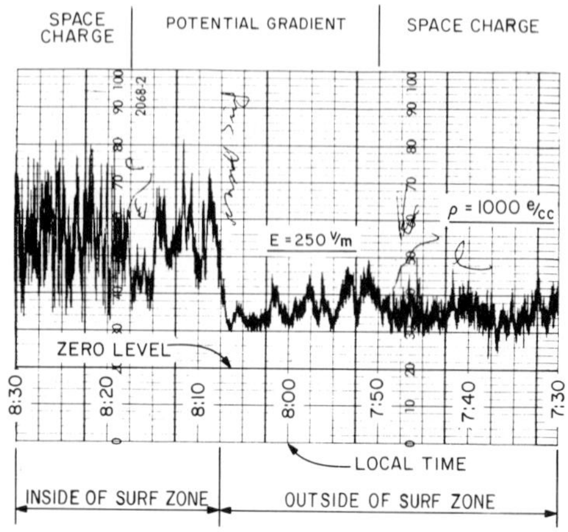

Fig. 14. An analog recording of the atmospheric electric field and the space charge density as measured from the instrumented fishing boat of Figure 13 as it moved from the open ocean environment into the protected area inside of the coral reef at Tent Bay, Barbados.

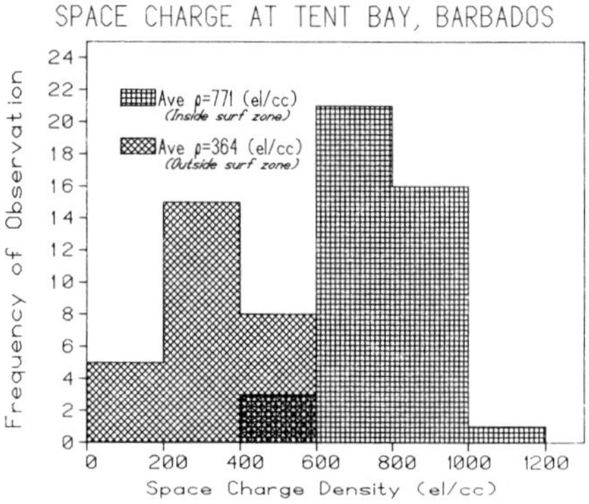

Fig. 15. A histogram showing the composite results of 10 minute averages of the atmospheric space charge measurements in and about the surf zone at Tent Bay, Barbados, 6-10 May 1969.

from the period between 6 and 10 May 1969, show a highly significant difference between the means of the two sets of data. These observations show that the marine surf zone is a strong source of positve space charge. The mechanism for the production of this charge is believed to be the natural charge of the jet drop

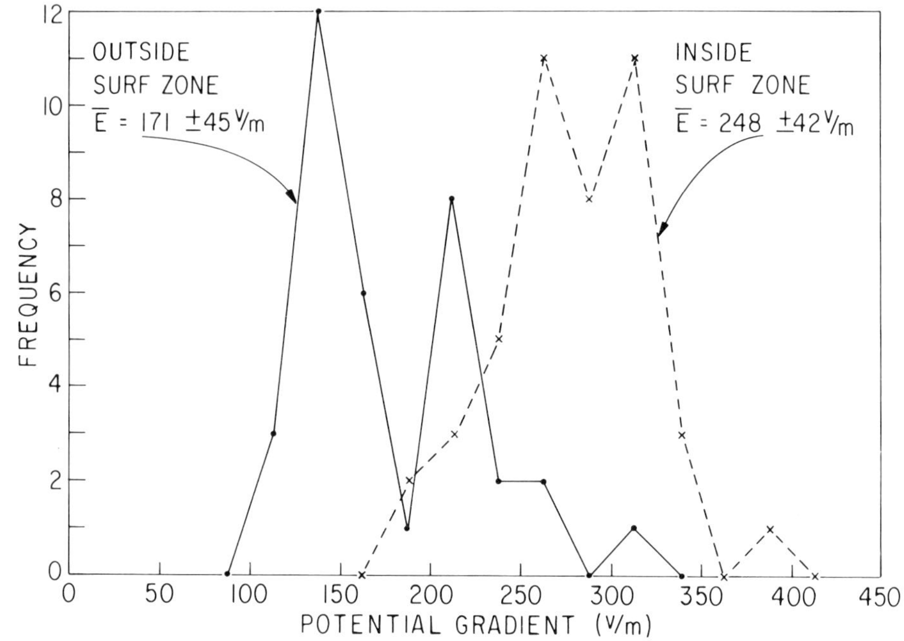

Fig. 16. A frequency of observation plot of atmospheric potential gradient measurements averaged over 10 minute intervals taken inside and outside the surf zone at Tent Bay, Barbados, 6-10 May, 1969.

produced by the bursting of air bubbles in the sea water.

The electric current production from the surf zone was estimated to be 4×10^{-11} amperes per square meter which is a factor of 20 higher in magnitude and opposite in direction to the normal air-earth currents. This convergence of currents results in the space charge build-up in the boundary layer near the surf zone and contributes to an increase in space charge down wind of the source of a least a factor of two.

Figure 16 shows the frequency of occurrence distribution of the potential gradient measurements which are separated into measurements made outside and inside the surf zone. The difference between the two mean potential gradients was 77 volts per meter. These results could only be produced by an additional overhead columnar atmospheric charge of about 430,000 elementary charges per square centimeter above the inside measurement site. This charge produced in the surf zone is mixed by eddy diffusion and if we assume it has a spatial distribution of the form:

$$q(z) = f(0)\exp(-z^2/2s^2)$$

where q(z) is the concentration of space charge at level z and s is the scale height, it can be fitted to the mean surface values of measured space charge density and potential gradient differences to arrive at a scale height of approximately 13 meters for the surf originated space charge.

Surf Electrification - Fresh Water

An interesting alternative to making electrical measurements over the sea is to make them over a large body of fresh water such as Lake Superior in North America. Gathman and Hoppel (1970a) performed a set of experiments in 1968 on the Keweenaw Peninsula, an arm of the State of Michigan, which extends out into

Lake Superior, to observe the differences in the electrification processes between fresh and sea water. The natural charging process of jet drops is strongly affected by the salinity of the water. Laboratory observations of the jet drop process show that bubbling air through distilled water produces a net negative atmospheric space charge, (Blanchard, 1966). This together with observations in the vicinity of water falls of negative space charge (Pierce and Whitson, 1965) indicates that measurements near fresh water lakes should be very different from measurements near the sea. This is because the polarity of the charge produced by the bubble mechanism in fresh water is different from that produced by the electrode effect. As the electrode effect space charge is normally positive in polarity, the ejection of negative fresh water jet drops into the boundary layer over fresh water will decrease the net space charge, not increase it as we observed over the ocean.

This process is illustrated in Figure 17 which shows the relationship between wind speed, breaker height, and space charge observed downwind of the surf zone along the shoreline of Lake Superior, plotted as a function of time. This representation shows variations which can be interpreted as interactions between the electrode effect, the fresh water whitecap process over the open water and the fresh water surf and/or spray. During calm periods where there was no breaker activity the data show high positive space charge which is evidence of the electrode effect in operation over the lake. This process does not depend on the chemistry of the water but is a function of the earth's electric field, the aerosol content and ion production in the air itself, and the stability of the air over the very cold waters of the lake. The lower space charge value during the rain event before noon on the 17th illustrates the effect of splashing in fresh water. Negative charge (either from fresh water jet drops produced by the bursting of bubbles introduced into the water by the falling rain drops or from the induction charging of the splashed drops themselves) is added to the positive space charge of the electrode effect reducing the net space charge value.

The high breakers during mid-day on the 18th, even with a relative low local wind, produced enough negative charge to counter the electrode effect's positive space charge and to thus produce a net negative charge in the boundary layer. On the other hand, medium positive space charge densities were observed from early morning on the 19th until midnight of the 20th because in that coastal location there were no breakers during this interval even though the wind speed varied between 3 and 10 knots. The final two days show high wind speeds, high breakers and the resultant high, negative, space charge density.

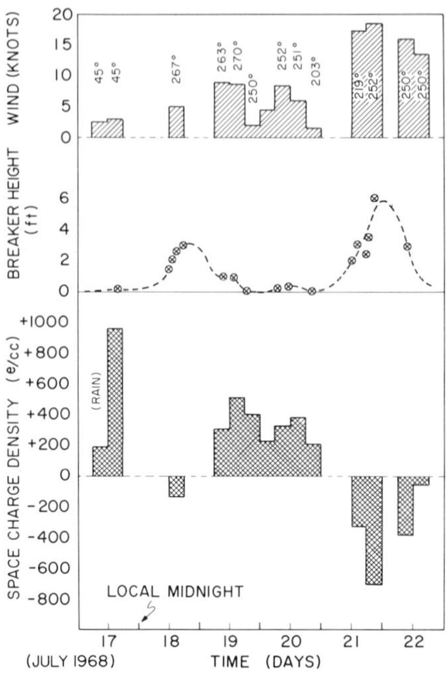

Fig. 17. Six hour averages of space charge density, breaker heights and wind vectors for a six day period at Eagle Harbor, Michigan, July 1968.

We may look at these data in other ways. Figure 18 is a plot of the measured space charge density on the beach of Lake Superior plotted as a function of breaker height. This plot shows the significant increase in negative space charge produced by the increasing breaker heights over the fresh water lake. This is an in-

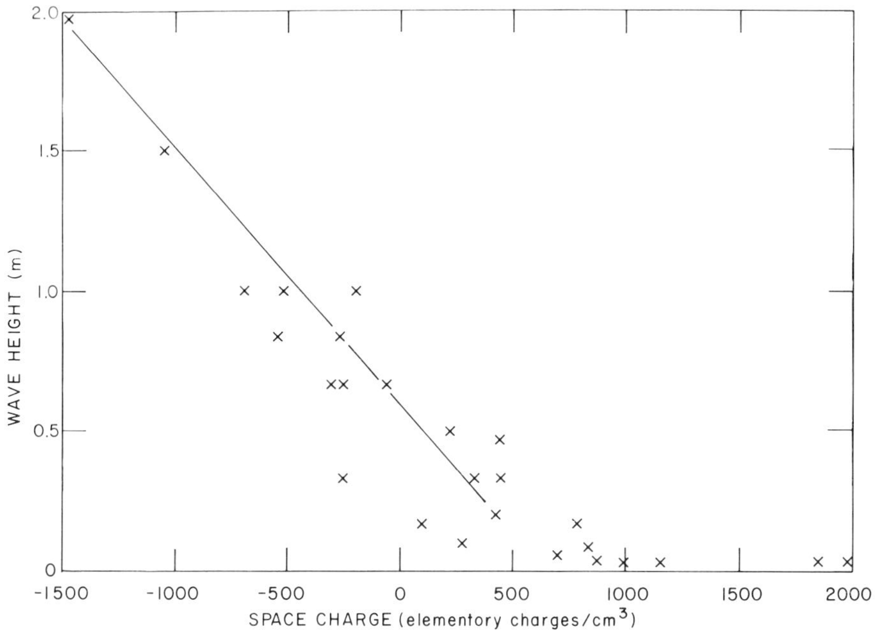

Fig. 18. Space charge density as a function of breaker height showing negative white water electrification with high breakers at Eagle Harbor, Michigan, July 1968.

dication that the fresh water surf and the breaking of air bubbles in fresh water produce negative space charge. There is no way to distinguish here however between the negative space charge from the jet droplets and the charge induced on the splashing drops and on the droplets blown off of the wave crest.

The efficiency of the Faraday cage used to measure space charge density on the shore of Lake Superior depended on natural ventilation of the cage to bring the charged particles from over the water into the interior of the cage. At very low wind speeds, a correction had to be made to account for the problem of diffusion of the atmospheric ions to the wires of the cage. Figure 19 is a plot of such corrected low wind speed data obtained during the non-raining electrode effect periods, plotted against stability. Stability was determined by calculating of Richardson's number obtained from profile measurements of the meteorological parameters on towers mounted in the water. This plot shows how the positive space charge is strongly dependent on atmospheric stability. This behaviour is of course another manifestation of the response to turbulence of the electrode effect, shown above in Figures 5 and 9.

Summary

There are three electrical processes operating over large water surfaces which have been shown to be dominant in their own separate wind speed domains. The electrode effect dominates at the lowest wind speeds. In this process the surface produces an imbalance of electrical carriers. The polarity of the space charge associated with the electrode effect depends only on the direction of the earth's electric field at the sea surface. If we exclude the relatively few cases of heavy local weather phenomenon such as thunderstorms, the electric field is always of a direction to produce positive

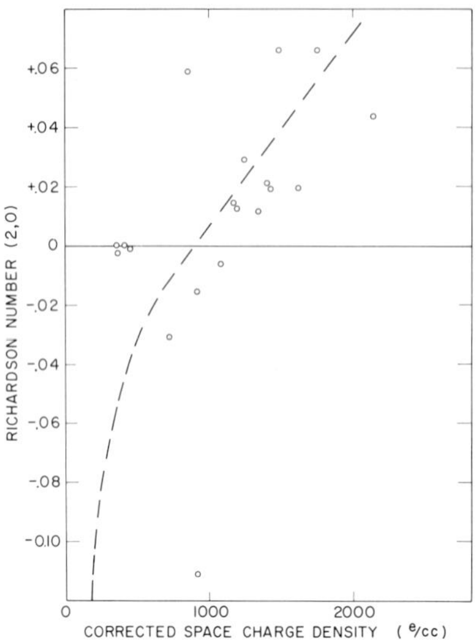

Fig. 19. Space charge density as a function of Richardson's number under electrode effect conditions at Eagle Harbor, Michigan, July 1968.

space charge near the air-sea interface. In an aerosol free environment, this space charge would be carried exclusively on small ions, but when larger aerosol are present these small ions diffuse to the larger particles so that in reality, the space charge is carried on both types of particles. The space charge decreases with increasing altitude and at any one fixed altitude and electric field level, the space charge decreases in magnitude with increasing mixing activity.

Electric charge is also produced by the bursting of small air bubbles at the air–water surface. The small jet drops ejected from the ocean's surface carry a natural positive charge. These jet drops are the predominant source of space charge in the intermediate range of wind speeds, starting from the onset of white water formation and continuing until this source is overridden by the inductively charged spray drops at higher sea states. The jet drop process is magnified in the area of surf activity with the charge polarity being reversed in fresh water environments.

Spray droplets are charged by the inductive process under the action of the Earth's electric field. This charge will depend on the magnitude and direction of the earth's field, but for most of the time will be small and negative in polarity. The resultant effect on the space charge in the marine boundary layer will be a function of the total number of spray droplets produced. Spray droplets are prominent at the highest wind speeds in a well developed sea. Although the general qualitative characteristic of this flux is known, a detailed quantitative study of the induced charge on these droplets is needed.

The features of the space charge near the sea surface described above reflect relationships which may be of value in understanding the complex relationships between meteorology and the specific aerosol and turbulent characteristics of the marine boundary layer. Some of these applications are suggested in Table II. These applications have to do with the recognition that certain processes at the air-water interface produce distinctive charge characteristics which can be used to identify the specific process whereby the droplets originated, given an understanding of the total atmospheric electrical environment. Properly designed sensitive electrical detectors measuring macrophysical properties of sampled air may be more practical to operate in hostile environments than arrays of micro physical and micro meteorological devices.

Table II

APPLICATIONS

Electrode Effect Perturbation
*Atmospheric turbulence - low wind

Space Charge Density - Polarity
*Sensitive indicator to distinguish between direct and indirect droplets

Space Charge Density - Magnitude
*Indication of surf activity
*Aerosol flux in high wind situations
*Indicator of onset of aerosol production

REFERENCES

Behacker, M., 1910: Beiträge zur Kenntnis der atmosphärischer Elektrizität XLI; Zur Berechnung des Erdfeldes under der Voraussetzung homogener Ionisierung der Atmosphäre, *S. B. Akad. Wien*, **119**, 675-684.

Blanchard, D. C., 1955: Electrified droplets from the bursting of bubbles at an air-sea water interface. *Nature* **175**, 334-336.

—, 1958: Electrically charged drops from bubbles in sea water and their meteorological significance. *J. Meteor.* **15**, 383-395.

—, 1963: Electrification of the atmosphere by particles from bubbles in the sea. *Prog. Oceanogr.* **1**, 71-202.

—, 1966: Positive space charge from the sea. *J. Atmos. Sci.* **23**, 507-515.

Gathman, S. G., 1972a: DC High-voltage follower electrometer, *US patent* no. 3,644,828.

—, 1972b: A field mill for tethered balloons, *Rev. Sci. Inst.* **43**, 1751-1754.

Gathman, S. G. and W. A. Hoppel, 1970a: Electrification processes over Lake Superior. *J. Geophys. Res.* **75**, 1041-1048.

—, 1970b: Surf electrification. *J. Geophys. Res.* **75**, 4525-4529.

Gathman, S. G. and E. M. Trent, 1968: Space charge over the open ocean. *J. Atmos. Sci.* **25**, 1075-1079.

Hoppel, W. A., 1967: Theory of the electrode effect. *J. Atmos. Terr. Phys.* **29**, 709-721.

—, 1968: The ions of the troposphere: their interactions with aerosols and the geoelectric field, Ph.D. thesis, The Catholic University, Washington, D. C.

Hoppel, W. A. and S. G. Gathman, 1971: Determination of eddy diffusion coefficients from atmospheric electrical measurements. *J. Geophys. Res.* **76**, 1467-1477.

—, 1972: Experimental determination of the eddy diffusion coefficient over the open ocean from atmospheric electric measurements. *J. Phys. Oceanogr.* **2**, 248-54.

Monahan, E. C., D. E. Spiel, and K. L. Davidson, 1983: Model of marine aerosol generation via whitecaps and wave disruption. Preprint vol. of 9th Conf. Aerospace and Aeronautical Meteorology, June 6-9, 1983, Omaha, Nebr. (A. M. S.), 147-152.

Mühleisen, R., 1961: Electrode effect measurements above the sea. *J. Atmos. Terr. Phys.* **20**, 79-81.

—, 1962: Die luftelektrischen Verhältnisse in Küstenaerosol 2. *Archiv. Meteorol., Geophys., Bioklimatol.*, **B 12**, 435-446.

—, 1969: Der Elektrodeneffekt beim luftelektrischen Feld über dem Meer. *Meteor. Rund.* **22**, 175-177.

Pierce, E. T. and A. L. Whitson, 1965: Atmospheric electricity and the waterfalls of Yosemite valley. *J. Atmos. Sci.* **22**, 314-319.

Schweidler, E. V., 1909: Über den Einfluss des Standortes auf Messungen der Aerstreuung and Leitfähigkeit der Atmosphäre. *Phys. Z.* **10**, 847-849.

Swann, W. F. G., 1913: The atmospheric potential gradient, and a theory as to the cause of its connection with other phenomena in atmospheric electricity, together with certain conclusions as to the expression for the electric force between two parallel charged plates. *J. Mag. Atmos. Elect.* **18**, 163-184.

SATELLITE MEASUREMENTS OF AEROSOLS OVER OCEAN SURFACES

M. GRIGGS
Science Applications, Inc.,
11526 Sorrento Valley Road,
San Diego, California 92121.

The ability to measure tropospheric aerosols over ocean surfaces has been demonstrated using several different satellite sensors. Landsat data originally showed that a linear relationship exists between the upwelling visible radiance and the aerosol optical thickness (about 90% of this thickness is generally in the lowest 3 km of the atmosphere). Similar relationships have also been found for sensors on GOES, NOAA-5 and NOAA-6 satellites. The linear relationship has been shown theoretically to vary with the aerosol properties, such as size distribution and refractive index, although the Landsat data obtained at San Diego showed little variability in the relationship. To investigate the general applicability of the technique to different locations, a global-scale ground truth experiment was conducted in 1980 with the AVHRR sensor on NOAA-6 to determine the relationship at eleven ocean sites around the globe. The data show good agreement at the sites analyzed to date, and suggest that the technique can be used routinely on a global basis. A preliminary analysis of AVHRR Channel 1 and Channel 2 radiances suggests that useful information on the aerosol size distribution may also be obtained from satellite observations. The data for an overpass at San Diego in the presence of many whitecaps were analyzed to infer the whitecap coverage; the result showed good agreement with previously determined relationships between whitecap coverage and wind speed.

Introduction

Tropospheric aerosols play an important role in environmental quality on local, regional, and global scales. The local and regional aerosols impact mainly on the ambient air quality, and changes in the global background levels of aerosols may affect our climate. In addition, the nature and distribution of aerosols over the oceans affects the use of electro-optical communications and weapons systems by the Navy. This paper discusses a satellite technique to measure aerosols over the ocean. This method relates the upwelling visible radiance measured by the satellite to the aerosol optical thickness of the atmosphere. Since 60% of the aerosols are typically in the lowest 1 km, and 90% in the lowest 3 km, it is clear that the satellite measurements will provide information of considerable importance to Naval operations. The technique, originally based on Landsat data (Griggs, 1975), has since been applied to data from other satellites (Griggs, 1979).

The relationship between the radiance and the aerosol optical thickness varies with several parameters such as the aerosol size distribution and refractive index, and the ocean surface reflectivity. However, the Landsat data obtained at San Diego showed little variability in the relationship, except on one particular occasion when there was a high whitecap coverage. This result is discussed in detail.

To investigate the possible variability of the radiance-aerosol optical thickness relationship around the globe, a ground-truth experiment (Griggs, 1983) was conducted in 1980 with the NOAA-6 satellite at eleven locations. The results of this experiment are used to demonstrate that the technique has global application. In addition, the preliminary results of a method using the AVHRR Channels 1 (0.65 μm) and 2 (0.86 μm) radiance data to infer aerosol size distribution are presented.

E. C. Monahan and G. Mac Niocaill (eds.), Oceanic Whitecaps, 245–250.

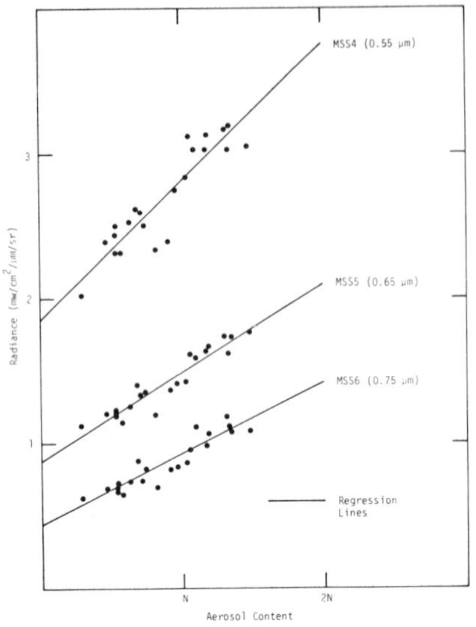

Fig. 1. Landsat radiance vs aerosol content at San Diego for MSS4, MSS5 and MSS6. (Normalized to nadir viewing and a sun zenith angle of 63°).

Background

The relationship (Griggs, 1975) between the upwelling visible radiance and the atmospheric aerosol content measured by Landsat 2 over the ocean at San Diego are shown in Fig. 1, and demonstrate the feasibility of measuring the aerosol content over oceans. The aerosol content is defined in terms of the Elterman (1965) model vertical aerosol optical thickness; i.e., the aerosol content is given by the ratio of measured aerosol optical thickness at wavelength λ to the model aerosol optical thickness at wavelength λ, multiplied by N. In the results reported here, the wavelength is always 0.5 μm, and the model aerosol optical thickness is 0.213 (to the base e). In order to use this technique to routinely map global aerosol distributions it is preferable to use a satellite such as NOAA-6 to obtain daily coverage. The AVHRR Channel 1 on NOAA-6 has a spectral bandpass very similar to that of MSS5, with the same bandwidth of 0.1 μm, and centered at 0.645 μm in comparison with a center of 0.65 μm for MSS5. Thus, the radiance values measured with the AVHRR Channel 1 may be directly compared with those of MSS5. The AVHRR radiance data are recorded in a 10-bit digital system and are more sensitive to aerosol changes than the 7-bit MSS5 data.

The Landsat data are based on nadir viewing, whereas the AVHRR scans up to 60° from the nadir; thus, the MSS5 relationship in Fig. 1 cannot be used directly with AVHRR radiances to infer the aerosol content. To use the MSS5 relationship, theoretical calculations have been performed to relate the nadir radiance to the radiance at other viewing angles, as a function of sun angle and aerosol content. These calculations, made with the Dave and Gazdag (1970) scattering code, have then been incorporated into a table-look-up algorithm so that the AVHRR radiance measurement together with the scan angle and sun angle can be used to infer the atmospheric aerosol content. The calculations use aerosol parameters such that the theory reproduces the linear regression found for the MSS5 data in Fig. 1.

In 1980, ground truth measurements of aerosol content at the time of the NOAA-6 overpasses (approximately 0730 l.s.t.) were made with hand-held sunphotometers at eleven locations in close proximity to the ocean. One hundred and sixty-two useful coincidences between ground truth measurements and the NOAA-6 overpass were obtained. Unfortunately the sunphotometer malfunctioned at one site, and the data from five other sites have not yet been analyzed. The results available to date are shown in Fig. 2 for data obtained when the sun zenith angle (θ_0) was less than 70°. At larger zenith angles, the table-look-up code is expected to introduce uncertainties in the predicted aerosol content due to the use of a flat earth model in the calculations.

In comparing the results for the different sites in Fig. 2, it appears that there is no significant global variation of aerosol properties that would preclude using this technique on a routine basis.

Aerosol size distribution from AHVRR data

A preliminary investigation has been made into the use of AVHRR Channels 1 and 2 to infer

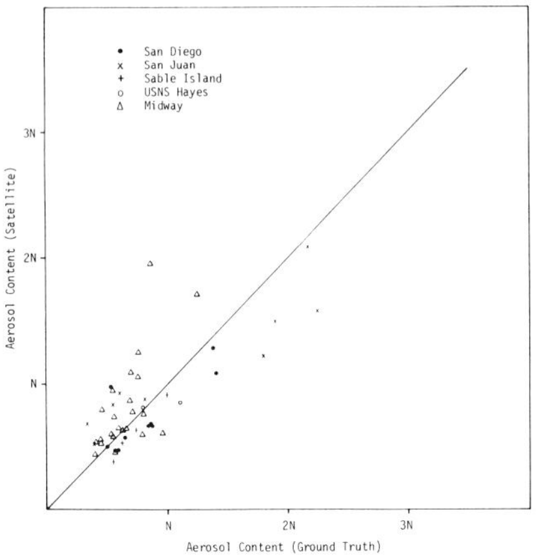

Fig. 2. Comparison of AVHRR and ground-truth measurements of aerosol content (for $\theta_0 < 70°$).

the value of ν in the Junge size distribution:

$$dn(r) = Cr^{-\nu} d \log r \text{ (cm}^{-3}\text{)} \tag{1}$$

where n(r) is the number of particles with radius r, and C is a constant depending on the number of particles per unit volume.

The value of ν is inferred from the satellite data by comparing the Channel 1 and 2 radiances (the Channel 2 radiance is corrected for water vapor absorption) with model calculations of these radiances, made as a function of N and ν for many sets of sun and view angles. The values of N and ν are determined so that the model radiances agree with the measured radiances in each AVHRR Channel for the given sun and view angles at the time of the measurement. A two-channel table look-up code, similar to the Channel 1 code was developed, covering ν values between 0 and 7.0; typical ν values in the atmosphere range between 2.0 and 5.0.

The value of ν can be determined from sunphotometer measurements since the spectral variation of the aerosol optical thickness, $\tau(\lambda)$, is related to ν, to a close approximation, by:

$$\tau(\lambda) = B\lambda^{-(\nu-2)} \tag{2}$$

where B is a constant.

In the 1980 experiment, multispectral sunphotometer measurements were made only at Barbados and on the USNS Hayes in the Atlantic just off the Virginia coast. The ground truth measurements of $\tau(\lambda)$ included observations at 0.5 μm and 0.86 μm so that a value of ν could be obtained by twice applying Eq. (2) using these wavelengths, and is directly comparable with the value of ν inferred from the AVHRR data at 0.65 μm and 0.86 μm.

Unfortunately, Barbados was the site, mentioned above, at which the sunphotometer malfunctioned. However, analysis of the data suggests that the error produced was a systematic one in both channels (0.5 μm and 0.86 μm). Since both channels are used, the determination of ν from Eq. (2) is not strongly affected, whereas the estimate of the aerosol content using just one channel (0.5 μm) is quite uncertain.

The comparison of the satellite and ground truth values of ν is shown in Fig. 3. The agreement is remarkably good considering the uncertainties in the ground truth data.

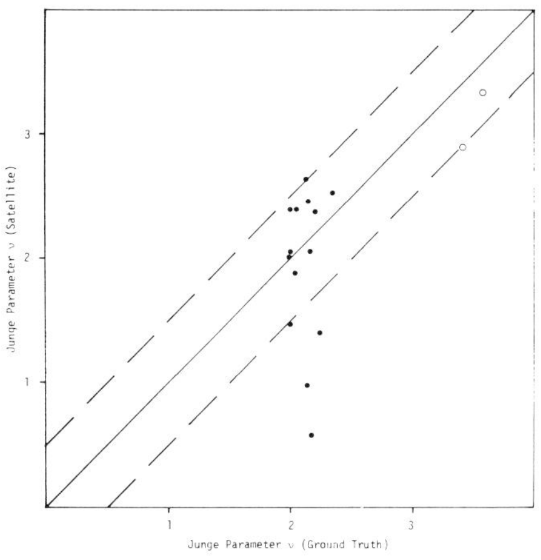

Fig. 3. Comparison of satellite and ground truth measurements of the Junge parameter (● Barbados; o USNS *Hayes*) for $\theta_0 < 70°$.

A satellite estimate of whitecap coverage

The presence of whitecaps on the ocean changes the reflectance of the ocean surface, and thus will affect the accuracy of the satellite technique used to measure atmospheric aerosols. The only documented case of whitecaps influencing the method was at San Diego on March 1, 1977 when ground truth measurements were taken at the time of a Landsat 2 overpass. The sky was clear, and a large coverage of whitecaps were noted by the ground truth observer. The local Weather Service recorded a mean wind speed of 7.7 m/s with gusts to 15 m/s. The ground-truth value of the aerosol content was 0.68 N and the Landsat 2 MSS5 radiance was 17% higher than expected based on the regression line shown in Fig. 1. This measured radiance corresponds to a surface albedo of 0.026 compared to the value of 0.015 used for the albedo in the model that fits the regression line.

It is assumed that this enhanced albedo is due entirely to whitecaps without any significant contribution from sunglitter. The Landsat observation was 2° from the nadir in the direction of the sun. Based on calculations by Wald and Monget (1983), using the Cox and Munk (1954) model, the glitter albedo is less than 0.001 under these conditions, and is negligible in comparison to the whitecap albedo.

The albedo (A) of the whitecap covered ocean can be represented approximately by

$$A = A_0 (1-W) + A_F W \tag{3}$$

where A_0 is the albedo of the calm ocean
A_F is the albedo of the whitecap foam
W is the fraction of the ocean covered by whitecap foam

If it is assumed that $A_F = 0.5$ [e.g., Wald and Monget (1983)] and $A_0 = 0.015$, then Eq. (3) yields W = 0.023. This value of W for a windspeed of 7.7 m/s compares well with previous work, based on the analysis of ocean photographs, reported by Monahan and Ó Muircheartaigh (1981) as can be seen in Fig. 4. The satellite inferred value of W is a little higher than would be predicted by these authors' least squares fit to the photographic data but falls in the region of the cold water data obtained in the 1980 JASIN experiment. It is to be noted that the San Diego water temperature was about 15°C, which is about the same as experienced during the JASIN experiment. Of course, even better agreement with the previous results can be achieved by choosing a

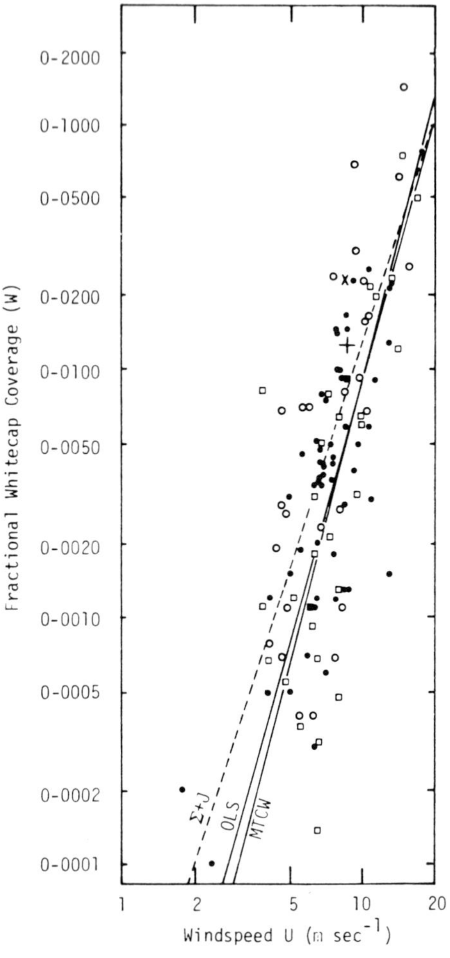

Fig. 4. Satellite inferred value of W versus windspeed compared with previous photographic results reported by Monahan and Ó Muircheartaigh (1981). (X is for $A_F = 0.5$; + is for $A_F = 0.8$).

larger value of A_F in Eq. (3), as can be seen from the $A_F = 0.8$ point in Fig. 4.

Effect of whitecaps on satellite measurement of aerosols

Using the relationship between whitecap coverage and windspeed shown in Fig. 4, and Eq. (3), the effective ocean albedo as a function of windspeed can be deduced. This albedo can then be used to determine the error in the aerosol content from calculations of radiance as a function of albedo (Griggs, 1978). This error in the aerosol content as a function of windspeed and whitecap coverage is shown in Fig. 5 for nadir viewing and a sun zenith angle of 65°.

From the data published by Monahan et al (1982) on the global and seasonal distribution of whitecaps it is concluded that whitecaps may be a problem for the satellite technique in winter months at higher latitudes. In these circumstances where the whitecap coverage is greater than 1%, errors in the aerosol content of greater than 0.15 N result. However, it is believed that these whitecap occurrences would generally be under cloud cover when the satellite technique cannot be used anyway.

Conclusions

Ground truth measurements at several ocean sites around the globe have been compared with NOAA-6 AVHRR measurements of the atmospheric aerosol content, and show good agreement for the sites analyzed to date, suggesting that the satellite technique to measure aerosols can be used routinely on a global basis. It also appears that the AVHRR Channel 1 and Channel 2 radiances can be used to provide useful information on the aerosol size distribution.

The presence of whitecaps on the ocean produces an error in the aerosol content inferred by the satellite technique, but is not significant at windspeeds below about 10 m/s. Analysis of a documented case of whitecaps affecting the technique showed that the whitecap coverage inferred from the satellite radiances for the measured windspeed showed good agreement with other methods of relating whitecap coverage to windspeed.

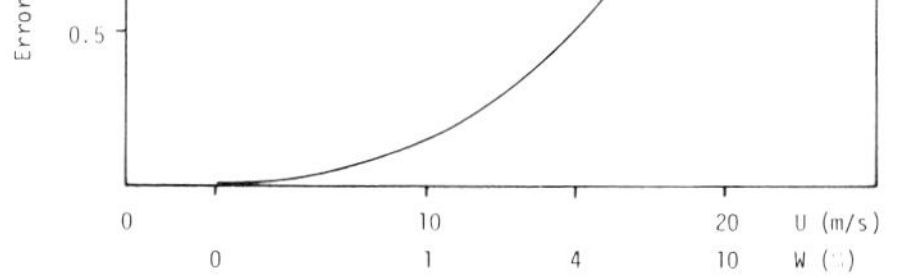

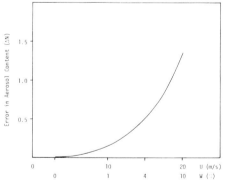

Fig. 5. Error in N due to whitecaps for nadir viewing and sun zenith angle of 63°. U (m/s) is the windspeed, W (%) is the whitecap coverage.

ACKNOWLEDGEMENTS

This research has been sponsored by ONR Contract N00014-77-C-0489, by NOAA Contracts MO-A01-78-00-4092 and NA-83-SAC-00106, and by NEPRF Contract N00228-82C-6199.

REFERENCES

Cox, C., and W. Munk (1954). Statistics of the sea surface derived from sun glitter. *J. Mar. Res.* 13, 198–212.

Dave, J. V., and J. Gazdag (1970). A modified Fourier transform method for multiple scattering calculations in the plane parallel Mie atmosphere. *Appl. Opt.* 9, 1457-1466.

Elterman, L. (1965). Atmospheric optics. *Handbook of Geophysics and Space Environments* (S. Valley, ed.) McGraw-Hill, 7-1 to 7-2.

Griggs, M. (1975). Measurements of atmospheric aerosol optical thickness over water using ERTS-1 Data. *J. Air Pollut. Control Assoc.* 25, 622-626.

Griggs, M. (1978). Determination of aerosol content in the atmosphere from Landsat data. Science Applications. Inc. Report No. SAI-78-525-LJ.

Griggs, M. (1979). Satellite observations of atmospheric aerosols during the EOMET cruise. *J. Atmos. Sci.* 36, 695-698.

Griggs, M. (1983). Satellite measurements of tropospheric aerosols. *Adv. Space Res.* 2, 109-118.

Monahan, E. C. and I. Ó Muircheartaigh (1983). Optimal power-law description of oceanic whitecap coverage dependence on wind speed. *J. Phys. Oceanogr.* 10, 2094-2099.

Monahan, E. C., M. C. Spillane, P. A. Bowyer, D. M. Doyle and J. J. Taper (1982). Whitecaps and the marine atmosphere, Report No. 4. ONR Grant N00014-78-G-0052.

Wald, L. and J. M. Monget (1983). Remote sensing of the sea-state using the 0.8–1.1 μm spectral band. *Int. J. Remote Sensing, Vol.* 4, No. 2, 433–446.

REMOTE SENSING SIGNATURES OF WHITECAPS

PETER KOEPKE
Meteorolog. Inst. Univ. München

In the solar spectral range, whitecaps are isotropic reflectors with a reflectance of about 55%. So even observation directions outside the sunglint, those often used for remote sensing, are affected by whitecaps. Due to the variation of the area and the reflectance of individual whitecaps with age, the optical effect of the whitecaps taken as the product of the percentage area covered with whitecaps and the spectral reflectance of dense foam must be reduced by an efficiency factor of about 0.4. Chlorophyll retrieval algorithms are not significantly affected by whitecaps, if they are based on radiance differences. Determination of the atmospheric turbidity using satellite measured radiances from cloud-free pixels over water is successful, but whitecaps are the most perturbing parameter. In the microwave region, the emissivity of whitecaps is higher than the emissivity of the water surface without foam. Therefore, the variation of measured brightness temperatures which thus results from the alteration in the amount of whitecaps is used to determine the wind speed near the surface.

1. Introduction

Each radiation field in an ocean-atmosphere system is determined by the optically active parameters of the system. It is completely described by the radiative transfer theory. So, in the solar spectral range the spectral radiance in a specific viewing direction, at a given height in or above the atmosphere, is a function of: the solar irradiance; the solar azimuth and zenith angles; the cloud properties; and the optical properties of the aerosol particles, of the atmospheric gases and of the ocean surface and consequently the optical properties of whitecaps. In the microwave region the spectral radiance (or the brightness temperature) in a given viewing direction, and for both polarization states, is a function of the same parameters, together with the vertical temperature profile. The influence of the aerosol particles can be neglected in the microwave region, but at extreme wavelengths the radiance is also a function of the galactic noise and the solar irradiance. If all the parameters are known, the radiances can be calculated via the radiative transfer equation.

The approach taken in remote sensing work is just the opposite: radiances are known from measurements, but the optical parameters are unknown. The parameter I am interested in, the so called 'wanted quantity', may be the area covered with whitecaps (e.g. Monahan, 1971), may be the underlight (to give information on the amount of chlorophyll, e.g. Gordon et al., 1980), may be the atmospheric turbidity (e.g. Koepke and Quenzel, 1982) or it may be the wind speed near the surface (e.g. Wentz et al., 1981). Because the radiances simultaneously depend on all the different optically acting parameters, in each case the parameters apart from the wanted one are the so called 'perturbing quantities'. Their actual values usually are known only within certain limits and therefore result in an equivalent uncertainty in the inverted value of the wanted quantity.

Fortunately, the influence of the different parameters on the radiances depends on the observation direction, on the direction of the sun and on the wavelength. These three quantities can be chosen and thus remote sensing of the wanted quantity can be optimized. Moreover,

E. C. Monahan and G. Mac Niocaill (eds.), Oceanic Whitecaps, 251–260.

simultaneous measurement at different wavelengths often allows the determination of different optical parameters simultaneously.

2. Reflectance of an ocean with whitecaps in the solar spectral range

In the solar spectral range the reflective character of an ocean surface is the result of three contributions: the reflectance due to whitecaps, the underlight and, most importantly, the specular reflectance of the water surface without foam.

In Equation 1, the reflectance factor of the ocean R_{oc}, which is angle and wavelength dependent, is given as the sum of the three components.

$$R_{oc} = W \cdot f_{ef} \cdot R_f + + (1 - W \cdot f_{ef} \cdot R_f) \cdot R_u + (1 - W) \cdot R_s \tag{1}$$

The first term on the right describes the influence of the whitecaps. The subscript f stands for foam. W multiplied by 100 gives the percentage area covered with whitecaps. The values given in the literature for W as a function of the windspeed are determined from photos of the ocean surface showing foam patches (e.g. Blanchard, 1971; Monahan, 1971; Toba and Chaen, 1973; Ross and Cardone, 1974; Wu, 1979; Monahan and Ó Muircheartaigh, 1980). R_f is the reflectance of whitecaps, measured by Whitlock et al. (1982). Usually in the literature the optical influence of whitecaps is described by the product $W \cdot R_f$ (e.g. Gordon and Jacobs, 1977; Koepke and Quenzel, 1979; Tassan, 1981). I introduce an additional factor, which I call the 'efficiency factor' f_{ef}, to take into account the increase of the area of an individual whitecap with increasing age and the simultaneous decrease of its reflectance. This factor will be discussed later in more detail.

The influence of the underlight is taken into account in the second term on the right of Eq. 1. Though the underlight is light scattered inside the water, it can be understood and described as a reflectance (e.g. Morel, 1980). In Eq. 1 the reflectance R_u is a reflectance just above the water surface which can be calculated simply from the reflectance just beneath the surface (e.g. Whitlock et al., 1981). The contribution due to underlight is weighted by the factor in brackets, to restrict it to the area where the absence of whitecaps allows the emergence of light at the water surface. This factor is based on the assumption that the reflectance of whitecaps is the same for light coming from above or below.

The third term of the reflectance factor of the ocean surface describes the specular reflection at the water surface without foam R_s. This component can be calculated for a flat surface with the Fresnel formula and therefore depends on the refractive index of the water, and, of course, on the angle of incidence and reflection. For the usually rough ocean, the slope of the waves reduces and broadens the glint (e.g. Saunders, 1967; Plass et al., 1977), an effect which must be taken into account in R_s, e.g. with data after Cox and Munk (1954). Of course, specular reflection is only possible at those parts of the surface which are free of foam. So again a corresponding weighting factor must be taken into account.

If the area covered with foam is very small, the weighting factors may be neglected and the reflection function of the ocean can be understood as the sum of the reflectance due to whitecaps, underlight and specular reflectance. The uncertainty due to this approximation is especially small, if, rather than considering angle dependent values of the reflectance factors, one considers albedo values which describe the reflection of the hemispherical fluxes.

Fig. 1 illustrates the angular dependency of the components of the reflectance of the ocean surface. It shows the reflectance factor, which gives the radiance reflected into a specific direction in multiples of the radiance which would be reflected by an isotropic Lambert reflector into this direction. The direction of the reflected radiance is given by its zenith angle ϑ_r, the abscissa in Fig. 1, and the azimuth φ_r, relative to the direction of the incident radiance, whose zenith angle is given by ϑ_i. The whitecaps are usually assumed to be isotropic reflectors, which seems to be correct from visual inspection. At the time of writing, I had access to no study which discussed an anisotropic reflection function for whitecaps. So the R_f value is

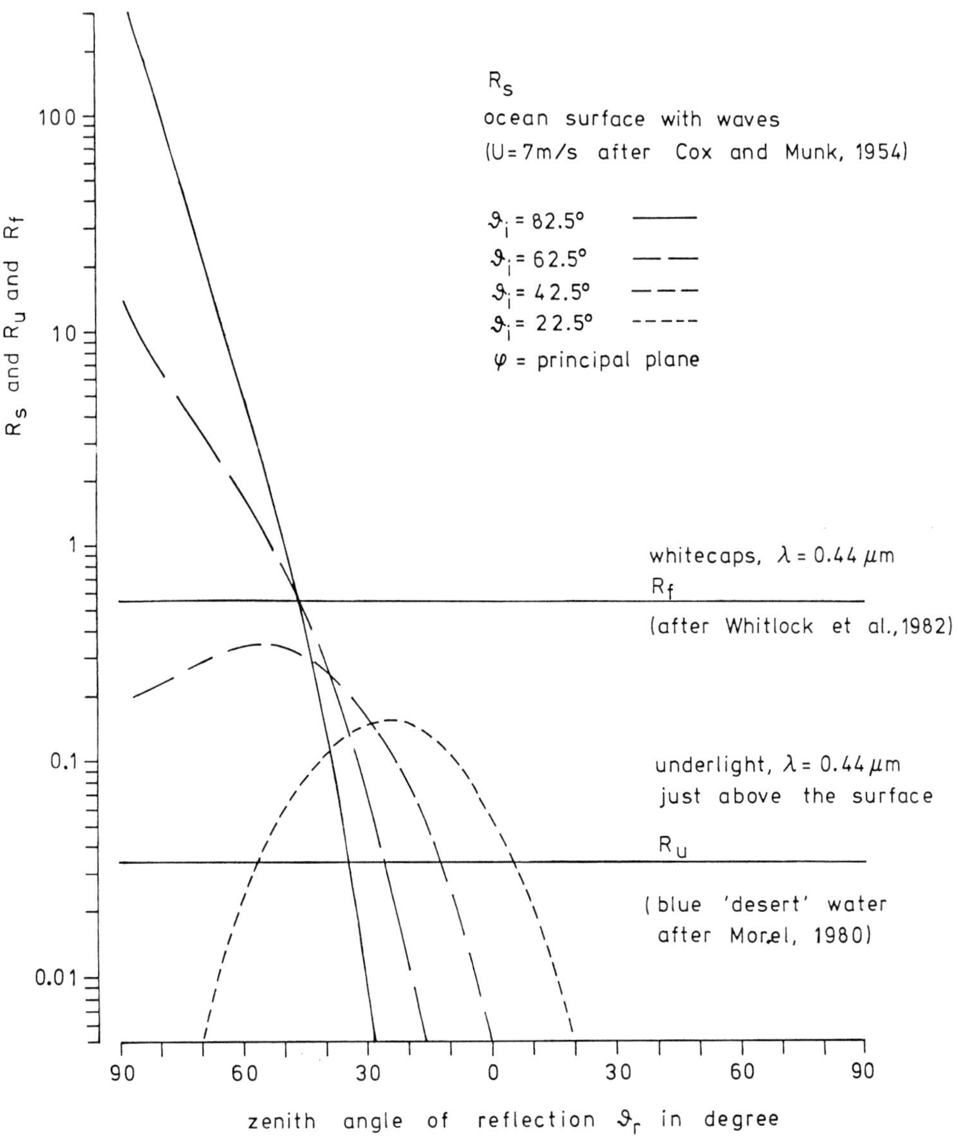

Fig. 1. Angle dependent reflectance factors of the components of an ocean surface: biconical reflectance R_s of the surface with waves appropriate for 7 m/s windspeed but without foam, isotropic hemispherical reflectance factor of whitecaps R_f, and isotropic hemispherical reflectance factor just above the ocean surface due to underlight R_u. ϑ = zenith angle, φ = azimuth angle.

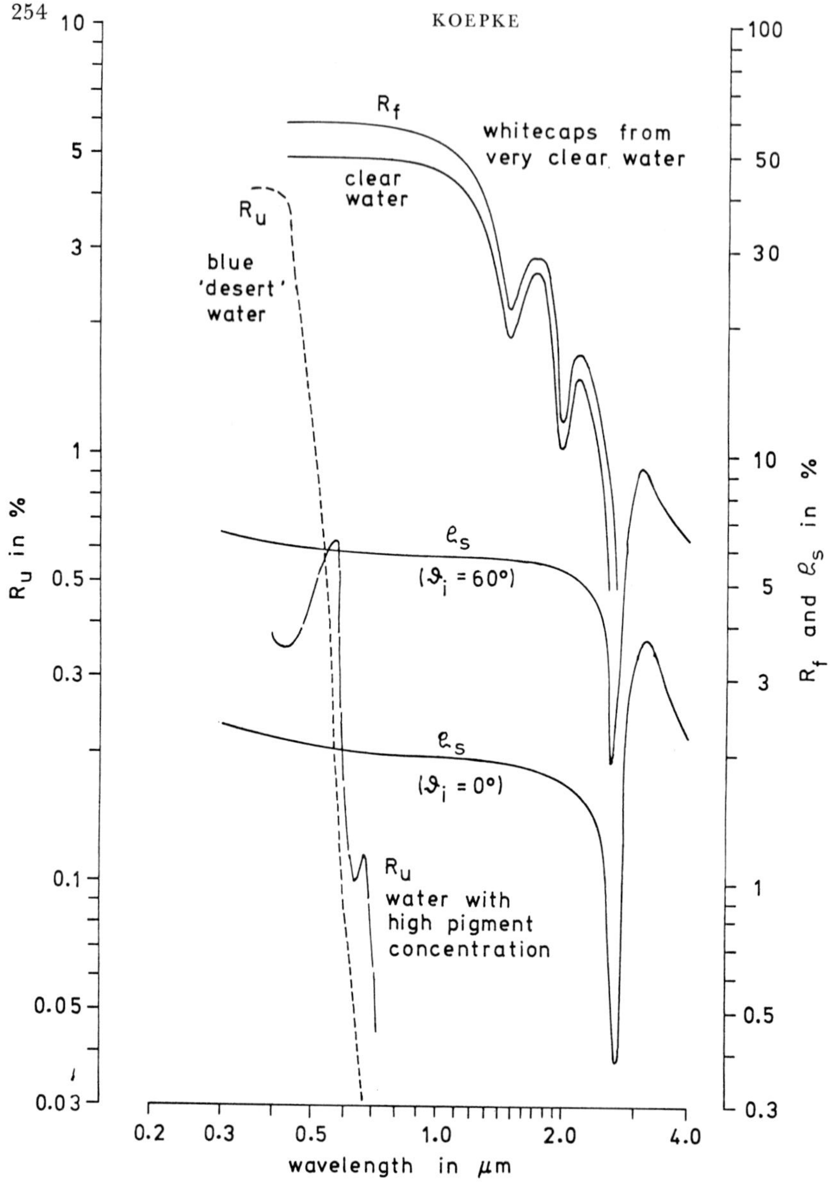

Fig. 2. Spectral reflectance of the components of an ocean surface hemispherical reflectance factor for directional incidence ρ_s of a flat surface, calculated with the Fresnel formula from the refractive index (Irvine and Pollack, 1968), reflectance of foam R_f from clear and very clear water measured by Whitlock (1982) (both right ordinate); and reflectance R_u just above the ocean surface due to underlight for two different water types measured by Morel (1980) (left ordinate).

independent of the angles of incidence and reflection, which leads to a horizontal line in Fig. 1. The reflectance due to the underlight R_u also can be assumed to be isotropic due to the multiple scattering processes inside the water, and due to the spreading out of the radiance field emerging upward through the usually rough water surface (e.g. Morel, 1980; Gordon et al., 1980). The specular reflection of the foam free surface R_s, however, shows high values around the specular angle and low or no reflection outside this region. Due to the Fresnel formula, the magnitude of the reflectance values (also in the case of waves) depends on the angle of incidence: flat incident radiation, coming from large zenith angles, is strongly reflected, while steep incoming radiance penetrates the water and is reflected only weakly.

Thus, the ocean looks dark in the directions outside the sunglint not so much due to its low albedo, but due to its strong anisotropy. Of course, specular reflectance takes place not only in the sunglint, but in all directions, always reflecting that part of the sky which is under the specular angle.

The spectral features of the reflectance of the ocean surface components are shown in Fig. 2. The reflectance of water without foam, R_s, is given for a flat ocean for two angles of incidence, nadir direction (low reflection) and a zenith angle of 60° (higher reflection). The right ordinate is valid for this part of Fig. 2. The values are calculated with the Fresnel formula from the spectral refractive index of water after Irvine and Pollack (1968). In the visible and near infrared part of the spectrum up to 2 μm, the part which is used for remote sensing and which is essential in energy questions, the reflection of clear water is nearly independent of wavelength. So a mean refractive index of 1.33 can be used. Towards longer wavelengths one can see a variation in R_s due to the presence of a strong absorption band.

The underlight is highly wavelength dependent due to the spectral properties of the material suspended in the water and due to the spectral absorption of the water itself. Examples of measured R_u-values after Morel (1980) are shown for dark blue water and water with high pigment concentration. The differences between these curves provide a basis for the determination from satellite remote sensing of chlorophyll in the water (e.g. Gordon et al., 1980). Since the R_u-values are low, they have been multiplied by a factor of 10 for better comparison with the other quantities. Therefore the left ordinate is valid for R_u. Due to the increasing absorption, the underlight decreases towards longer wavelengths. The ocean water can be assumed to be black at wavelengths longer than 0.7 μm. But with high sediment load, there will be underlight even at wavelengths above 1.0 μm (e.g. Morel, 1980; Whitlock, et al., 1981).

The spectral reflectance of whitecaps was measured by Whitlock et al. (1982) in the laboratory. The reflectances, R_f, presented in Fig. 2 for thick whitecaps from clear and very clear water, have nearly constant values of about 55% in the visible spectral range and show a decrease towards longer wavelengths due to the absorption of the liquid water.

Consequently, the remote sensing of whitecaps is most suitably done at wavelengths in the near infrared in directions outside the sunglint. Under these conditions a good contrast will exist between the whitecaps and the surrounding water. An additional advantage of these longer wavelengths, compared to the wavelengths in the visible region, is the decreased optical depth of the atmosphere, resulting in lower path radiances, in higher illumination due to the direct sun and in lower sky radiance, the reflection of which cannot be avoided. Of course, in this spectral range the absorption due to atmospheric water vapor must be taken into account (e.g. Koepke and Quenzel, 1978).

3. *The optical effectiveness of whitecaps*

As mentioned above, the optical influence of whitecaps is usually described by the product $W \cdot R_f$, with each quantity coming from different publications.

The R_f-values used so far lie between .4 and 1.0 (e.g. Payne, 1972; Maul and Gordon, 1975; Gordon and Jacobs, 1977; Koepke and Quenzel, 1979; Quenzel and Kaestner, 1980; Tassan, 1981), independent of wavelength. Now, after the publication of Whitlock et al., (1982), measured spectral R_f values can be used.

The W-values are determined by different authors (cited in Section 2) with the photographic method where more or less 'the outline of the white area was traced and the area was measured' (Toba and Chaen, 1973). Consequently, in determining the W-values, whitecaps of different age are taken into account, and the areas that contribute to the W-values have reflectances that do not correspond to the R_f values measured for dense foam.

From series of photos taken at 1-second intervals at the research platform *Nordsee* I determined the increase of the area and the decrease of the reflectance of individual whitecaps during their life (Koepke, 1984). From these data, under the assumption of a maximum reflectance R_f of 55% (Whitlock et al., 1982) I determined an 'effective reflectance' R_{ef}, the reflectance of the whitecaps weighted with the area of whitecaps both integrated over the time (Eq. 2).

$$R_{ef} = \frac{\int_0^{t_{id}} a(t) \cdot r(t) \cdot dt}{\int_0^{t_{id}} a(t) \cdot dt} \qquad (2)$$

a(t) is the area of individual whitecaps and r(t) is their mean reflectance value, both as a function of time. Due to the decrease of their reflectance the whitecaps vanish. At an age t_{id} they will no longer be identified and so will not be considered in the W-values. Since in a single photo whitecaps of each age are present with the same probability, the effective reflectance from Equation 2 is the appropriate reflectance to associate with the whitecap areas determined from such photos. The effective reflectance depends not only on the high variability of different whitecaps and the subjective judgement which enters in the analysis of the data (e.g. Blanchard, 1971), but also on the criteria which are used to get W. Consequently the values of R_{ef} have a high uncertainty.

Besides, foam patches (whitecaps) and foam streaks must be distinguished. The foam streaks, compared to patches, have a longer life time, cover a relatively constant area and have low reflectance values.

The resulting values are (Koepke, 1984):
effective reflectance of whitecaps (foam patches,

$$R_{ef,fp} = (22 \pm 8)\,\%$$

and effective reflectance of foam streaks,

$$R_{ef,fs} = (10 \pm 4)\,\%$$

Among other things, the life time of foam depends on the water temperature as has been demonstrated by laboratory measurements (Miyake and Abe, 1948). This is also noticeable in the different data sets discussed by Monahan and Ó Muircheartaigh (1980). So, at temperatures other than the one that pertained during the *Nordsee* measurements (about 16°C), the values of $R_{ef,fp}$ and $R_{ef,fs}$ may be slightly different.

The reduction of the reflectance of whitecaps with age, due to the increase of their area and the thinning of the foam, can be assumed to be independent of the wavelength. Therefore, to describe the effective reflectance, I introduce an 'efficiency factor' $f_{ef} = R_{ef}/55\%$ which allows the use of the spectral reflectance values of dense foam layers given by Whitlock, et al. (1982). So the effective optical influence of whitecaps as used in Eq. 1, is given by the product $W \cdot f_{ef} \cdot R_f$ with W the wind speed dependent coverage of whitecaps, f_{ef} the efficiency factor and R_f their spectral reflectance.

The efficiency factors are:
for foam patches $f_{ef,fp} = 0.4 \pm 0.15$
and foam streaks $f_{ef,fs} = 0.18 \pm 0.07$

Of course the f_{ef}-values show the same high uncertainty as the values of the effective reflectance. But it is evident that the optical influence of whitecaps in the solar spectral range, their remote sensing signature in this range, is lower than that derived from combining the somewhat inconsistent data available in the literature so far. The relatively low influence of whitecaps is confirmed by extensive measurements by Payne (1972), who found that in the albedo of the sea surface 'effects of whitecaps are not noticeable at wind speeds up to 30 kt'.

4. Whitecaps as a perturbing parameter

Single whitecaps cannot be detected from satellite borne radiometers, due to the spatial resolution, which is not better than 30m in available satellite systems (Thematic Mapper, Landsat 4).

But, of course, increasing the amount of whitecaps leads to an increase of the radiances at the top of the atmosphere. The relative increase of these radiances due to whitecaps depends not only on the amount of whitecaps and the wavelength, but also on the atmospheric turbidity, on the position of the sun and on the viewing direction. These other factors taken together determine the path radiance, which will mask the surface, and the attenuation of the radiance coming from the surface.

Fig. 3 shows the relative increase of the spectral radiance at the satellite, ΔL_{SAT} as a function of the area covered with whitecaps. In the sunglint ($\vartheta_{sat} = \vartheta_0$; $\varphi = 0°$) the radiances are high, but the relative increase due to whitecaps is small, since the whitecaps are highly reflecting but reduce the area where specular reflectance takes place. The opposite behaviour is shown at the nadir direction, which is representative of the directions used in remote sensing. The radiances are relatively low, but the influence of whitecaps is high. For both directions, the influence of the efficiency factor is shown. With a value of $f_{ef} = 0.4$ an increase of the whitecaps to 2% leads to an increase of the nadir radiance of about 5%.

Chlorophyll detection from satellite usually is based on a comparison of radiances at wavelengths with high and low chlorophyll absorption (e.g. Hojerslev, 1981). If the sea surface is covered with whitecaps, the radiances at both wavelengths will increase. Consequently the uncertainty due to whitecaps in the chlorophyll determination is negligible, if an algorithm is used which applies radiance differences (Tassan, 1981). The necessary supposition that the whitecaps have the same reflectance at both wavelengths, is a valid one (Whitlock et al., 1982). If radiance ratios are used to determine chlorophyll, the uncertainty due to the actual unknown amount of whitecaps becomes much greater (Tassan, 1981). Nevertheless, even the ratio results will become more acceptable if an efficiency factor is used to describe the foam influence.

Remote sensing of the atmospheric aerosol optical depth in cloudfree pixels over water surfaces using radiances at one wavelength is possible and has been done successfully (e.g. Kästner et al., 1983). In this determination technique, the whitecaps are one of the most essential perturbing parameters (Köpke and Quenzel, 1983). But again, the effective perturbation of the radiances due to whitecaps will be smaller than was concluded in that paper, because no efficiency factor was taken into account at that time.

5. Effect of whitecaps in the atmospheric window and in the microwave range

The existence of an atmospheric window at about 11 μm makes possible the determination of the sea surface temperature from space (e.g. Smith et al., 1970; Maul, 1981; McClain, 1981). The radiance to be measured at the satellite depends, among other things, on the emissivity of the sea surface and on the skin temperature (that is the temperature of the emitting uppermost microlayer of less than 20 μm thickness) which is usually lower than the bucket temperature (e.g. Hasse, 1963). The emissivity of the water is high, about 0.99 for nadir direction (e.g. Saunders, 1968). So whitecaps are not noticeable because of their higher emissivity. They could influence the measured radiance due to a skin temperature, different from that of the foam free water. But this effect will not be great, because the skin temperature of the foam free water is itself also a function of the windspeed.

In the microwave range, the emissivity of the calm sea surface is much lower than that of foam. Therefore, the brightness temperature of an ocean surface will increase with increasing amounts of whitecaps. Fig. 4 shows the spectral emissivity of sea water over the microwave wavelength band after Gloersen et al. (1981). Added in the figure is the emissivity of foam after Droppleman (1970), which is not a function of the wavelength, but depends on the ratio of foam depth to wavelength.

The emissisivity of the foam free water surface is not only a function of the wavelength, but also of the water temperature and the salinity. Between 0.88 and 3.2 cm there is a wave-

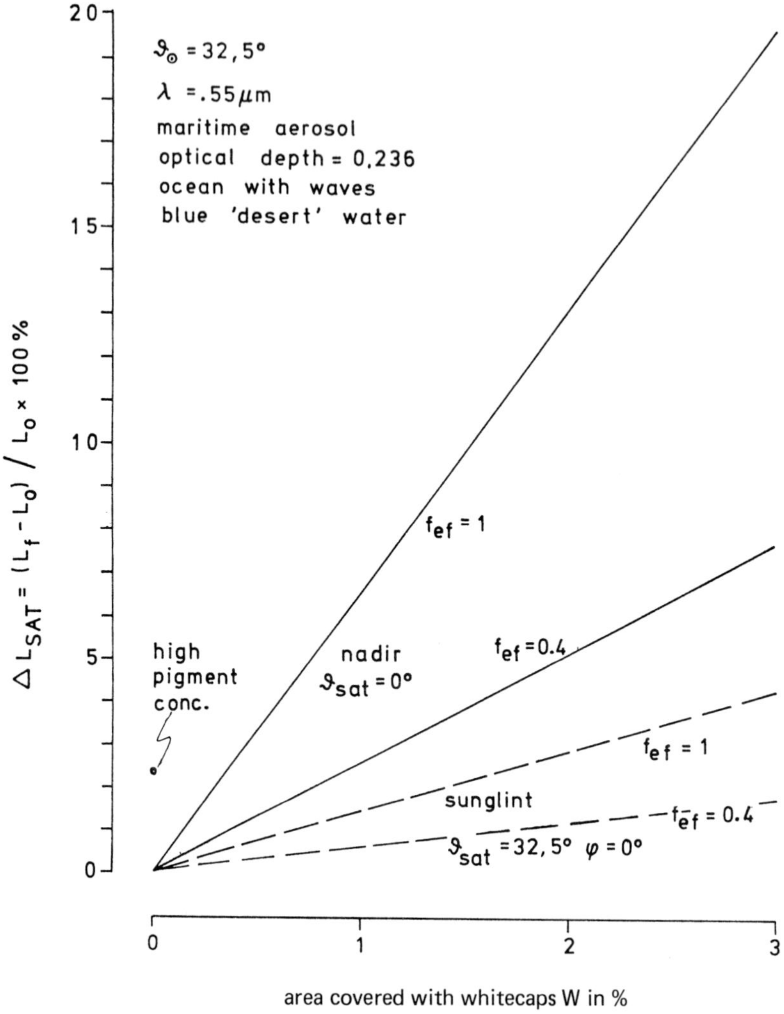

Fig. 3. Relative increase of spectral radiances at the top of the atmosphere calculated as a function of the area covered with whitecaps. f_{ef} = efficiency factor (see text), ϑ = zenith angle of sun and satellite.

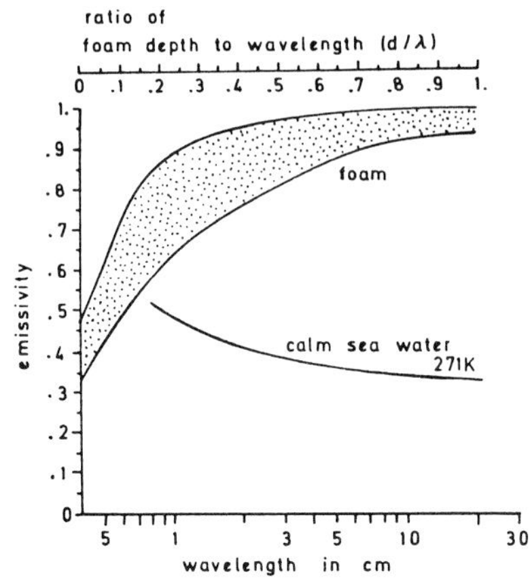

Fig. 4. Spectral emissivity of sea water as a function of microwave wavelength (after Gloersen et al., 1981; lower abscissa) and variation of emissivity of foam with foam depth as a function of the ratio d/λ (after Droppleman, 1970; upper abscissa).

length range where the brightness temperature is only slightly dependent on the physical temperature of the water (e.g. Edgerton and Trexler, 1969) and therefore correlates well with the emissivity and thus with the amount of whitecaps.

The modification of the radiances owing to atmospheric absorption and emission is highly variable due to varying temperature profiles, gas amounts, clouds and precipitation in the atmosphere. But since this modification depends on wavelength, remote sensing of the ocean-atmosphere system in the microwave region is made with different wavelengths simultaneously. The Scanning Multi-Channel Microwave Radiometer (SMMR) (Gloersen and Barath, 1977; Njoku et al., 1980) uses channels set at 0.8, 1.4, 1.7, 2.7 and 4.6 cm wavelength.

With retrieval techniques (GOASEX, 1979; Wilheit and Chang, 1979), making use of the brightness temperatures at all these wavelengths and for horizontal and vertical polarizations, it has been possible to determine the sea-surface temperature, cloud droplet concentration, rainfall rates, the atmospheric water vapor content and the near surface winds. These winds are derived from the variation of the brightness temperature due to whitecaps and the correlation between the amount of whitecaps and the windspeed. To become free of the problem of the effectiveness of the whitecaps, usually a direct correlation between brightness temperature and wind speed is used in the microwave region (e.g. Williams, 1969; Wentz et al., 1981).

6. *Conclusion*

Whitecaps are, with respect to remote sensing, an essential parameter of the ocean-atmosphere system. Their effective optical properties must be taken into account.

This work was partly sponsored by the German Minister for Research and Technology (BMFT) under grant MF-0235, which is gratefully acknowledged.

REFERENCES

Blanchard, D. C. (1971). Whitecaps at sea. *J. Atmos. Sci.* 28, 645.

Cox, C. and W. Munk (1954). Statistics of the sea surface derived from sun glitter. *J. Mar. Res.* 13, 198–227.

Droppleman, J. D. (1970). Apparent microwave emissivity of sea foam. *J. Geophys. Res.* 75, 696–698.

Edgerton, A. D. and D. T. Trexler (1969). Oceanographic application of remote sensing with passive microwave techniques. *Proc. 6th Symp. on Remote Sensing*, Univ. of Mich., 767–788.

Gloersen, P., W. J. Campbell and D. Cavalieri (1981). Global maps of sea ice concentration, age and surface temperature derived from Nimbus-7 scanning multichannel microwave radiometer data: A case study, in: *Oceanography from Space* (J. Gower, ed.), Plenum Press, New York, 777–783.

Gloersen, P. and F. T. Barath (1977). A scanning multichannel microwave radiometer for Nimbus-G and Seasat-A. *IEEE J. Oceanic Eng., OE-2*, 172–178.

GOASEX Workshop Report (1979), SMMR panel report, Jet Propulsion Lab., Pasadena, CA

Gordon, H. R. and M. M. Jacobs (1977). Albedo of the ocean-atmosphere system: influence of sea foam. *Appl. Opt.* 16, 2257–2260.

Gordon, H. R., D. K. Clark, J. L. Mueller and W. A. Hovis (1980). Phytoplankton pigments from the Nimbus-7 Coastal Zone Color Scanner: Comparison with surface measurements. *Science* **210**, 63-66.

Hasse, L. (1963). On the cooling of the sea surface by evaporation and heat exchange. *Tellus* **15**, 363-366.

Hojerslev, N. K. (1981). Assessment of some suggested algorithms on sea colour and surface chlorophyll, in: *Oceanography from Space* (J. Gower, ed.), Plenum Press, New York, 347-353.

Irvine, W. M. and J. B. Pollack (1968). Infrared optical properties of water and ice spheres. *Icarus* **8**, 324-360.

Kästner, M., P. Köpke and H. Quenzel (1983). Monitoring of Saharan dust over the Atlantic using Meteosat-VIS-data. *Adv. Space Res.* **2**, 119-121.

Koepke, P. (1984). Effective reflectance of oceanic whitecaps. *Appl. Opt.* **23**, 1816-1824.

Koepke, P. and H. Quenzel (1978). Water vapor: spectral transmission at wave-lengths between 0.7μm and 1μm. *Appl. Opt.* **17**, 2114-2118.

Koepke, P. and H. Quenzel (1979). Turbidity of the atmosphere determined from satellite: Calculation of optimum viewing geometry. *J. Geophys. Res.* **84**, 7846-7856.

Köpke, P. and H. Quenzel (1982). Most suitable conditions for aerosol monitoring from space. *Adv. in Space Res.* **2**, 29-32.

Maul, G. A. and H. R. Gordon (1975). On the use of the Earth Resources Technology Satellite (LANDSAT-1) in oceanography. *Rem. Sensing of Environment* **4**, 95-128.

Maul, G. A. (1981). Application of GOES visible-infrared data to quantifying mesoscale ocean surface temperatures, *J. Geophys. Res.* **86**, 8007-8021.

McClain, E. P. (1981). Multiple atmospheric-window techniques for satellite-derived sea surface temperatures, in: *Oceanography from Space* (J. Gower, ed.), Plenum Press, New York, 73-85.

Miyake, Y. and T. Abe (1948). A study on the foaming of sea water. *J. Marine Res.* **7**, 67-73.

Monahan, E. C. (1971). Oceanic whitecaps. *J. Phys. Oceanogr.* **1**, 139-144.

Monahan, E. C. and I. G. Ó Muircheartaigh (1980). Optimal Power-law description of oceanic whitecap coverage dependence on wind speed. *J. Phys. Oceanogr.* **10**, 2094-2099.

Morel, A. (1980). In-water and remote measurements of ocean color. *Bound.-Layer Met.* **18**, 177-201.

Njoku, E. G., J. M. Stacey, and F. T. Barath, (1980). The Seasat scanning multichannel microwave radiometer (SMMR): Instrument description and performance. *IEEE J. Oc. Eng. OE-5*, 100-115.

Payne, R. E. (1972). Albedo of the sea surface. *J. Atmos. Sci.* **29**, 959-970.

Plass, G. N., G. W. Kattawar and J. A. Guinn (1977). Isophotes of sunlight glitter on a wind-ruffled sea. *Appl. Opt.* **16**, 643-653.

Quenzel, H. and M. Kaestner (1980). Optical properties of the atmosphere: calculated variability and application to satellite remote sensing of phytoplankton. *Appl. Opt.* **19**, 1338-1344.

Ross, D. B. and V. Cardone (1974). Observations of oceanic whitecaps and their relation to remote measurements of surface wind speed. *J. Geophys. Res.* **79**, 444-452.

Saunders, P. M. (1967). Shadowing on the ocean and the existence of the horizon. *J. Geophys. Res.* **72**, 4643-4649.

Saunders, P. M. (1968). Radiance of sea and sky in the infrared window 800-1200 cm^{-1}. *J. Opt. Soc. America* **58**, 645-652.

Smith, W. L., P. K. Rao, R. Koffler and W. P. Curtis (1970). The determination of sea surface temperature from satellite high resolution infrared window radiation measurements. *Month. Weather Rev.* **98**, 604-611.

Tassan, S. (1981). The influence of wind in the remote sensing of chlorophyll in the sea, in: *Oceanography from Space* (J. Gower, ed.), Plenum Press, New York, 371-375.

Toba, Y. and M. Chaen (1973). Quantitative expression of the breaking of wind waves on the sea surface. *Records of oceanogr. works in Japan* **12**, 1-11.

Wentz, F. J., E. J. Christensen and K. A. Richardson (1981). Dependence of sea-surface microwave emissivity on friction velocity as derived from SMMR/-SASS, in: *Oceanography from Space* (J. Gower, ed.), Plenum Press, New York, 741-749.

Whitlock, C. H., L. P. Poole, J. W. Usry, W. M. Houghton, W. G. Witte, W. D. Morris and E. A. Gurganus (1981). Comparison of reflectance with backscatter and absorption parameters for turbid waters. *Appl. Opt.* **20**, 517-522.

Whitlock, L. H., D. S. Bartlett and E. A. Gurganus (1982). Sea foam reflectance and influence on optimum wavelength for remote sensing of ocean aerosols. *Geophys. Res. Letters* **9**, 719-722.

Wilheit, T. T. and A. T. C. Chang (1979). An algorithm for retrieval of ocean surface and atmospheric parameters from the observations of the scanning multichannel microwave radiometer (SMMR), NASA Tech Memo, 80277, Goddard Space Flight Center.

Williams, G. F., Jr. (1969). Microwave radiometry of the ocean and the possibility of marine wind velocity determination from satellite observations. *J. Geophys. Res.* **74**, 4591-4594.

Wu, J. (1979). Oceanic whitecaps and sea state. *J. Phys. Oceanogr.* **9**, 1064-1068.

THE INFLUENCE OF WHITECAPS ON THE ALBEDO OF THE SEA SURFACE

P. J. STABENO & E. C. MONAHAN
University College, Galway

The influence of whitecaps on the total albedo of the sea surface is significant over broad reaches of the world's oceans. The sea surface albedo due to whitecaps is estimated by combining the monthly North Atlantic charts of the calculated fractional whitecap coverage (Spillane, 1982) with estimates of the albedo of an individual whitecap. This latter quantity, which is determined from an extremely simple model for a whitecap consisting of uniform bubble layers, is about 50%, which is consistent with the localized whitecap albedo mentioned by Payne (1972) and Whitlock, et al. (1982). During the winter months the whitecap contribution to the albedo for regions of the North Atlantic amounts to more than 15% of the observed average monthly albedo as reported by Payne (1972) and Hummel and Reck (1979).

1. Introduction

The albedo is the ratio of the upward to downward irradiance just above the sea surface, and is dependent upon both atmospheric and oceanic conditions. In a cloudless sky, the sea surface albedo is probably most sensitive to variations in the zenith angle of the sun. To determine the reflectance of light from the water's surface theoretically, Fresnel's equations may be used. These equations are only directly applicable under certain limited conditions, i.e. for direct radiation striking a smooth surface. Under these conditions the albedo varies from a reflectance of 2% when the sun is directly overhead ($\theta=0°$), to total reflectance when the sun approaches the horizon ($\theta=90°$). Actual conditions do not necessarily conform to the ideal necessary for the direct application of Fresnel's laws. Both cloud cover and particles in the atmosphere cause the sunlight to be scattered and thus the light becomes diffuse. For diffuse light both theoretical results (obtained by the integration of Fresnel's equations over the angles of incidence) and data from observations have been published (Burt, 1953; Nunez, et al., 1972; Kondratyev, 1972). The variation in the albedo under such natural conditions is, of course, dependent upon the thickness of the cloud cover. Observations for highly diffuse light give an albedo of between 7% and 10% (Kondratyev, 1969 and 1972), with only a weak dependence upon the zenith angle of the sun, while theoretical calculations indicate a reflectance of 6.6% (Burt, 1953).

In addition to the dependence of the albedo on these atmospheric conditions there are also the variations due to the sea surface. The albedo of a roughened sea, with non-breaking waves, varies from that of a flat surface. For zenith angles near 0° the reflectance increases, and when it is nearer to 90° it decreases (Kondratyev 1969; Nunez, et al., 1972), when the sea surface is roughened. The opacity of the water also has an effect. With a decrease in water transparency there is an increase in the albedo (Kondratyev, 1972). It is only to be expected that whitecaps would also influence the total albedo. The exact nature of the effect, however, is open to question. From data collected by Payne (1972) it appears that an increase in the windstress (and hence in the whitecap cover) decreases the albedo. A plot in Neumann and Peirson (1966), however indicates just the opposite, namely that an increase in whitecap

E. C. Monahan and G. Mac Niocaill (eds.), Oceanic Whitecaps, 261–266.

cover increases the reflectance. A similar result was observed by Griggs and mentioned at this workshop. From a measurement of the sea surface albedo obtained via satellite in an instance where the wind speed at the surface was known, he calculated what portion of the sea surface was covered with whitecaps. The resulting data point compared well with the whitecap coverage at that windspeed reported by Monahan (1982). From the results of a laboratory experiment (Whitlock, et al., 1982) it is clear that the presence of even a single layer of bubbles increases the reflectance. If the number of layers of bubbles increases from one to twenty then the localized albedo increases from 10% to approximately 50%. This latter value is the percentage that is generally accepted for the reflectance of whitecaps (Payne 1972; Whitlock, et al., 1982).

2. *The Whitecap Model*

A simple model of bubble rafts in the ocean should give some indication of the effect of a whitecap upon the reflectance of light. Calculations using Fresnel's equations show that the presence of a bubble (either on the surface or below it) does not increase the albedo of the sea surface near the bubble significantly. The high reflectance of whitecaps must be due, in large part, to the thickness of the bubble rafts. Thus, it is expected that the albedo of the whitecap declines as the bubbles burst, the raft thins out, and the whitecap decays. The dependence of the albedo upon the number of layers of bubbles is clearly seen in the data of Whitlock, et al. (1982). The decrease in the albedo of a whitecap as it ages was examined in a poster by Koepke at this symposium. A similar behavior is observed in snow. As the depth of the snow increases so does the surface albedo until the snow reaches a depth of 12.7 centimeters (Hummel and Reck, 1979). The albedo now is near its maximum and any further increase in the depth of the snow causes a negligible change in the albedo.

In developing a first model simplicity is desired. The interface between two contiguous bubbles is planar. A group of bubbles compressed together would therefore have internal walls each of which is a plane. If this is carried to an extreme and all the bubbles are taken to be the same size, then a possible (if improbable) configuration is that they stack together as blocks (figure 1a). Consider a light beam passing through the vertical walls of the bubble: a portion of this beam is transmitted, while the remainder is reflected. However, both parts of the beam subsequently intersect the horizontal walls at the same angle (α), as shown in figure 1a. Therefore, in this simple model the vertical walls play no part in the determination of the albedo. This leads to the geometry shown in figure 1b. Using Fresnel's laws of reflectance at

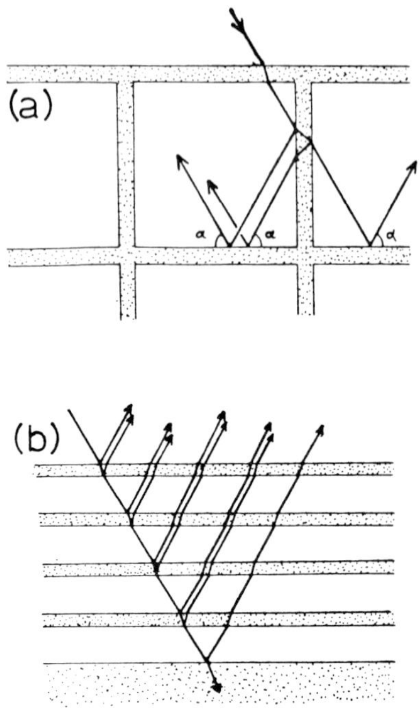

Fig. 1. a) A schematic representation of the bubble structure. The beam of light intersects the surface of the whitecap and each of the horizontal air-water interfaces at an angle α.

b) The modified structure neglecting the vertical walls. There are four layers of bubbles and hence nine interfaces from which reflectance is possible. The beam undergoes at most one reflection.

each air-water and water-air interface the albedo, A, can be calculated. It is expected that the major contribution to the albedo of the bubble raft would be from that part of the beam that is reflected from only one interface. The total albedo is expressed as

$$A = A_1 + A_3 + A_5 + \ldots \qquad (1)$$

where each A_i is the albedo of the beam that is reflected from i interfaces. Shown in figure 1 is that portion of the beam that is reflected only once. From examination of the model it is evident that:

(1a)

$$A_1 = R \sum_{n=0}^{2N} (1-R)^{2n}$$

(1b)

$$A_3 = R^3 (1-R)^2 \left\{ \frac{(N+1)(N+2)}{2} (1-R)^{2N} + \sum_{n=0}^{N-1} \frac{(n+1)(n+2)}{2} \left[(1-R)^{2n} + (1-R)^{2(2N-n)} \right] \right\}$$

(1c)

$$A_5 = R^5 (1-R)^2 \sum_{n=0}^{6N-3} B_N(n) (1-R)^{2n}$$

where N is the number of layers of bubbles and R is the reflectance as calculated by Fresnel's equations. The coefficients of the last series are given in table 1. These series lend themselves to a limiting solution as N, the number of layers, approaches infinity. Thus:

$$A_\infty = R \left\{ \left[1-(1-R)^2 \right]^{-1} + R^2 (1-R)^2 \left[1-(1-R)^2 \right]^{-3} + R^4 (1-R)^2 \left[1+(1-R)^2 \right] \left[1-(1-R)^2 \right]^{-5} \right\} + o(R^7) \qquad (2)$$

which is the maximum albedo of the whitecap.

The results of the calculations carried out using equations 1 and 2 are displayed in figure 2. Presented along with the theoretical calculations are the data from Whitlock, et al. (1982). The index of refraction for pure water as a function of wavelength is used in calculating the albedo in the model. It is clear from figure 1 that the reflectance of one layer of bubbles in the experiment is considerably greater than that predicted by the model. There are several possible explanations for this. Obviously, the model is extremely simple. The orientation of the walls of the bubbles of a whitecap are in reality random. Also, in analysing the experimental data, no reflectance from the surface of the container, nor any back scattering of light, was considered. Another possible cause of the discrepancy is that in the experiment soap was added to the water which would change the index of refraction.

The solid circles in figure 2 show the measured reflectance for multiple layers of bubbles (Whitlock, et al, 1982). The decrease in the reflectance observed in the experiment with increasing wavelength is due mainly to the higher absorption rate of light at longer wave-

Table 1: $B_N(n)$ Coefficients in Equation 1c.

N \ n	1	2	3	4	5	6	7	8	9	10	11	12	13	14	15	16	17
1	1	6	5	1													
2	1	6	20	50	60	53	35	15	5	1							
3	1	6	20	50	105	196	245	255	232	185	126	70	35	15	5	1	
4	1	6	20	50	105	196	336	540	672	735	735	681	585	462	330	210	126
5	1	6	20	50	105	196	336	540	825	1210	1485	1653	1720	1695	1590	1420	1203

N \ n	18	19	20	21	22	23	24	25	26	27	28
1											
2											
3											
4	70	35	15	5	1						
5	960	715	495	330	210	126	70	35	15	5	1

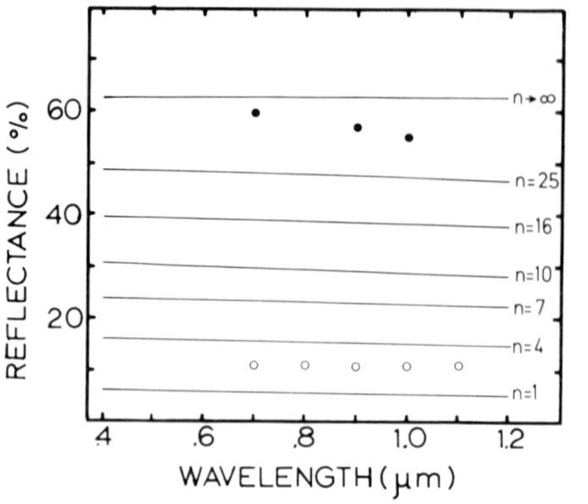

Fig. 2. The reflectance of a whitecap as a function of wavelength is shown here. The results of the laboratory experiment (Whitlock et al., 1982) are indicated by circles. The open ones represent a single layer of bubbles and the solid, multiple layers. The lines are the prediction of the model for various number of layers of bubbles.

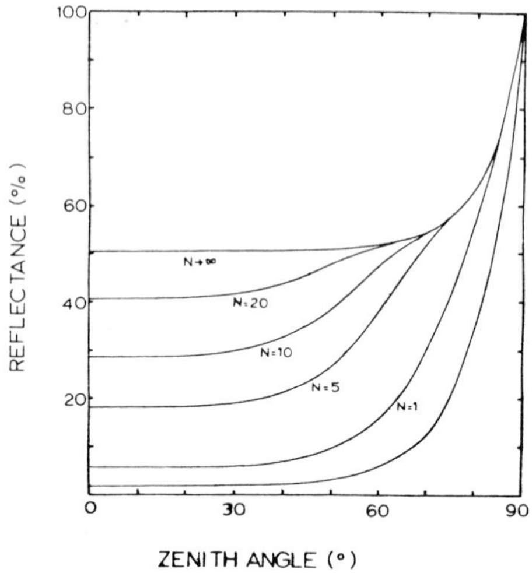

Fig. 3. The percentage of light reflected from a whitecap as calculated by the model. N is the number of layers. A refractive index of 1.33 was used in the calculation.

lengths. The slight decrease in the reflectance predicted by the model is due solely to the decrease of the index of refraction with increasing wavelength. There is a sufficient depth of bubbles in the experiment so that the bottom of the tank was not visible, but the number of layers of bubbles in the raft is not explicitly given in the article. If a bubble size of 0.4 centimeters is taken, there would be 25 layers in the 10 centimeters of foam. The model then would predict a reflectance of just less than 0.5, which is comparable to the experimentally determined reflectance of between 0.5 and 0.6.

As can be seen in figure 3, for angles less than 50° there is a very weak dependence of the albedo of the whitecap upon the zenith angle. Also as the number of layers increases, the reflectance's dependence upon the zenith angle decreases. As $N \rightarrow \infty$ the reflectance is virtually independent of the zenith angle for all angles less than 75°. Hence, the choice of a reflectance of 50% regardless of the angle of incidence is not unreasonable.

Although the model follows the trend of the data from the laboratory experiment reasonably well, it does not immediately follow that it would also model whitecaps in the open ocean. There, a whitecap may last only 10 to 15 seconds and individual bubbles probably have a life time of less than two seconds once they surface (Zheng, et al., 1983). A more complex model with various sizes of bubbles and a disordered orientation of interfaces would be more realistic. There is always a considerable number of subsurface bubbles that have not combined with the raft which may also play a role and hence should be considered.

3. *Observations and Conclusions*

If the albedo as calculated by Payne (1972), or Hummel and Reck (1979), for the North Atlantic is considered along with the fractional whitecap coverage calculated by Spillane (1982) then the relative influence of whitecaps upon the total albedo can be determined. An individual whitecap is assumed to have an albedo of 50%. Using the monthly averaged albedo in the North Atlantic as calculated by Payne (1972), a plot of the percentage of sea surface albedo that is due solely to whitecaps can be generated (figure 4). The values for each latitude are integrated over longitude for every

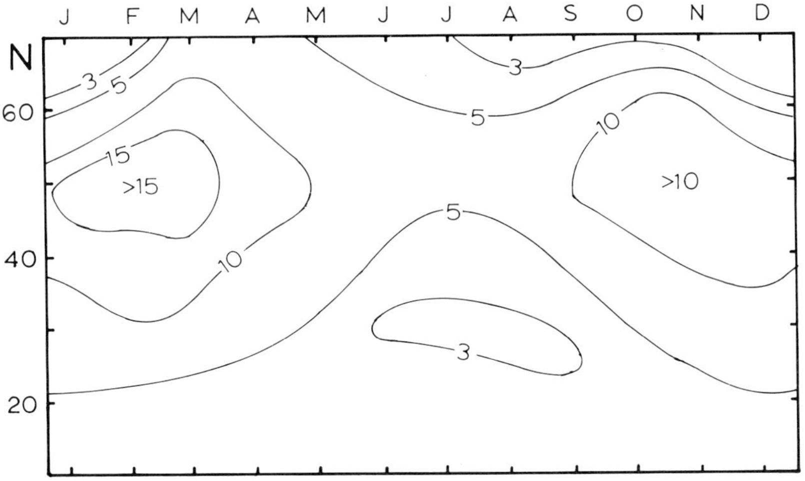

Fig. 4. Contours of the percentage of the total albedo due to whitecaps in the North Atlantic. The contours were calculated by assuming that 50% of the light incident upon the whitecaps is reflected. The percentage of the sea surface covered by whitecaps is obtained from Spillane (1982) and the total albedo is given in Payne (1982).

month of the year. As can be seen in this figure, the maximum influence of whitecaps on the albedo is during the winter months at about 50° N. At this time and location there is a maximum whitecap coverage of 4%. The relatively high coverage by whitecaps of the sea surface is enough to make the whitecap contribution to the albedo proportionally high even at a time of the year when other factors tend to increase the total albedo. Further north the relative influence of the whitecaps becomes smaller, because the sun is so low in the sky that the reflectance off even a bubble-less ocean surface is very high and in this calculation the dependence of the localized albedo of a whitecap on the altitude of the sun has been neglected. To the south of 50° N the wind speed decreases. This decrease in the windstress causes a decrease in the whitecap coverage, and in the whitecap contribution to the albedo, that dominates the smaller decrease in the total albedo due to the higher elevation of the sun. The percentage of the total albedo due to the whitecaps hence decreases. During the summer the proportion of the albedo due to the reflectance of light from the whitecaps is minimal since the occurrance of storms decreases so dramatically.

A contour plot for just the winter months (January, February and March) of the influence of whitecaps on the total albedo in the North Atlantic is shown in figure 5. The total albedo is obtained from data given by Humel and Reck (1979). The maximum influence is in the central North Atlantic, which is to be expected since the strongest winds are there. The slight variations between figure 4 and 5 are due to the use of different data sets.

From this simple calculation it is clear that whitecaps can play a significant role in increasing the albedo of the sea surface in regions of high winds.

ACKNOWLEDGEMENTS

The study of whitecaps at University College, Galway, is being funded by the US Office of Naval Research, via Grant N-00014-78-G-0052.

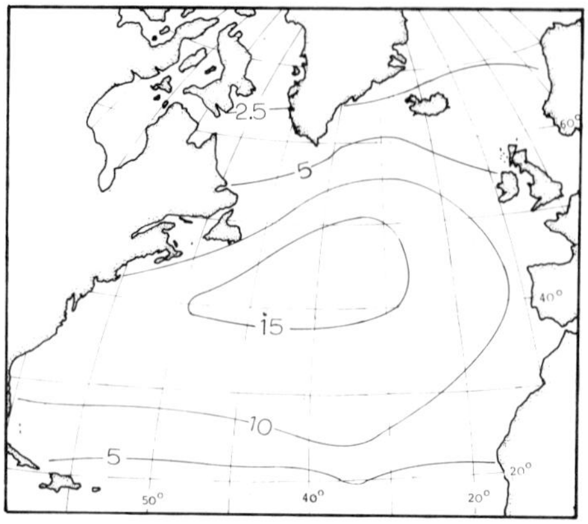

Fig. 5. A chart of the north Atlantic showing the influence of whitecaps on the total albedo (in percentage) during the winter months (January, February and March). The amount of whitecap coverage is taken from Spillane (1982) and the total albedo is from Hummel and Reck (1979). A local reflectance for the whitecaps of 0.5 is used.

REFERENCES

Burt, Wayne V., 1953. A note on the reflection of diffuse radiation by the sea surface. *Trans. AGU* **34**, 199-200.

Burt, Wayne V., 1954. Albedo over wind-roughened water. *J. of Met.* **11**, 283-90.

Hummel, J. R., and R. A. Reck, 1979. A global surface albedo model. *J. App. Met.* **18**, 239-253.

Kondratyev, K. Y., Radiation Processes in the Atmosphere, 1972.

Neumann, G., and W. J. Pierson, Jr., 1966. Principles of Physical Oceanography. Prentice-Hall, Inc.

Nunez, M., J. A. Davies and P. J. Robinson, 1972. Surface albedo at a tower site in Lake Ontario. *Bound. Layer Met.* 3, 77-86.

Payne, R. E., 1972. Albedo of the sea surface. *J. of Atmos. Sciences* **29**, 959-970.

Powel, W. M., and G. L. Clarke, 1936. The reflection and absorption of daylight at the surface of the ocean. *J. of Optical Soc. Am.* **26**, 111-120.

Spillane, M. C., 1982. Estimation of global whitecap coverage from ship observations. Chapter 5 in: Monahan, E. C., M. C. Spillane, P. A. Bowyer, D. M. Doyle, and J. J. Taper, *Whitecaps and the Marine Atmosphere, Report No. 4*, University College, Galway.

Whitelock, C. H., D. S. Bartlett and E. A. Gurganus, 1982. Sea foam reflectance and influence on optimum wave-length for remote sensing of ocean aerosols. *Geo. Res. Letters* **9**, 719-722.

Zheng, Q. A., Y.-H. v. Klemas and L. Hsu, 1983. Laboratory measurement of water surface bubble life time. *J. of Geophys. Res.* **88**, 701-706.

ABSTRACTS

AN ATTEMPT TO DETERMINE THE SPACE CHARGE PRODUCED BY A SINGLE WHITECAP UNDER LABORATORY CONDITIONS

PETER A. BOWYER
University College, Galway,
Ireland

The Whitecap Simulation Facility located in UCG has been used to generate repeatable whitecaps under controlled conditions over the last 4-5 years. Amongst other things, the electric charge transferred into the atmosphere has been measured using an Obolensky filter. The charge was found to vary with (1) time after splash of measurement; (2) intensity of splash (controlled by the height of water behind the tank gates before a splash), and (3) water temperature. These results can be combined with global results relating whitecaps and windspeed to produce a new estimate of the contribution of the ocean to the global electrical circuit. This ranges from 160A in January to 220A in June.

ACOUSTIC PROPAGATION IN LIQUID CONTAINING GAS-BUBBLES: EFFECT OF THE BUBBLES' SIZE AND DISTRIBUTION

MARTINE BRIQUET
Sintra Alcatel, Arcueil,
Cedex, France

Gas bubbles in a liquid disturb the propagation of acoustic waves and this phenomenon is being thoroughly studied, especially in underwater acoustic.

Some authors have given theoretical approaches: some other investigators have measured experimentally the bubbles' size and have given a statistical representation. Among them, Davids and Thurston (1950), Laird and Kendig (1952) have demonstrated that the Poisson distribution gives a good fit to the observed data.

The principal purpose of the present investigation was to display the relation existing between bubble characteristics - size, number of bubbles per cm^3, bubble volume ratio - and disturbance on the propagation of sound waves, such as sound velocity variation and attenuation. For this purpose, we have used the preceding distributions and developed two other new ones: the first has the same number of bubbles, the second is defined by a same bubble volume ratio. Results of sound velocity and attenuation are shown for each distribution. The general theory presented by Carstensen and Foldy in 1947 has been applied to the propagation of sound through a bubble screen of different thickness. Both transmission loss through the screen and reflection from the screen are given for each distribution. The absorption and reflection capacities of the bubble screen are investigated.

N. Davids & E. Thurston, JASA **22**(1), 1950, pp. 20-23
D. Laird & P. Kendig, JASA **24**(1), 1952, pp. 29-32
E. Carstensen & L. Foldy, JASA **19**(3), 1947, pp. 481-501.

THE GENERATION, TRANSPORT, AND DEPOSITION OF MARINE AEROSOLS: A TURBULENCE MODELING STUDY

STEPHEN D. BURK
Naval Environment Prediction
Research Facility, Monterey, CA

A numerical planetary boundary layer (PBL) model is used to address problems involving the generation, transport, and deposition of large and giant-sized sea-salt aerosol. The second order closure turbulence model discussed by Burk (1977, 1980) provides the core PBL model to which conservation equations describing turbulent aerosol transport are added. The surface aerosol generation rate, which enters the equation set as a lower boundary condition, is taken from the production flux expressions developed by Monahan. Turbulent aerosol transport is handled with the second order closure formulation, while sedimentation is computed with Kasten's expression for equilibrium aerosol size. Dry deposition fluxes are computed as complicated functions of Stokes settling speed and the rate of turbulent supply of particles to the surface layer.

A series of progressively more complex numerical experiments are performed to study the dynamics of the aerosol size distribution throughout the PBL. In the first set of experiments we begin with an aerosol-free, neutral PBL in which the surface layer wind is either Beaufort 3, 5, or 7. The Monahan aerosol production expression, which is a function of the surface stress and particle radius, is then activated and aerosols are diffused throughout the PBL, undergo sedimentation, and are deposited at the surface. Integration proceeds until a quasi-steady aerosol distribution is attained for each experiment at the different Beaufort wind forces. These results can then be compared with the classic Woodcock measurements.

In another set of experiments, we begin with the equilibrium distributions described above, and then require the wind speed to undergo either a ramp function increase or decrease during a three-hour period. Once the wind is altered, a great many feedbacks on the character of the aerosol distribution occur. The turbulence intensity is altered, the surface aerosol generation flux changes, and even the deposition velocity is affected. We then examine the dynamic behaviour of the temporally changing size distribution.

The final numerical experiment examines the vertical distribution of aerosols in a tradewind PBL with particular emphasis on results in the high humidity region immediately beneath the tradewind inversion.

MOMENTUM FLUX IN WIND WAVES

L. CAVALERI
S. ZECCHETTO
I.S.D.G.M.-CNR
Venice, Italy

Sparse measurements from different authors (Shonting 1964; Yefinov and Khristoforov, 1971; Cavaleri, Ewing and Smith, 1978) have put in evidence the occasional presence of a departure of the kinematics components of a wind wave from the linear theory relationships. The departure is such as to imply the presence of a strong vertical flux of horizontal momentum in the downwards direction.

A full set of measurements, throughout a whole storm, including growing and subsequent decay, has been obtained in the Northern Adriatic Sea taking advantage of the oceanographic tower of CNR, placed in the open sea on 16 m of depth. The measurements include surface elevation, three-dimensional particle velocity field and underwater pressure, all measured on the same vertical and at a choosable depth.

The analysis has shown the downward vertical momentum flux to be strictly associated with the presence of highly nonlinear and breaking waves, and consequently to the presence of an active wind. The analysis of records obtained during swell conditions shows no presence of momentum transfer, with a possible indication of a flux in the opposite direction.

FURTHER EXPERIMENTS WITH A LABORATORY BREAKING WAVE MODEL

RAMON CIPRIANO
Atmospheric Sciences Research Center
SUNY, Albany, New York

A laboratory breaking wave model is used to elucidate the mechanisms by which bursting bubbles contribute to the production of the marine aerosol. The production of jet and film drops is strongly dependent on bubble size, and their relative importance depends critically on the shape of the whitecap bubble size distribution. Earlier experiments involving measurements of both bubble and aerosol spectra produced by the model suggested that the large end of the bubble spectrum dominates the small end of the droplet spectrum, and conversely. In these experiments, the bubble spectrum reaching the simulation tank surface was measured at different positions with respect to the whitecap center. This was not the case for droplet spectra: the simulation tank was covered by an enclosure to confine the whitecap-produced aerosol, so that aerosol measurements represented an integration, over the entire tank surface, of droplets from bubble spectra which varied greatly with position in the whitecap. The above inference that large bubbles dominate small droplet production was based partly on other laboratory experiments involving droplet production from individually-bursting bubbles at a quiescent water surface, and clearly such results must cautiously be applied to the chaotic situation at the whitecap surface. The simulation tank experiments have now been modified so that the flux of droplets ejected from the water surface can be measured precisely with respect to position from the whitecap center, as was the flux of bubbles reaching the water surface. The results of these experiments are completely in accord with the original hypothesis.

In all of the above whitecap simulations, the 'wave height' was held constant. Experiments are now in progress to determine the effect of this parameter on the shape of the bubble spectrum.

WHITECAPS, 10-m WINDSPEED AND MARINE AEROSOL INTER-RELATIONSHIPS AS OBSERVED DURING THE 1980 STREX EXPERIMENT

DAVID M. DOYLE
Department of Oceanography,
University College, Galway,
Ireland

The influence of 10m elevation windspeed (U) on oceanic whitecap coverage (W) and the concomitant production of marine aerosol particles, as measured during the 1980 STRX experiment, is evaluated.

Asuming a power law dependence for W(U) of the form $W = \alpha U^{\lambda}$ and using an ordinary least squares fitting technique a λ value of 2.21 was obtained. Thermal stability conditions were predominantly unstable. The relationship between aerosol particles with radii greater than 8μm (N8+) and W was also investigated. Assuming a power law dependence of the form $N8+ = CW^{\gamma}$ and using the same fitting technique a γ value of 0.71 was obtained. This figure compares favourably with previously obtained values of this quantity. The enhancement of the dependence of aerosol concentration upon whitecap cover with increasing droplet radius did not occur.

The various inter-relationships, as considered in light of ambient meteorological conditions, are compared with previously published data sets. Comparisons with aerosol measurements as obtained by the author at an island field station are also made.

AEROSOL MEASUREMENTS AT A REMOTE COASTAL SITE

H. J. EXTON and M. H. SMITH
Physics Dept., UMIST, Manchester M60, England,
and
R. R. ALLAN
Space Dept., RAE, Farnborough, England

Measurements of aerosol parameters have been made on the Hebridean island of South Uist off the north-western coast of Scotland during the years 1980-81. Spectra characterization is based on measurements with PMS light-scattering probes. An ASASP (0.1-1.5μm radius) and CSASP (0.25-16μm) have been operated from a 10m tower situated a few metres from the high water mark of a gently sloping sandy beach. At 2m height, next to the high water mark, an ASASP, FSSP (0.25-23.5μm) and an OAP (5-150μm) have also been operated. Meteorological sampling, consisting of wind speed and direction, and wet and dry bulb temperatures, ran continuously with the probes; additional routine observations of cloud type and height, possible pollution sources and other useful information was recorded by hand.

The spectral volume distributions indicate two aerosol components in the 0.1-16μm band: the number concentration of particles in the 0.1-0.25μm band can generally be related to air mass history, whilst the integrated 0.25-16μm concentrations are strongly windspeed dependent. If both maritime and continental conditions are included, then the small particle number concentration correlates well with the radon count, a frequently used air mass indicator. Superimposed on the continentally influenced small particle count is a windspeed related component, providing evidence of submicron, sea-produced particles. The relationship of small ($<0.25\mu$m) particle concentration and windspeed is similar to the large ($>0.25\mu$m) particle windspeed relationship, which in turn compares very favourably with the results of Woodcock at cloud-base, and those of Lovett near the sea-surface. X-ray and scanning electron microscope analysis of recent Junge-Jänicke slides has provided further evidence of the origin of the sampled particles in the windspeed-related whitecapping phenomenon.

New data of radii up to 150μm suggest a third family of sea-originating particles in the radius band 10-150μm peaking at about 50μm. Although the beach-based experiments mean that these particles are produced in large quantities by the surf zone region, in conditions of strong wind close to the sea surface, similar spectral distributions are to be expected.

BUBBLE COALESCENCE IN SEA- AND FRESHWATER: REQUISITES FOR AN EXPLANATION

HERMANN GUCINSKI
Anne Arundel Community College
Arnold, MD 21012

The persistence of visible bubbles in whitecaps and breaking seas at sea is longer than that in freshwater during similar wind and sea conditions. One factor is undoubtedly the stabilising presence of available organic matter much of which is surface active and spontaneously absorbs to bubble interfaces. Laboratory experiments where the surfactant effects were eliminated or minimized still appear to show a residual effect - there is a nonlinear but increasing reduction of bubble coalescence with increasing ionic strengths due to the addition of electrolytes. 3-1 and 2-2 electrolytes such as $AlCl_3$ and $MgSO_4$ have a greater coalescence reducing effect than 1-1 electrolytes such as NaCl or KCl.

No generally accepted theory of coalescence exists. Invoking viscosity as retarding the thinning of the liquid film separating two approaching bubbles accounts for many observations, but fails for KCl which appears to reduce viscosity in the range where coalescence reduction is also observed. Viscosity theories also rely in shared bubble surface segments not always observed during experiments. Theories involving repulsive effects from diffuse double layers, such as the DLVO theory, are attractive, but present theoretical difficulties at very close approach distances - less than about 2 nm - and at high ionic strengths. Moreover, available data on jet-drop charge from bursting bubbles and surface potential measurements as a function of ionic strength and concentration do not relate well to the observed shape of the coalescence versus concentration curve.

Nevertheless, electrostatic repulsion effects are difficult to dismiss because of their evident contribution in similar models such as micellar and emulsion stability, behaviour in lipid vesicles. Coalescence theory that invokes the diffuse double layer and consequent electrostatic repulsion must include the effect of changes in the immediate ion vicinity by the water of hydration, both in terms of ordering, i.e. entropy effects, and in terms of screening, i.e. partial charge neutralization. Treating the bubble interface as consisting of a layer of dipoles produced by charge separation due to hydration, preferential surface adsorption - or more correctly, surface rejection - due to water structuring at interfaces, remaining repulsive energies can be sufficient to contribute to the reduction of bubble coalescence with increasing ionic strengths.

THE DISTRIBUTION OF AEROSOL OVER SEA AND ICE

AUSTIN W. HOGAN
Atmospheric Sciences Research Center,
SUNY, Albany, New York.

The southern hemisphere ocean is characterized by strong westerly winds and immense waves. These winds and waves are the result of the family of cyclones along the polar front. The region poleward of the polar front has much calmer seas, but aerosol production at this relatively calm sea surface is sufficient to produce an enhanced aerosol concentration in the lower kilometer of the atmosphere, around the periphery of the Antarctic continent.

Measurement of aerosol concentration above the cloud tops (at about the 400 m b level) while flying over members of this family of cyclones at 45 to 60S shows five to tenfold increases in aerosol concentration when compared to clear air along the same flight track. Surface aerosol measurements at the south pole over 1000 km from open water, show similar enhancement in aerosol concentration when the remnants of these storms penetrate to the interior of Antarctica.

The number of aerosol particles measured at the cloud tops, and during south polar storms, was greater than the number measured over the southern hemisphere ocean by several observers. The size of the particles is in general smaller. This diminution of size may be indicative of nucleus regeneration in the cloud and precipitation process.

THE COMPLEX REFRACTIVE INDEX OF MARINE AEROSOL CONSTITUENTS

S. G. JENNINGS
Dept. of Experimental Physics
University College, Galway,
Ireland

Values of the real and imaginary index of refraction of the marine aerosol from the visible to middle infrared wavelengths are reviewed. Wide variation in the measured values of the imaginary index of refraction of marine aerosol constituents is noted. The effect of uncertainty of the complex refractive index of marine aerosol particulate on the extinction and absorption of radiation is examined. Bi-modal lognormal size distributions based on measured marine aerosol size distributions are used in the analysis. A small effect ($\leqslant 20\%$) on extinction due to uncertainty in refractive index is found at visible and near IR wavelengths. Changes in

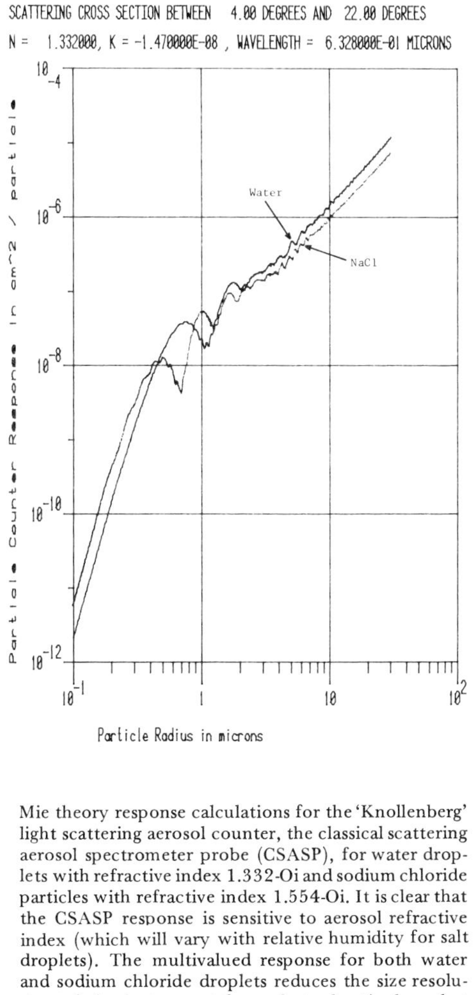

Mie theory response calculations for the 'Knollenberg' light scattering aerosol counter, the classical scattering aerosol spectrometer probe (CSASP), for water droplets with refractive index 1.332-Oi and sodium chloride particles with refractive index 1.554-Oi. It is clear that the CSASP response is sensitive to aerosol refractive index (which will vary with relative humidity for salt droplets). The multivalued response for both water and sodium chloride droplets reduces the size resolution of the instrument from that advertised so that channels must be regrouped to avoid regions of multivalued response. It can be seen that size resolution is severely degraded for droplet radius $\leqslant 2.0 \mu m$ using the CSASP.

extinction by as much as a factor of 3 and 50% are caused by variation in refractive index in the 8-12μm and 3-5μm wavelength bands. The absorption coefficient is severely affected (by over 2 orders of magnitude at visible and 3-5μm wavelength bands and up to a factor of 20 in the 8-12μm band) through uncertainty in the imaginary index of the marine aerosol.

Sizing uncertainty of commercially available 'Knollenberg' light-scattering aerosol counters: CSASP and ASASP (frequently used for marine aerosol sizing) due to uncertainty in refractive index of marine aerosol constituents, is assessed. There is virtually no size resolution for particles with radius $\geqslant 0.5 \mu m$ using the ASASP, which precludes use of its range 0. Size resolution of marine aerosol constituents NaCl and $(NH_4)_2SO_4$ is severely degraded for particle radius $\leqslant 1.5 \mu m$ using the classical scattering aerosol spectrometer probe.

WHITECAP COVERAGE MEASUREMENTS USING AN AIRBORNE MULTI-SPECTRAL SCANNER

BRYAN R. KERMAN
Boundary-Layer Research Division
STEVEN PETEHERYCH
Aerospace Meteorology Division
Atmospheric Environment Service
Ontario, Canada
HAROLD H. ZWICK
Moniteq Ltd.
Ontario, Canada

Whitecaps are associated with many significant environmental effects. The bubbling occurring with the

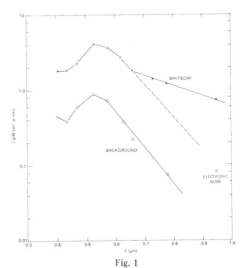

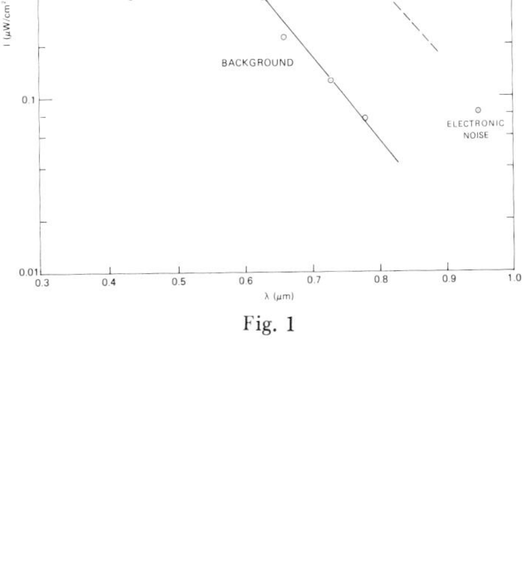

Fig. 1

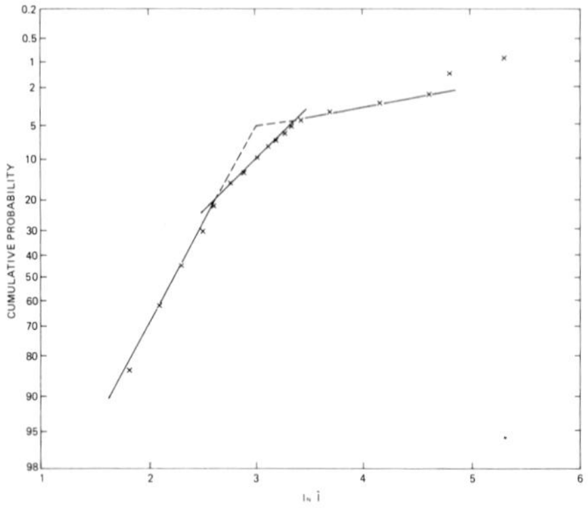

Fig. 2.

breaking wave is associated with a net exchange of gas between the atmosphere and ocean, the production of sea salt aerosols, an electrical charge exchange, chemical fractionation, a flux of trace metals, organics, bacteria and viruses. As well, bubbles near the surface are associated with the generation, scattering and propagation of sound underwater.

The common dominant effect in each of these processes is the areal extent of the breaking waves. Classically measurements of whitecap coverage have involved tedious and semi-quantitative analysis of the most distinct photographic images of breaking waves. In this study we describe a preliminary set of flights of a multispectral scanner in aircraft operated by the Canada Centre for Remote Sensing over Lake Ontario. Aspects of the spectral signature of the background and of the whitecaps will be discussed. A method for delineating background, glint and whitecaps based on the cumulative probability of the logarithm of the red-weighted radiance will be presented.

OCEANIC WHITECAPS: THEIR EFFECTIVE REFLECTANCE

PETER KOEPKE
Meteorolog. Inst. Univ. München

The aim of this article is the determination of an 'effective reflectance of whitecaps' which allows the use of the W(U) values from the literature to describe the optical influence of whitecaps in the solar spectral range.

The total foam reflectance, $R_{f,tot}$, naturally is the sum of the optical influence of all the individual whitecaps. This optical influence is given by the product of the area of each individual whitecap with its corresponding reflectance. However, these individual data in general are not available. So, usually a fixed R_f-value between 0.5 and 1.0 independent of wavelength or a wavelength dependent R_f, after Whitlock et al. (1982), is combined with W(U) to describe the optical influence of whitecaps (e.g. Maul and Gordon, 1975; Gordon and Jacobs, 1977; Koepke and Quenzel, 1981; Tassan, 1981). These rather high foam reflectance values are of the right order of magnitude for fresh, dense foam, but not for realistic foam on water surfaces which is typically comprised of patches of any age and consequently very different, mostly lower, reflectance.

The area of an individual whitecap increases with its age, while its reflectance decreases. Since in the W(U)-values determined from photos whitecaps of different age are taken into consideration, the combination of W with R_f-values valid for dense, fresh foam will give too high $R_{f,tot}$-values. Consequently, a lower 'effective reflectance' R_{ef} must be used.

$$R_{f,tot} = W \cdot R_{ef}$$

Even better is the use of an 'efficiency factor' f_{ef}, which makes possible the combination of spectral R_f-values, as measured by Whitlock et al. (1982), with the W-values depending on wind speed.

$$R_{f,tot}(\lambda) = W \cdot f_{ef} \cdot R_f(\lambda)$$

If an ocean surface is analysed as in the W(U)-determination, it can be assumed that whitecaps of any age exist with the same probability. Consequently, the effective reflectance of all the whitecaps in the area can be identified with the effective reflectance of an average whitecap, taking into account its total life time. However, in the W(U)-determination whitecaps are not taken into account if they have decayed to a reflectance less than the threshold reflectance for 'white', which occurs at an age t_{id} for the average whitecap. It follows that the integration to determine R_{ef} ends at the time t_{id}, since the values of the effective reflectance will be used in combination with W(U)-values.

$$R_{ef} = \frac{\int_0^{t_{id}} a(t) \cdot r(t) \cdot dt}{\int_0^{t_{id}} a(t) \cdot dt}$$

The normalized area of whitecaps, a(t), and their reflectance as a function of time, r(t), are determined from series of about 10 photos, taken in steps of one second.

The photos were made from the upper deck (30 m height) of the research platform *Nordsee* in the German Bight at 50°43'N and 7°10'E, between August 22 and September 21, 1978, at wind speeds between 8.5 and 15 m/s. The areas were determined with the method used in the W(U)-determination. The reflectances were based on the analysis of the film density as described by Austin and Moran (1974). As maximum reflectance a value of 55% was used since this is in agreement with the values measured by Whitlock et al. (1982) and calculated by Stabeno and Monahan (1983) for the visible spectral region for the foam on the water surface. Since this value represents the mean value for the total area of the fresh, dense foam patches, it is also in agreement with the value found by Austin and Moran (1974).

The reduction of the effective reflectance of whitecaps due to expansion of their area and thinning of the foam can be assumed to be independent of the wavelength. Thus the effective reflectance determined in the visible can be extrapolated to other spectral regions or wavelengths by use of an 'efficiency factor' f_{ef}, which is defined as the ratio of the effective reflectance to the reflectance of dense foam.

$$f_{ef} = \frac{R_{ef}}{R_f} = \frac{R_{ef}}{0.55}$$

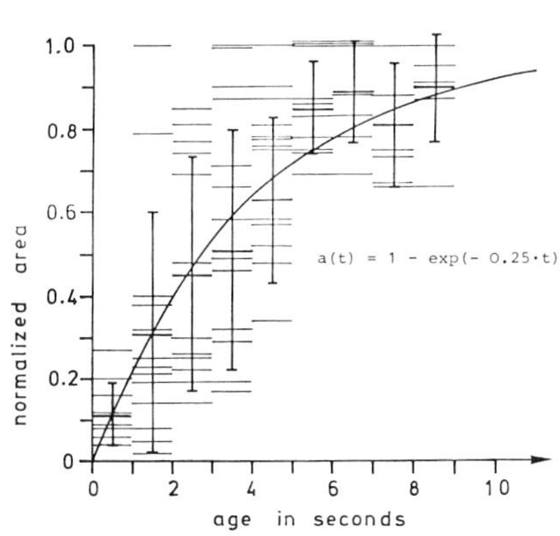

Fig. 1. Normalized area a(t) of whitecaps (foam patches) as a function of their age. Wind speed between 7.5 and 8.5 ms^{-1}, water temperature between 15 and 16°C.

Fig. 1 shows the normalized area a(t) of individual foam patches as a function of their age, together with the mean values and the standard deviations. The high standard deviations are the result of the high variability

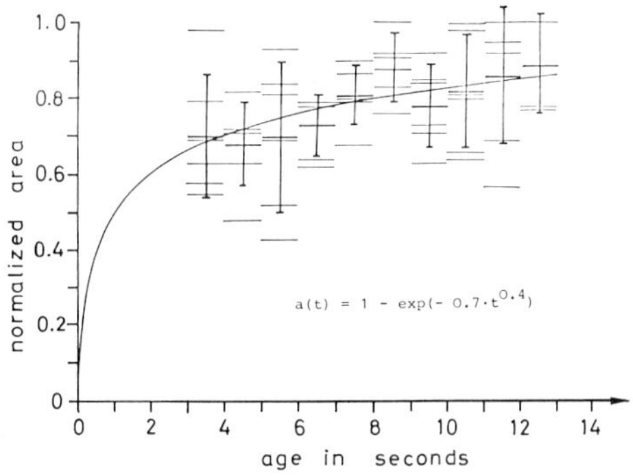

Fig. 2. Normalized area a(t) of foam streaks. Wind speed between 14 and 15 ms^{-1}, water temperature between 15 and 16°C.

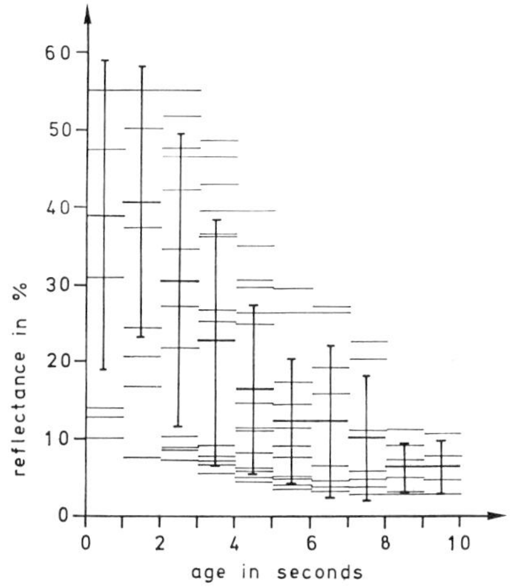

Fig. 3. Reflectance r(t) in the visible spectral range of whitecaps (foam patches) as a function of their age. Maximum reflectance of fresh dense foam R_f = 55% after Whitlock et al. (1982). Other data as in Fig. 1.

of the individual whitecaps and of the relatively low number of series analysed, but also of the subjective judgement used to determine the values. The uncertainty in the age of each foam patch at the time of each photograph is ±0.5 s. In Fig. 2 are to be seen the a(t)-values determined for foam streaks. They start with an assumed age of 3.5 seconds. The reflectance of individual whitecaps and foam streaks is shown in Fig. 3 and

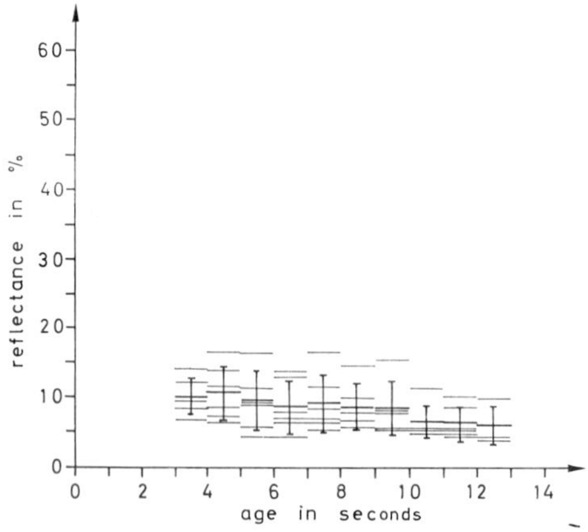

Fig. 4. Reflectance r(t) of foam streaks. Other data in Fig. 2.

Fig. 4 as a function of time, together with the mean values and the standard deviations. Again the high standard deviations are due to the variability of the reflectance of individual whitecaps, taken together with the uncertainty of the subjective judgement.

Resulting average values are:

for foam patches (whitecaps) $R_{ef,fp} = (22 \pm 8)\%$
$f_{ef,fp} = 0.4 \pm 0.15$

and for foam streaks $R_{ef,fs} = (10 \pm 4)\%$
$f_{ef,fs} = 0.18 \pm 0.07$

The values are given with high uncertainty to take into account the uncertainty of the individual data as well as the uncertainty due to the lack of knowledge of the t_{id}-value used in the determination of W(U). A more detailed description of the method and the results as a function of t_{id} and wind speed is given by Koepke (1984). Comparison of radiances and albedo values measured as a function of the amount of whitecaps (Payne, 1972; Austin and Moran, 1974; Maul and Gordon, 1975; Viollier et al., 1981) with calculated values, taking into account the effective reflectance, results in a very good agreement (Koepke, 1984).

The effective reflectance of whitecaps is more than a factor of 2 lower than reflectance values presently used in remote sensing and radiation budget studies. Consequently, the optical influence of oceanic whitecaps can be inferred to be essentially less than it has up to now been assumed to be.

Austin, R. W. and S. Moran (1974): Reflectance of whitecaps, foam and spray. In: *Ocean Color Analysis*, Univ. California, VIS-Lab., San Diego, S10 Ref. 74-10.

Gordon, H. R. and M. M. Jacobs (1977): Albedo of the ocean-atmosphere system: Influence of sea foam. *Appl. Opt.* **16**, 2257-2260.

Koepke, P. (1984): Effective reflectance of oceanic whitecaps. Submitted to *Appl. Opt.*

Koepke, P. and H. Quenzel (1981): Turbidity of the atmosphere determined from satellite: calculation of optimum wavelength. *J. Geophys. Res.* **86**, 9801-9805, and **87**, 7350.

Maul, G. A. and H. R. Gordon (1975): On the use of the earth resources technology satellite (Landsat-1) in optical oceanography. *Remote Sens. Envir.* **4**, 95-128.

Payne, R. E. (1972): Albedo of the sea surface. *J. Atm. Sci.* **29**, 959-970.

Stabeno, P. J. and E. C. Monahan (1983): The influence of whitecaps on the albedo of the sea surface. In: *Whitecaps and the marine atmosphere*, Report No. 5, University College, Galway, Ireland.

Tassan, S. (1981): The influence of wind in the remote sensing of chlorophyll in the sea. In: *Oceanography from Space* (J. Gower, ed.) Plenum Press, New York, 371-375.

Viollier, M., N. Baussart and P. Y. Deschamps (1981): Preliminary results of CZCS Nimbus-7 experiment for ocean colour remote sensing: Observation of the Ligurian Sea. In: *Oceanography from Space* (J. Gower, ed.) Plenum Press, New York, 387-393.

Whitlock, L. H., D. S. Bartlett and E. A. Gurganus (1982): Sea foam reflectance and influence on optimum wavelength for remote sensing of ocean aerosols. *Geophys. Res. Letters* **9**, 719-722.

AEROSOL POPULATIONS IN THE MARINE ATMOSPHERE

EUGENE J. MACK
Arvin/Calspan Advanced Technology Center
Buffalo, New York 14225

Since 1975 Calspan has acquired considerable data describing marine boundary-layer aerosol properties at sea and in maritime locales of North America and Europe. The data show that the marine aerosol population varies considerably in composition, both temporally and spatially, and does not necessarily comprise primarily sea salt particles. At sea, the aerosol burden is dependent on RH, winds and sea state; but a continental/anthropogenically-derived component to the marine aerosol population is generally always observed, even in remote marine areas. In coastal regions, dramatic changes in the aerosol population occur with wind shifts or airmass changes. As a result of compositional differences, response of the aerosol to fluctuations in relative humidity is expected to differ from one locale or airmass to another. Therefore, aerosol size spectra alone are not sufficient for the prediction of visibility or the potential performance of EO systems under changing humidity conditions.

This paper discusses these aerosol data in the context of meteorologic, oceanographic and geographic scenarios. More recent data, obtained on the western Mediterranean and at Atlantic coastal sites of the US, and correlations with sun photometry are emphasized.

This work was sponsored in part by the Office of Naval Research (Code 465), The Naval Environmental Prediction Research Facility, The Air Force Geophysics Laboratory, the Naval Air Systems Command (AIR 370), the Army's Atmospheric Sciences Laboratory (WSMR), and Calspan IR&D.

THE INFLUENCE OF FETCH ON WHITECAP COVERAGE AS DEDUCED FROM THE ALTE WESER LIGHT-STATION OBSERVER'S LOG

E. C. MONAHAN and C. F. MONAHAN
University College, Galway
Ireland

Some 1500 visual observations of whitecapping as logged at the Alte Weser Light-station offshore from Bremerhaven in the period from September 1970 to May 1972 were made available by Dr H. Gienapp of the Deutsches Hydrographisches Institut. Each observation included estimates of the fraction of the waves with whitecaps (F), the average width of these whitecaps (B), and the wavelength of the swell (L). The mean whitecap coverage (W) for each observation was estimated using Eq. 1, and each observation was

$$W = FBL^{-1} \quad (1)$$

assigned a fetch category: infinite (azimuth 308°-340°), 131 observations; open North Sea (256°-308°, 340°-358°), 235 observations; limited (358°-053°, 247°-256°), 250 observations; or extreme limited (053°-247°), 831 observations.

Preliminary analysis of the 'infinite fetch' data yields a whitecap-wind power-law relationship (Eq. 2) with

$$W_I(U) = \alpha_I U^{3.25} \quad (2)$$

a wind-dependence similar to those obtained from the analysis of whitecap photographs (Monahan and Ó Muircheartaigh, 1980), while the analysis of the 'extreme limited fetch' data shows a significantly lower whitecap coverage at the lower wind speeds but comparable coverage at relatively high (15 ms^{-1}) winds (Eq. 3).

$$W_E = \alpha_E U^{4.16} \quad (3)$$

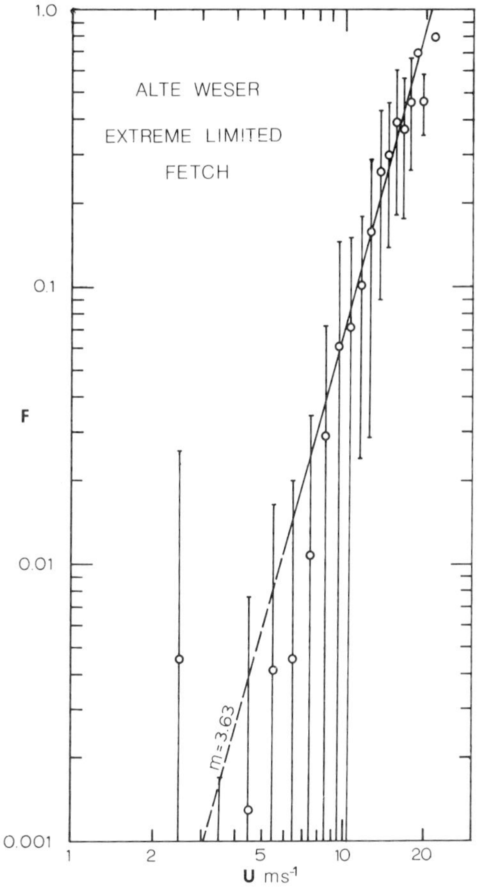

Fig. 1. The fraction of waves bearing whitecaps versus wind speed, for conditions of extreme limited fetch.

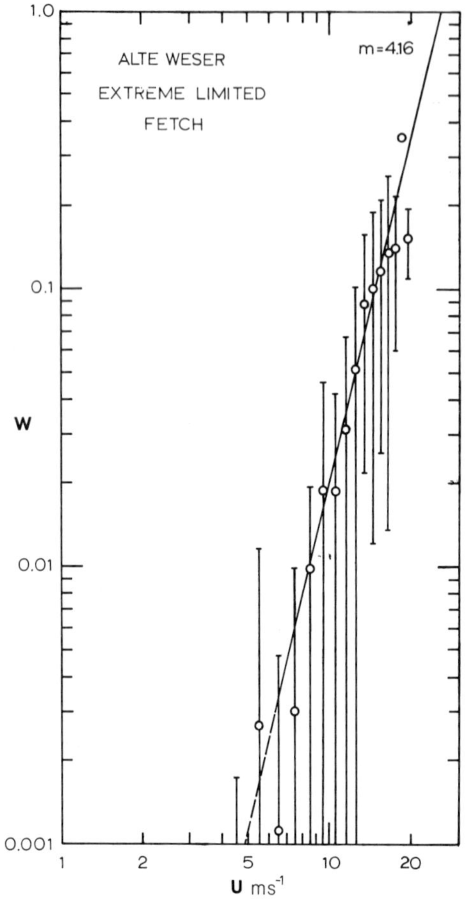

Fig. 2. The typical breadth of the whitecaps observed versus wind speed, for conditions of extreme limited fetch.

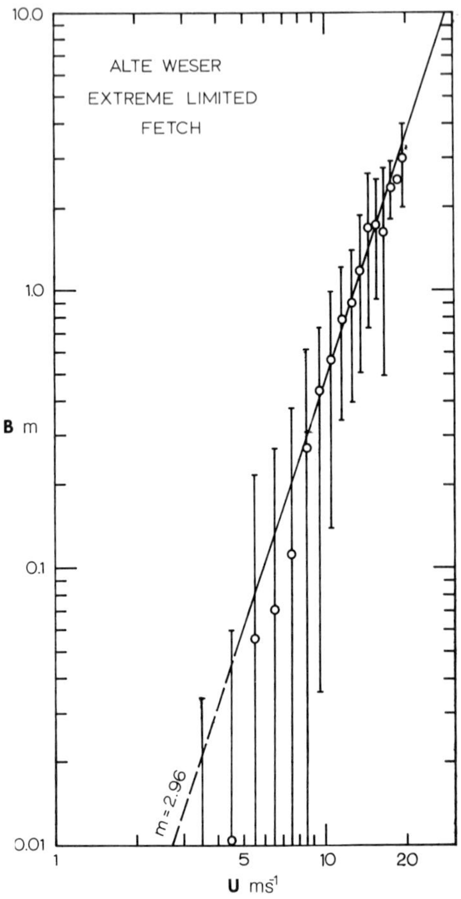

Fig. 3. Whitecap coverage versus wind speed, for extreme limited fetch conditions. Note strong dependence of whitecap coverage on wind speed: $W = \alpha U^{4.16}$

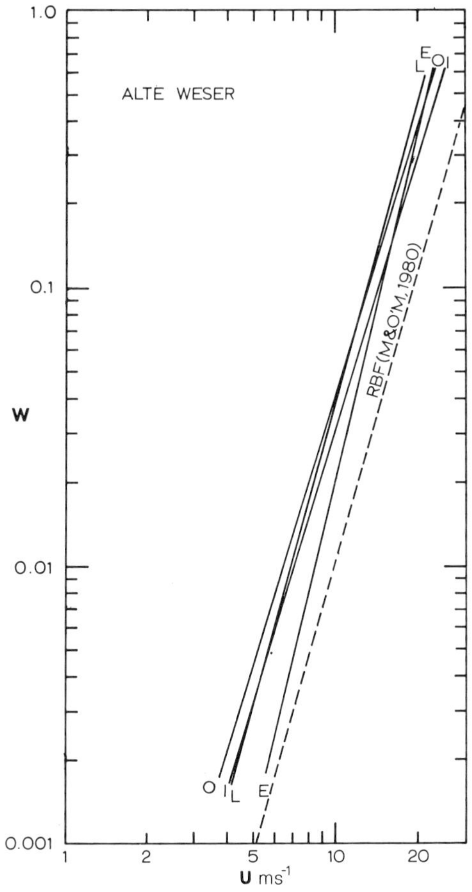

Fig. 4. Whitecap coverage versus wind speed, for various fetch conditions (I, infinite; O, open North Sea; L, limited; and E, extreme limited). Note that at relatively low wind speeds W_E is much less than other W's, but at high wind speeds, e.g. 15ms^{-1}, all W's are comparable. Note also that W_I and W_O show a similar wind dependence to that of the W_{RBF} expression of Monahan and O'Muircheartaigh (1980), but that at any wind speed the value of W_I (or W_O) is 3 to 4 times as great as the value of W_{RBF} at that wind speed. This discrepancy probably reflects the observers' tendency to exaggerate when making visual estimates of such quantities. (The W_{RBF} expression was obtained by applying the technique of robust biweight fitting to the W-values obtained from the analysis of photographic data.)

THE EFFECT OF STABILITY ON THE CONCENTRATION OF AEROSOL IN THE MARINE ATMOSPHERIC BOUNDARY LAYER

P. M. PARK and H. J. EXTON
Physics Dept., UMIST, Manchester, England

Measured concentrations of airborne aerosol produced at the sea surface might be expected to depend not only on production rates but also on mixing processes. In the boundary layer, turbulent motion will be generated by the aerodynamic friction of the underlying surface, and convective motion by conditional instability of the air, dependent on the temperature difference between the sea surface and the overlying air. This temperature difference will be seasonally and diurnally dependent, and also affected by the inflow of warmer or colder air.

Although aerosol concentrations at South Uist are strongly windspeed dependent, considerable data scatter does exist. In order to ascertain whether this scatter is caused by mixing processes, best fits to concentration/windspeed curves were used to normalize high wind speed concentrations to 6 ms^{-1}, just less than whitecapping initiation velocity.

Turbulent mixing heights were calculated from site windspeed and the calculated local roughness length. Convective mixing heights were estimated from radiosonde ascents; the ascent station being chosen according to the daily air trajectory. The technique used involved comparing those ascents with the dry and saturated adiabatic ascents of parcels of air at sea surface temperature and site mixing ratio. The higher of the two heights, turbulent or convective, was taken as the mixing height. Comparisons between calculated and observed mixing heights at a rural site suggest that this method is satisfactory.

South Uist volume spectra divide at 0.25μm radius, with larger particles locally produced, and smaller ones being mainly transported over long distances. Graphs were plotted of windspeed corrected, integrated large particle number concentration against estimated mixing height for each seasonal data set and for the data set as a whole. The results show a reasonable adherence to a relationship of the type

$$y = \frac{a}{x} + b.$$

For similar graphs using small particle numbers, far greater scatter results.

There is, therefore, evidence that mixing processes do affect the concentration of airborne maritime aerosol.

SODIUM CHLORIDE AND WATER TEMPERATURE EFFECTS ON BUBBLES

COLIN POUNDER
Department of Physics, UMIST,
Manchester, England

The dissimilar characteristics between fresh and sea water whitecaps as observed by Monahan were investigated by Scott (1975) who found that the presence of NaCl prevented the occurrence of bubble coalescence. Results are here reported of an investigation into the behaviour of pairs of bubbles. In distilled water a pair of bubbles has always been found to coalesce into a single bubble but in NaCl (aq) and in seawater this does not occur. Bubbles in saline water have been observed to push each other aside and break free to rise as two separate bubbles. It therefore seems reasonable to assert that in freshwater the number of bubbles for whitecapping is reduced by coalescence and vice versa in seawater. A hypothesis for the effect of NaCl is that some preferential ion attachment into the surface electrical double layer occurs and that therefrom a repulsion takes place at bubble surfaces.

Above: Bubbles in distilled water coalesce.
Below: The presence of NaCl prevents coalescence.

The effect of water temperature on the numbers and sizes of bubbles has been investigated, for distilled and sea waters, over a temperature range of 34°F to 85°F. It has been found that bubble size decreases with an increase in temperature and that numbers increase. There is a distinct difference in characteristics of numbers and size spectra between water types which may be attributed in part to the coalescence factor cited above.

Bubble density (Bd) and size (Bs) have been found to be related to water temperature according to the empirical relationships

$$Bd = 0.8T + 6.7$$
$$Bs = -0.7T + 95.5$$

where Bd = bubble density in number m^{-3}
Bs = bubble diameter in microns
T = water temperature in °C

I suggest that the discrepancy between Blanchard's observed bubble density and that reported by Johnson and Cooke may be accounted for by differences in water temperature.

The effect of NaCl on coalescence and of water temperature on bubble density and size will be reflected in the production of aerosol droplets since the factors of (i) Worthington jet diameter, affecting top jet droplet size (ii) jet drop height (iii) jet drop ejection speed, are all dependent on bubble size.

BUBBLE GENERATION BY SURFACE WAVE BREAKING

MING-YANG SU
ALBERT W. GREEN
Naval Ocean Research and Development
Activity, NSTL Station, Mississippi 39529

Physical processes of air entrainment and subsequent air bubble generation from breaking of steep surface gravity waves under controlled situations will be described and discussed in this paper.

It is generally known from actual field measurements that wind-wave breaking in the ocean contributes by far the most significant portion of air/gas/particle transfers across the water surface; and yet, the underlying physio-chemical processes are only poorly understood. To a large extent, this status might be attributed to the highly nonlinear complex wave breaking process which has defied full understanding despite a long history of both theoretical and experimental investigations.

Recently, a close interplay between experimental and theoretical studies has led to an important breakthrough on the physics of deep-water wave breaking as a consequence of a new type of three-dimensional wave instability and subsequent bifurcation (for details see

Su (1982), Su et al. (1982), McLean (1982), Saffman & Yuen (1980, 1982)). This breakthrough now provides, in turn, a sounder foundation for studying bubble generations resulting from wave breaking.

In this paper we shall present some experimental results conducted in a large-scale wave tank, which exhibits clearly the three-dimensional wave plunging and spilling that cause air entrainment and bubble generation/transport in the deep water. The rapid and complex physical processes involved are recorded by photographic techniques. These experimental results are then used to delineate general characteristics of the physical mechanisms involved.

Su, M. Y. (1982), J.F.M., **124**, 73-108.
Su, M. Y., et al. (1982), J.F.M., **124**, 45-72.
McLean, J. (1982), J.F.M., **114**, 315-330.
Saffman, P. G.. and H. C. Yuen (1980), J.F.M., **101**, 797-808.
Saffman, P. G. and H. C. Yuen (1982), J.F.M., **124**, 109-121.

BACTERIAL ENRICHMENTS IN THE AEROSOL FROM A LABORATORY BREAKING WAVE

LAWRENCE D. SYZDEK
Atmospheric Sciences Research Center
SUNY, Albany, New York

Enrichments of the bacteria *Serratia marinorubra* were determined for the aerosol produced from a model breaking wave. The drop-size distribution, and bacterial counts were determined from isolated zones above positions relative to the center of the bubble upwelling. Enrichments for drops from 2- to 20-μm diameter were about 100 to 20 respectively at the center of the bubble plume (CBP), and 30 to 7 respectively at 12 cm from CBP.

The flux of bacteria produced above the CBP, in the aerosol size range 2- to 6-μm diameter, was about three to five times that found at 12 cm from CBP. For drops 8- to 20-μm diameter the bacterial flux was about the same at both the CBP and 12 cm from CBP. This is consistent with the data showing a greater concentration of bubbles larger than 1mm diameter occurring at CBP. Bubbles in this size range produce predominantly film drops smaller than 20-μm diameter. At increasing distances from CBP jet drops may dominate the aerosol spectrum due to greater concentrations of bubbles smaller than 1mm diameter.

Bacterial enrichment in the aerosol from breaking waves is dependent upon the adsorption of cells to rising bubbles, the concentration in the surface microlayer, and the mechanism of drop formation. Film drops produced in the zone above the central bubble plume appear to be the dominant source for the biologically enriched aerosol produced from a breaking wave.

SUPPLEMENTARY BIBLIOGRAPHY

Abe, T., 1954. A study on the foaming of sea water (8th report). On the mechanism of decay of the foam layer of sea water (Part 3), [in Japanese], *Journal of the Oceanographical Society of Japan* **10**, No. 1, pp. 15-21.

Abe, T., T. Ono and N. Kishino, 1963. A fundamental study on the prevention of the salty damages due to the foaming of sea water (Preliminary Report), *Journal of the Oceanographical Society of Japan* **18**, pp. 185-192.

Aitken, J., 1881. On dust, fogs, and clouds, *Trans. Roy. Soc. Edinburgh* **30**, pp. 337-368.

Aliverti, G. and G. Lovera, 1950. Sui nuclei di condensazione di origine marittima, *Geofisica Pura e Applicata* **16**, pp. 133-135.

Andreas, E. L., R. M. Williams and C. A. Paulson, 1981. Observations of condensate profiles over Arctic leads with a Hot-Film anemometer. *Quarterly Journal of the Royal Meteorological Society* **107**, pp. 437-460.

Artyukhina, T. S., 1976. Surface oscillations in a jet falling from a cylindrical crest sill (on a dam). *Fluid Mechanics-Soviet Research* **5**, No. 3, pp. 67-75. (Original in *Izvestiya Vsesoyuznogo Nauchno-Issledovatel'skogo Instituta Gidrutekhniki* **104**, pp. 42-48, 1974).

Atkinson, L. P., 1973. Effect of air bubble solution on air-gas exchange, *J. Geophys. Res.* **78(6)**, pp. 962-968.

Bachhuber, C. and C. Sanford, 1974. The rise of small bubbles in water. *J. Appl. Phys.* **45**, pp. 2567-2569.

Banner, M. L. and W. K. Melville, 1976. On the separation of air flow over water waves, *Journal of Fluid Mechanics* **77**, pp. 825-842.

Bausum, H. T., S. A. Schaub, K. F. Kenyon and M. J. Small, 1982. Comparison of coliphage and bacterial aerosols at a waste-water spray irrigation site, *Appld. Environ. Microbiol.* **43**, pp. 28-38.

Baylor, E. R., M. B. Baylor, D. C. Blanchard, L. D. Syzdek and C. Appel, 1977a. Virus transfer from surf to wind, Science 198, pp. 575-580.

Baylor, E. R., V. Peters and M. B. Baylor, 1977b. Water to-air transfer of virus, *Science* **197**, pp. 763-764.

Beck, J. B., 1918. Observations on salt storms and the influence of salt and saline air upon animal and vegetable life, *Amer. J. Sci.* **1**, pp. 388-397.

Belot, Y., C. Capot and D. Gauthier, 1982. Transfer of Americium from sea water to atmosphere by bubble bursting, *Atmospheric Environment* **16**, pp. 1463-1466.

Berg, W. W., Jr., and J. W. Winchester, 1978. Aerosol chemistry of the marine atmosphere, in: *Chemical Oceanography, Vol. 7* (2nd edition), (J. P. Riley and R. Chester, eds.), pp. 173-231, Academic Press, New York.

Blanchard, D. C., 1971. The oceanic production of volatile cloud nuclei. *J. Atmos. Sci.* **28**, 811-812.

Blanchard, D. C., 1974. International symposium on the chemistry of sea/air particulate exchange processes: summary and recommendations. *Journal de Recherches Atmosphériques* **8**, 509-513.

Blanchard, D. C., 1975. Bubble scavenging and the water-to-air transfer of organic material in the sea. chapter 18 (pages 360-387) of *Applied Chemistry at Protein Interfaces*, editor Robert Baier, *Advances in Chemistry Series*, 145.

Blanchard, D. C., 1976. Nature's microtome: bursting bubbles at the surface of the sea. *Bulletin of the South Carolina Academy of Science* **38**, 38-53.

Blanchard, D. C., 1977. Comments on 'Airborne sea salt sedimentation measurements and a method of reproducing ambient sedimentation rates for the study of its effect on vegetation.' *Atmospheric Environment* **11**, 565-566.

Blanchard, D. C., 1978. Jet drop enrichment of bacteria, virus, and dissolved organic material. *Pure and Appld. Geophys.* **116**, 302-308.

Blanchard, D. C., and E. J. Hoffman, 1978. Control of jet-drop dynamics by organic material in seawater. *J. Geophys. Res.* **83**, 6187-6191.

Blanchard, D. C., and A. T. Spencer, 1957. Condensation nuclei in the vicinity of the island of Hawaii. *Tellus* **9**, 525-527.

Blanchard, D. C. and A. T. Spencer, 1964. Condensation nuclei and the crystallization of saline drops. *J.*

Atmos. Sci. 21, 182–186.
Blanchard, D. C., and A. T. Spencer, 1970. Experiments on the generation of raindrop-size distributions by drop breakup. *J. Atmos. Sci.* 27, 101–108.
Blanchard, D. C., and L. D. Syzdek, 1970. Mechanism for the water-to-air transfer and concentration of bacteria. *Science* 170, 626–628.
Blanchard, D. C., and L. D. Syzdek, 1972. Variation in Aitken and giant nuclei in marine air. *J. Phys. Oceanography*, 2, 255–262.
Blanchard, D. C., and L. D. Syzdek, 1972a. Concentration of bacteria in jet drops from bursting bubbles. *J. Geophys. Res.* 77, 5087–5099.
Blanchard, D. C., and L. D. Syzdek, 1973. Reply. *J. Phys. Oceanography* 3, 168.
Blanchard, D. C., and L. D. Syzdek, 1974. Bubble tube: apparatus for determining rate of collection of bacteria by an air bubble rising in water. *Limnology and Oceanography* 19(1), 133–138.
Blanchard, D. C., and L. D. Syzdek, 1974. Importance of bubble scavenging in the water-to-air transfer of organic material and bacteria. *Journal de Recherches Atmosphériques* 8, 529–540.
Blanchard, D. C., and L. D. Syzdek, 1975. Electrostatic collection of jet and film drops. *Limnology and Oceanography* 20, 762–774.
Blanchard, D. C., and L. D. Syzdek, 1978. Seven problems in bubble and jet drop researches. *Limnology and Oceanography* 23, 389–400.
Blanchard, D. C., and L. D. Syzdek, 1978. Reply to comment by F. MacIntyre. *Limnology and Oceanography* 23, 573.
Blanchard, D. C., and L. D. Syzdek, 1982. Water-to-air transfer and enrichment of bacteria in drops from bursting bubbles. *Applied and Environmental Microbiology* 43, 1001–1005.
Blanchard, D. C., L. D. Syzdek and M. E. Weber, 1981. Bubble scavenging of bacteria in freshwater quickly produces bacterial enrichment in airborne jet drops. *Limnology and Oceanography* 26, 961–964.
Bondur, V. G., and E. A. Sharkov, 1982. Statistical characteristics of foam formations on a disturbed sea surface [in Russian]. *Oceanology* 22, 372–379.
Bordonsky, G. S., I. B. Vasil'kova, V. M. Veselov, N. N. Vorsin, Yu. A. Militsky, V. G. Mirovsky, V. V. Nikitin, V. Yu. Raiser, Yu. B. Khapin, E. A. Sharkov, and V. S. Etkin, 1978. The spectral characteristics of microwave thermal emission from some foam formations. *Fizika Atmosfery i Okeana* 14, 656–663.
Bortkovskii, R. S., 1972. Mechanism of ocean-atmophere interaction during a storm, *Trudy No. 282, Glavnaya Geofizicheskaya Observatoriya, Leningrad*, 187–193.
Bortkovskii, R. S., 1975. Refinement of heat and moisture exchange of ocean and atmosphere during a storm, *Trudy No. 326, Glavnaya Geofizicheskaya Observatoriya, Leningrad*, 58–69 [in Russian].
Bortkovskiy, R. S., and D. F. Timanovskiy, 1982. On the microstructure of the breaking crest of wind waves 12v. *Atm. and Ocean Phys.* 18(3), 255–256 [Eng. translation].
Brierly, W. B., 1970. Bibliography on atmospheric (cyclic) sea-salts. Tech. Rpt. 70-63-ES, U.S. Army Natick Laboratories, Natick, MA, 01760.
Brocks, K., and L. Krügermeyer, 1972. The hydrodynamic roughness of the sea surface. In Gordon (ed.), *Studies in Physical Oceanography, Vol. 1*, pp. 75–92, Gordon and Breach, New York.
Broecker, W. S., and T. H. Peng, 1974. Gas exchange rates between air and sea. *Tellus* 26, 21–35.
Brutsaert, W. H., 1982. *Evaporation into the atmosphere, Theory, history and applications.* D. Reidel Pub. Co., pp 299.
Bunker, A. H., B. Haurwitz, J. S. Malkus, and H. Stommel, 1949. Vertical distribution of temperature and humidity over the Caribbean Sea. *Pap. Phys. Oceanogr. and Meteor.* 11(1), pp. 82.
Burger, S. R. and D. C. Blanchard, 1983. The persistence of air bubbles at a seawater surface. *Journal of Geophysical Research* 88, 7724–7726.
Byers, H. R., J. R. Sievers, and B. J. Tufts, 1957. Distribution in the atmosphere of certain particles capable of serving as condensation nuclei, in: *Artificial Stimulation of Rain*, pp. 47–72, Pergamon Press, New York.
Cardone, V. J., 1969. Specification of the wind distribution in the marine boundary layer for wave forecasting, Geophysical Sciences Laboratory, New York University, New York, *Technical Report 69-1*, pp. 1–131.
Carlucci, A. F., and P. M. Williams, 1965. Concentration of bacteria from sea water by bubble scavenging. *J. Cons. Perm. Int. Explor. Mer.* 30, 28–33.
Cato, D. H., 1978. Review of ambient noise in the ocean: Non-biological sources. *Bulletin of Australian Acoustical Society, Vol. 6*, 31–36.
Coantic, M., F. Ramanonjiarisca, P. Mestayer, F. Resch, and A. Favre, 1981. Wind tunnel simulation of small-scale ocean-atmosphere interactions. *Journal of Geophysical Research* 86, 6607–6626.
Cokelet, E. D., 1977. Breaking Waves. *Nature* 267, 769–774.
Crow, S. A., D. G. Ahearn, W. L. Cook, and A. W. Bourquin, 1975. Densities of bacteria and fungi in coastal surface films as determined by a membrane-absorption procedure. *Limnology and Oceanography* 20, 644–646.
Dalen, J., and A. Lovik, 1981. The influence of wind-induced bubbles on echo integration surveys. *Journal of the Acoustical Society of America* 69, 1653–1659.
Darrozès, J. S., and P. Ligneul, 1982. The production of drops by the bursting of an air bubble at an air-liquid interface, in: *Proc. 2nd Intl. Colloquium on Drops and Bubbles*, JPL Pub. 82-7, pp. 157–165, Jet Propulsion Lab., Calif. Inst. Tech., Pasadena, CA.
Davidson, K. L., and V. R. Noonkester, in press. Observations of the occurrence of encroachment within the marine atmospheric boundary layer (CEWCOM–

76). Second Conference of Coastal Meteorology, pp. 284-287.

Davidson, K. L., G. E. Schacher, C. W. Fairall, and A. K. Goroch, 1981. Verification of the bulk method for calculating overwater optical turbulence. *Applied Optics* **20**, 2919-2924.

Deacon, E. L., 1981. Sea-air gas transfer: The wind speed dependence. *Boundary Layer Meteorology* **21**, 31-37.

Dessens, H., 1946. Les noyaux de condensation de l'atmosphère. *C. R. Acad. Sci. Paris* 223, 915-917.

Dikinov, Kh. Zh., and V. D. Zholvdev, 1980. Calculations of the drag coefficient and wind stress at the ocean surface. *Oceanology* (English ed.) **20**, 550-554.

Dimmick, R. L., and A. B. Akers, 1969. *An Introduction to Experimental Aerobiology.* Wiley-Interscience, pp. 494.

Dimmick, R. L., H. Wolochow, and M. A. Chatigny, 1979. Evidence for more than one division of bacteria within airborne particles. *Appl. Environ. Microbiol.* **38**, 642-643.

Dobson, F., L. Hasse, and R. Davis (eds.), 1980: *Air-Sea Interaction, Instruments and Methods.* Plenum Press, New York. pp. 801.

Dondero, T. J., Jr., R. C. Rendtorff, G. F. Mallison, R. M. Weeks, J. S. Levy, E. W. Wong, and W. Schaffner, 1980. An outbreak of Legionnaires' Disease associated with a contaminated air-conditioning cooling tower. *N. Engl. J. Med.* **302**, 365-370.

Donelan, M. A., 1978. Whitecaps and momentum transfer, pp. 273-287 in: Favre, A., and K. Hasselmann (eds.), 1978, *Turbulent fluxes through the sea surface, wave dynamics and prediction.* NATO Conference Series: V, Air-Sea Interactions, Vol. 1, pp. 1-677.

Donelan, M. A., 1982. The dependence of the aerodynamic drag coefficient on wave parameters. Preprint volume, First International Conference on Meteorology and Air-Sea Interaction in the Coastal Zone, The Hague, The Netherlands, 10-14 May, 1982, 398-404.

Dunckel, M., L. Hasse, L. Krügermeyer, D. Schriever and J. Wucknitz, 1974. Turbulent fluxes of momentum, heat and water vapor in the atmospheric surface layer at sea during ATEX. *Boundary-Layer Meteorology* **6**, 81-106.

Dunckel, M., L. Hasse, L. Krügermeyer, D. Schriever and J. Wucknitz, 1974. Turbulent fluxes of momentum, heat and water vapor in the atmospheric surface layer at sea during ATEX. *Boundary-Layer Meteorology* **6**, 121-146.

Duce, R. A., 1982. SEAREX: a multi-institutional investigation of the sea/air exchange of pollutants and natural substances, in: *Marine Pollutant Transfer Processes* (M. Waldichuk, G. Kullenberg, and M. Orren, eds.), Elsevier Pub. Co., in press.

Duce, R. A., and E. J. Hoffman, 1976. Chemical fractionation at the air/sea interface. *Annual Rev. Earth and Planetary Sciences* **4**, 187-228.

Ellison, T. H., and J. S. Turner, 1959. Turbulent entrainment in stratified flows. *Journal of Fluid Mechanics* **6**, 432-448.

Emerson, S., 1975. Gas exchange in small Canadian shield lakes. *Limnology and Oceanography* **20**, 754-761.

Environment Canada, 1975. State of sea photographs for the Beaufort Wind Scale, Atmospheric Environment Service, Information Canada, Ottawa, pp. 1-51.

Eriksson, E., 1959. The yearly circulation of chloride and sulfur in nature; meteorological, geochemical and pedological implications. Part I. *Tellus* **11**, 375-403.

Eriksson, E., 1960. The yearly circulation of chloride and sulfur in nature; meteorological, geochemical and pedological implications. Part II. *Tellus* **12**, 63-109.

Fitzgerald, J. W., and R. E. Ruskin, 1976. A marine aerosol model for the North Atlantic. Naval Research Laboratory, Washington, D.C., Manuscript Report, pp. 1-8.

Fliermans, C. B., W. B. Cherry, L. H. Orrison, and L. Thacker, 1979. Isolation of *Legionella pneumophila* from nonepidemic-related aquatic habitats. *Appld. Environ. Microbiol.* 37, 1239-1242.

Fraizier, A., M. Masson, and J. C. Guary, 1977. Recherches préliminaires sur le rôle des aérosols dans le transport de certains radioéléments du milieu marin au milieu terrestre. *J. Rech. Atmos.* **11**, 49-60.

Frank, E. R., J. P. Lodge, Jr., and A. Goetz, 1972. Experimental sea salt profiles. *Journal of Geophysical Research* 77, No. 27, 5147-5151.

Garner, D. M., 1969. Vertical surface acceleration in a wind-generated sea. *Deutsche Hydrographische Zeitschrift* 22, 163-168.

Garrett, W. D., 1967a. Stabilization of air bubbles at the air-sea interface by surface-active material. *Deep-Sea Research* **14**, 661-672.

Garrett, W. D., 1967b. The organic chemical composition of the ocean surface. *Deep-Sea Research* **14**, 221-227.

Garrett, W. D., 1968. The influence of monomolecular surface films on the production of condensation nuclei from bubbled sea water. *J. Geophys. Res.* **73**, 5145-5150.

Garrett, W. D., 1981. Comment on 'Organic particle and aggregate formation resulting from the dissolution of bubbles in sea-water' (Johnson and Cooke). *Limnol. and Oceanogr.* **26**, 989-992.

Garretson, G. A., 1973. Bubble transport / theory with application to the upper ocean. *Journal of Fluid Mechanics* 59, 187-206.

Gathman, S. G., 1983. Optical properties of the marine aerosol as predicted by a basic version of the Navy aerosol model. Memorandum report 5157, Naval Research Laboratory, Washington, D.C., pp. 1-34.

Gaudin, A. M., N. S. Davis, and S. E. Bangs, 1962. Flotation of *Escherichia coli* with sodium chloride. *Biotech. and Bioengr.* **4**, 211-222.

Gerba, C. P., C. Wallis, and J. L. Melnick, 1975. Microbiological hazards of household toilets: droplet production and the fate of residual organisms. *Appld. Microbiology* **30**, 229-237.

GESAMP (Joint Group of Experts on the Scientific Aspects of Marine Pollution), 1980. Interchange of pollutants between the atmosphere and the oceans. GESAMP reports and studies No. 13, pp. 55, World Meteorological Organisation, Genf.

Ginsburg, A. I., A. M. Shutko, M. A. Antipychev, and A. G. Grankov, 1982. Microwave radiation from water surface as related to its temperature variations in the presence of ripples (A laboratory study), *IEEE Journal of Oceanic Engineering, OE-7*, 33-34.

Gordon, H. R., 1976. Radiative transfer: a technique for simulating the ocean in satellite remote sensing calculations. *Appl. Opt.* **15**, 1974-1979.

Goroch, A., S. Burk, and K. L. Davidson, 1979. Stability effects on aerosol size and height distributions. *Tellus* **31**, 1-6.

Gras, J. L., and G. P. Ayers, 1983. Marine aerosol at southern mid-latitudes. *Journal of Geophysical Research* **88**, 10661-10666.

Greenhow, M., 1983. Free-surface flows related to breaking waves. *J. Fluid Mech.* **134**, 259-275.

Gregory, P. H., 1961. *The Microbiology of the Atmosphere.* Interscience Publishers, Inc., New York, pp. 251.

Griesseier, H., 1952. Zur Reflexion der Strahlung an einer unbewegten Wasseroverflache. *Zeitschrift für Meteorologie* **6**, 53-57.

Grieves, R. B., and S. L. Wang, 1967. Foam separation of bacteria with a cationic surfactant. *Biotech. and Bioengr.* **9**, 187-194.

Gruft, H., J. Katz, and D. C. Blanchard, 1975. Postulated source of *Mycobacterium intracellulare* (Battey) infection. *Amer. J. Epidemiology. 102*, 311-318.

Hasse, Lutz, 1968. Zur Bestimmung der vertikalen Transporte von Impuls und fühlbarer Wärme in der wassernahen Luftschichte auf See. *Hamburger Geophys. Einzelschr.* **11**, 1-70. (English translation: 1970 Dept. Oceanogr., Oregon State Univ. Ref. No. 70-20, pp. 1-55).

Hasse, Lutz, 1971. The sea surface temperature deviation and the heat flow at the sea-air-interface. *Boundary-Layer Meteorology* **1**, 368-379.

Hasse, L., K. Brocks, M. Dunckel, and H. Gorner, 1966. Eddy flux measurement at sea. *Beiträge zur Physik der Atmosphäre* **39**, 2-4, pp. 254-257.

Hasse, L., M. Grünewald, J. Wucknitz, M. Dunckel, and D. Schriever, 1978. Profile derived turbulent fluxes in the surface layer under disturbed and undisturbed conditions during GATE. *Meteor. Forschungsergebnisse* **B 13**, 24-40.

Hasselmann, D. E., 1978. Wind-wave generation by energy and momentum flux to the forced components of a wave field. *J. Fluid. Mech.* **85**, 543-572.

Hasselmann, D. E., 1979. The high wavenumber instabilities of a Stokes wave. *J. Fluid. Mech.* **93**, 491-499.

Hasselmann, D. E., M. Dunckel, and J. A. Ewing, 1980. Directional wave spectra observed during JONSWAP 1973. *J. Phys. Oceanogr.* **10**, 1264-1280.

Hasselmann, K., 1974. On the spectral dissipation of ocean waves due to white-capping. *Boundary-Layer Meteorology* **6**, 107-127.

Hayami, S., and Y. Toba, 1958. Drop production by bursting of air bubbles on the sea surface (1) experiments at still sea water surface. *Journal of the Oceanographical Society of Japan* **14**, 145-50.

Heintzenberg, J., 1980. Particle size distribution and optical properties of arctic haze. *Tellus* **32**, 251-60.

Hejkal, T. W., P. A. LaRock, and J. W. Winchester, 1980. Water-to-air fractionation of bacteria. *Appld. Environ. Microbiology* **39**, 335-338.

Hickey, J. L. S., and P. C. Reist, 1975. Health significance of airborne micro-organisms from wastewater treatment processes. Part II: Health significance and alternatives for action. *J. Water Poll. Con. Fed.* **47**, 2758-2773.

Higgins, F. B., Jr., 1964. *Bacterial Aerosols from Bursting Bubbles.* Ph.D. thesis, Georgia Institute of Technology.

Hobbs, P. V., 1971. Simultaneous airborne measurements of cloud condensation nuclei and sodium-containing particles over the ocean. *Quart. J. Roy. Met. Soc.* **97**, 263-271.

Hofer, R., and E. G. Njoku, 1981. Regression techniques for oceanographic parameter retrieval using space-borne microwave radiometry. *IEEE Transactions on Geoscience and Remote Sensing, GE-19*, pp. 178-189.

Hofer, R., E. G. Njoku, and J. W. Waters, 1981. Microwave radiometric measurements of sea surface temperature from the Seasat satellite: first results. *Science* **212**, 1385-1387.

Hoffman, E. J., and R. A. Duce, 1976. Factors influencing the organic carbon content of marine aerosols: a laboratory study. *J. Geophys. Res.* **81**, 3667-3670.

Hoffman, E. J., and R. A. Duce, 1977. Organic carbon in marine atmospheric particulate matter: concentration and particle size distribution. *Geophys. Res. Letters* **4**, 449-452.

Hoffman, E. J., G. L. Hoffman, and R. A. Duce, 1974. Chemical fractionation of alkali and alkaline earth metals in atmospheric particulate matter over the North Atlantic. *J. de Rech. Atmos.* **8**, 675-688.

Hogan, A. W., 1975. Antarctic aerosols. *Journal of Applied Meteorology* **14**, 550-559.

Hogan, A. W., 1981. Aerosol measurements over and near the South Pacific Ocean and Ross Sea. *Journal of Applied Meteorology* **20**, 1111-1118.

Hogan, A. W., and V. Mohnen, 1979. On the global distribution of aerosols. *Science* **205**, 1373-1375.

Hogan, S. J., 1979. Some effects of surface tension on steep water waves. *Journal of Fluid Mechanics* **91**, 167-180.

Hogan, S. J., 1980. Some effects of surface tension on steep water waves. Part 2. *Journal of Fluid Mechanics* **96**, 417-445.

Hogan, S. J., 1981. Some effects of surface tension on steep water waves. Part 3. *Journal of Fluid Mechanics* **110**, 381-410.

Hogan, S. J., 1981. Relationships between integral properties of gravity-capillary interfacial waves. *Physics of Fluids* **24**, 774-775.

Hogan, S. J., 1983. Energy flux in capillary-gravity waves. *Physics of Fluids* **26**, 1206-1209.

Hogan, S. J., 1983. Subharmonic generation of deep-water capillary waves. *Physics of Fluids* **26**, 000-000.

Hollinger, J. P., 1970. Passive microwave measurements of the sea surface. *Journal of Geophysical Research* 75, No. 27, pp. 5209-5213.

Horrocks, W. H., 1907. Experiments made to determine the conditions under which 'specific' bacteria derived from sewage may be present in the air of ventilating pipes, drains, inspection chambers and sewers. *Proc. Roy. Soc. London* 79 (Ser. B), 255-266.

Hoyt, J. W., and J. J. Taylor, 1977. Waves on water jets. *Journal of Fluid Mechanics* **83**, 119-127 & 11 plates following p. 128.

Hoyt, J. W., J. J. Taylor, and C. D. Runge, 1974. The structure of jets of water and polymer solution in air. *Journal of Fluid Mechanics* **63**, 635-640 and 14 plates following p. 640.

Hsu, S. A., and T. Whelan, 1976. Transport of atmospheric sea salt in coastal zone. *Environmental Science and Technology* **10**, 281-283.

Hunter, K. A., and P. S. Liss, 1981. Organic sea surface films, in: Marine Organic Chemistry, Chapter 9, pp. 259-298 (E. K. Duursma and R. Dawson, eds.), Elsevier Scientific Pub. Co., pp. 522.

Iribarne, J. V., D. Corr, B. Y. H. Liu, and D. Y. H. Pui, 1977. On the hypothesis of particle fragmentation during evaporation. *Atmos. Environ.* **11**, 639-642.

Jacobs, W. C., 1937. Preliminary report on a study of atmospheric chlorides. *Mon. Wea. Rev.* **65**, 147-151.

Jarvis, N. L., 1972. Effect of various salts on the surface potential of the water-air interface. *Journal of Geophysical Research* 77, No. 27, p. 5177.

Jeffrey, D. J., 1982. Quasi-stationary approximations for the size distribution of aerosols. *Journal of the Atmospheric Sciences* **38**, 2440-2443.

Jennings, S. J., 1983. Extinction and liquid water content of fog at visible wavelengths. *Applied Optics* **22**, 2514-2515.

Johnson, B. D., R. M. Gershey, R. C. Cooke, and W. H. Sutcliffe, 1982. A theoretical model for bubble formation at a frit surface in a shear field. *Separation Science and Technology* **17**, 1027-1039.

Johnson, B. D., R. M. Gershey, R. C. Cooke, and W. H. Sutcliffe, 1982. A device for the production of small bubbles in seawater. *Limnology and Oceanography* **27**, 369-373.

Josberger, E. G., 1980. The effect of bubbles released from a melting ice wall on the melt-driven convection in salt water. *Journal of Physical Oceanography* **10**, 474-477.

Kanwisher, J., 1963. Effect of wind on CO_2 exchange across the sea surface. *Journal of Geophysical Research* **68**, 3921-3927.

Kawai, S., 1981. Visualization of airflow separation over wind-wave crests under moderate wind. *Boundary-Layer Meteorology* **21**, 93-104.

Kazakov, A. L., and V. N. Lykosov, 1980. Parameterization of heat and moisture exchange during storms with application to problems of atmosphere-ocean interaction. *Meteorologiya i Gidrologiya* **8**, 58-64.

Keith, C. H., and A. B. Arons, 1954. The growth of sea-salt particles by condensation of atmospheric water vapor. *Journal of Meteorology* **11**, 173-184.

Kennedy, R. M., and R. L. Snyder, 1983. On the formation of whitecaps by a threshold mechanism. Part II: Monte Carlo experiments. *Journal of Physical Oceanography* **13**, 1493-1504.

Kerman, B. R., 1983. Distribution of bubbles near the ocean surface. *Cornell Gas Transfer Symposium*, Preprint, pp. 1-10.

Kerman, B. R., 1984. A model of interfacial gas transfer for a well-roughened sea, pp. 311-320 in: *Gas Transfer at Water Surfaces* (W. Brutsaert and G. H. Jirka, eds.), Reidel Publishing Co., Dordrecht.

Kerman, B. R., D. L. Evans, D. R. Watts, and D. Halpern, 1983. Wind dependence of underwater ambient noise. *Boundary-Layer Meteorology* **26**, 105-114.

Kinsman, B., 1969. Who put the wind speeds in Admiral Beaufort's force scale? *Oceans* **2**, No. 2, pp. 18-25.

Kitaigorodskii, A. A., and M. A. Donelan, 1983. Wind-wave effects on gas transfer. Study 83-340, National Water Research Institute, Environment Canada, pp. 1-27. (Text of paper presented at Int. Symp. on Gas Transfer at Water Surfaces, Cornell U., Ithaca, N.Y.).

Kjeldsen, S. P., M. Lystad, and D. Myrhaug, 1981. Forecast of breaking waves on the Norwegian continental shelf. Norwegian Meteorological Institute and Norwegian Hydrodynamic Laboratories, Trondheim, Norway, pp. 1-136.

Kjeldsen, S. P., and D. Myrhaug, 1982. Kinematics and dynamics of breaking waves - Main Report. Report STF 60A78100, Vassdrags- og Havnelaboratoriet, Trondheim, Norway, pp. 1-218.

Kondratyev, K. Ya., R. M. Welch, S. K. Cox, V. S. Grishechkin, V. A. Ivanov, M. A. Prokofyev, V. F. Zhvalev, and O. B. Vasilyev, 1981. Determination of vertical profiles of aerosol size spectra from aircraft radiative flux measurements. 1: Retrieval of spherical size distributions. *Journal of Geophysical Research* **86**, 9783-9793.

Kopelevich, O. V., and E. M. Mezhericher, 1980. On estimating the fraction of foam coverage of sea surface from spectral values of visible radiance. *Oceanology* (Eng. Trans.) **20**, 30-34.

Krügermeyer, L., 1976. Vertical transports of momentum, sensible and latent heat from profiles at the tropical Atlantic during ATEX. *"Meteor" Forschungsergebnisse* **B 11**, 51-77.

Ledbetter, J. O., L. M. Hauck, and R. Reynolds, 1973. Health hazards from wastewater treatment practices. *Environmental Letters* **4**, 225-232.

Leetma, A., and A. F. Bunker, 1978. Updated charts of the mean annual wind stress, convergences in the Ekman Layers, and Sverdrup transports in the North Atlantic. *Journal of Marine Research* **36**, No. 2, 311-322.

Ling, S. C., and T. W. Kao, 1976. Parameterization of the moisture and heat transfer process over the ocean under whitecap sea states. *Journal of Physical Oceanography* **6**, 306-315.

Lodge, J. P., 1954. Analysis of micron-sized particles. *Analytical Chemistry* **26**, 1829-1831.

Lodge, J. P., 1955. A study of sea-salt particles over Puerto Rico. *J. Meteor.* **12**, 493-499.

Lodge, J. P., H. F. Ross, N. K. Sumida, and B. Tufts, 1956. Analysis of micron-sized particles: Determination of particle size. *Analytical Chemistry* **28**, 423-424.

Longuet-Higgins, M. S., 1973. A model of flow separation at a free surface. *J. Fluid Mech.* **57**, 129-148.

Longuet-Higgins, M. S., 1974. On the mass, momentum, energy and circulation of a solitary wave. *Proc. Roy. Soc. Lond.* **A 337**, 1-13.

Longuet-Higgins, M. S., 1979. The almost-highest wave: A simple approximation. *Journal of Fluid Mechanics* **94**, 269-273.

Longuet-Higgins, M. S., 1979. The trajectories of particles in steep, symmetric gravity waves. *Journal of Fluid Mechanics* **94**, 497-517.

Longuet-Higgins, M. S., 1980. Modulation of the amplitude of steep wind waves. *Journal of Fluid Mechanics* **99**, 705-713.

Longuet-Higgins, M. S., 1980. On the forming of sharp corners at a free surface. *Proc. Roy. Soc. Lond.* **A 371**, 453-478.

Longuet-Higgins, M. S., 1981. Trajectories of particles at the surface of steep solitary waves. *Journal of Fluid Mechanics* **110**, 239-247.

Longuet-Higgins, M. S., 1981. On the overturning of gravity waves. *Proc. R. Soc. Lond.* **A 376**, 377-400.

Longuet-Higgins, M. S., 1982. Parametric solutions for breaking waves. *Journal of Fluid Mechanics* **121**, 403-424.

Longuet-Higgins, M. S., 1983. Bubbles, breaking waves and hyperbolic jets at a free surface. *Journal of Fluid Mechanics* **127**, 103-121.

Longuet-Higgins, M. S., and E. D. Cokelet, 1976. The deformation of steep surface waves on water. I. A numerical method of computation. *Proc. R. Soc. Lond.* **A 350**, 1-26.

Longuet-Higgins, M. S., and E. D. Cokelet, 1978. The deformation of steep surface waves on water. II. Growth of normal-mode instabilities. *Proc. R. Soc. Lond.* **A 364**, 1-28.

Longuet-Higgins, M. S., and J. D. Fenton, 1974. On the mass, momentum, energy and circulation of a solitary wave. II. *Proc. R. Soc. Lond.* **A 340**, 471-493.

Longuet-Higgins, M. S., and M. J. H. Fox, 1977. Theory of the almost-highest wave: The inner solution. *Journal of Fluid Mechanics* **80**, 721-741.

Longuet-Higgins, M. S., and N. D. Smith, 1983. Measurement of breaking waves by a surface jump meter. *Journal of Geophysical Research* **88**, 9823-9831.

Longuet-Higgins, M. S. and R. W. Stewart, 1960. Changes in the form of short gravity waves on long waves and tidal currents. *Journal of Fluid Mechanics* **8**, 565-583.

Longuet-Higgins, M. S., and J. S. Turner, 1974. An 'entraining plume' model of a spilling breaker. *Journal of Fluid Mechanics* **63**, 1-20.

Lovik, A., 1980. Acoustic measurements of the gas bubble spectrum in water, in: *Cavitation and Inhomogeneities in Underwater Acoustics* (ed. Lauterborn), Springer-Verlag, Berlin, pp. 211-218.

McDonald, R. L., C. K. Unni, and R. A. Duce, 1982. Estimation of atmospheric sea salt dry deposition: Wind speed and particle size dependence. *Journal of Geophysical Research* **87**, 1246-1250.

Macha, J. M., and D. J. Norton, 1981. Boundary-layer flow at the air-water interface with spray entrainment. *Journal of Hydronautics* **15**, 55-61.

MacIntyre, F., 1968. Bubbles: a boundary-layer 'microtome' for micron-thick samples of a liquid surface. *J. Phys. Chem.* **72**, 589-592.

MacIntyre, F., 1974. Chemical fractionation and sea-surface microlayer processes, in: *The Sea (Marine Chemistry)*, Vol. 5, pp. 245-299. (E. D. Goldberg ed.), John Wiley and Sons, New York.

MacIntyre, F., 1978. Additional problems in bubble and jet drop research. *Limnol. and Oceanogr.* **23**, 571-573.

Markson, R., J. Sedlacek, and C. W. Fairall, 1981. Turbulent transport of electric charge in the marine atmospheric boundary layer. *Journal of Geophysical Research* **86**, pp. 12, 115-121.

Mason, B. J., 1971. *The Physics of Clouds.* Oxford University Press, pp. 671.

Medwin, H., 1965. Design and use of an acoustic spectrometer for the detection of particulate matter and bubbles in the sea. *Proceedings, 5e Congrès International d'Acoustique*, Liège, 7-14 September, 1965, J27, pp. 1-4.

Medwin, H., 1974. Acoustic fluctuations due to microbubbles in the near-surface ocean. *J. Acoustical Soc. Am.* **56**, 1100-1104.

Medwin, H., 1977. Acoustical determinations of bubble-size spectra. *Journal of the Acoustical Society of America* **62**, 1041-1044.

Medwin, H., J. Fitzgerald, and G. Rautmann, 1975. Acoustic miniprobing for ocean microstructure and bubbles. *Journal of Geophysical Research* **80**, 405-413.

Metnieks, A. L., 1958. The size spectrum of large and giant sea-salt nuclei under maritime conditions. *Geophysical Bulletin, No. 15*, pp. 1-45. Dublin Institute for Advanced Studies.

Monahan, E. C., 1970. Reply. *Journal of Atmospheric Sciences* 27, No. 8, 1220-1221.

Monahan, E. C., 1973. Comments on 'Variations in Aitken and giant nuclei in marine air'. *Journal of Physical Oceanography* 3, No. 1, 167-168.

Monahan, E. C., 1980. Sea state via satellite measurement of whitecap coverage, pp. 39-42, in S. M. P. McKenna Lawlor, ed., *Irish Participation in Space Science*, Proceedings of a symposium held in the Royal Irish Academy, 23 May 1980, R.I.A., occasional series.

Monahan, E. C., 1982. Comment on 'Bubble and aerosol spectra produced by a laboratory "breaking wave"', by R. J. Cipriano and D. C. Blanchard. *Journal of Geophysical Research* 87, 5865-5867.

Monahan, E. C., 1982. Whitecapping, a manifestation of air-sea interaction with implications for remote sensing, pp. 113-131, in: *Processes in Marine Remote Sensing*, J. Vernberg and F. Diemer, eds., Belle W. Baruch Library of Marine Science No. 12, University of South Carolina Press.

Monahan, E. C., 1983. Positive charge flux from the world ocean resulting from the bursting of whitecap bubbles, pp. 85-87, in: *Proceedings in Atmospheric Electricity*, L. H. Ruhnke and J. Latham, eds., Deepak Publishing, Hampton, Virginia.

Monahan, E. C., and I. Ó Muircheartaigh, 1981. Improved statement of the relationship between surface wind speed and oceanic whitecap coverage as required for the interpretation of satellite data, pp. 751-755, in: *Oceanography from Space*, J. F. R. Gower, ed., Plenum Pub.

Monahan, E. C., and I. G. Ó Muircheartaigh, 1981. The JASIN whitecap coverage observations, and the resulting inferred wind dependence of the drag coefficient. JASIN Meteorological Workshop, Wormley, 29 June -2 July, 1981. Published in 1982 in *JASIN News* 25, (I.O.S., Wormley), pp. 3-5, 9.

Monahan, E. C., and I. G. Ó Muircheartaigh, 1982. Reply. *Journal of Physical Oceanography* 12, 751-752.

Monahan, E. C., and M. C. Spillane, 1984. The role of oceanic whitecaps in air-sea gas exchange, pp. 495-503, in: *Gas Transfer at Water Surfaces*, W. Brutsaert and G. H. Jirka, eds., Reidel Publishing Co., Dordrecht.

Moore, D. J., and B. J. Mason, 1954. The concentration, size distribution and production rate of large salt nuclei over the oceans. *Quarterly Journal of the Royal Meteorological Society* 80, 583-590.

Moore, R. K., and A. K. Fung, 1979. Radar determination of winds at sea. *Proceedings of the IEEE* 67, No. 11, 1504-1521.

Morelli, J., T. Marchal, L. Girard-Reydet, B. Remy, A. Dutot, P. Perros, and P. Carlier, 1981. Résultats préliminaires d'une étude de la composition chimique de l'aérosol côtier, pp. 418-427, in: *Proceedings of the Second European Symposium on Physico-Chemical Behaviour of Atmospheric Pollutants*, Varese, Italy, B. Versino and H. Ott, eds., Reidel Publishing Co., Dordrecht.

Muir, M. S., 1977. Atmospheric electric space charge generated by the surf. *Journal of Atmospheric and Terrestrial Physics* 39, 1341-1346.

Neiburger, M., 1948. The reflection of diffuse radiation by the sea surface. *Transactions, American Geophysical Union*, 29, 647-652.

New, A. L., 1983. A class of elliptical free-surface flows. *J. Fluid. Mech.* 130, 219-239.

Nifuku, M., B. Vonnegut, and D. C. Blanchard, 1977. Charged drops produced by bursting of bubbles at the surface of organic liquids. *J. Electrostatics* 2, 279-282.

Noonkester, V. R., 1980. Offshore aerosol spectra and humidity relations near Southern California. *Proceedings, Second Conference on Coastal Meteorology*, A.M.S., Boston, Massachusetts, pp. 113-120.

Norkrans, B., and F. Sörensson, 1977. On the marine lipid surface microlayer-bacterial accumulation in model systems. *Botanica Marina* 20, 473-478.

Novarini, J. C., and D. R. Bruno, 1982. Effects of the sub-surface bubble layer on sound propagation. *Journal of the Acoustical Society of America* 72, 510-514.

O'Connor, T. C., 1966. Condensation nuclei in maritime air. *Journal de Recherches Atmosphériques* 2, 2nd year, No. 2-3, pp. 181-184.

Okuda, K., S. Kawai, M. Tokuda and Y. Toba, 1976. Detailed observation of the wind-exerted surface flow by use of flow visualization methods. *Journal of the Oceanographical Society of Japan* 32, 53-64.

Paterson, M. P., and K. T. Spillane, 1969. Surface films and the production of sea-salt aerosol. *Quart. J. Roy. Met. Soc.* 95, 526-534.

Patterson, E. M., C. S. Kiang, A. C. Delany, A. F. Wartburg, A. C. D. Leslie, and B. J. Huebert, 1980. Global measurements of aerosols in remote continental and marine regions: Concentration, size distributions and optical properties. *Journal of Geophysical Research* 85, No. C12, 7361-7376.

Peng, T. H., T. Takahashi, and W. S. Broecker, 1974. Surface radon measurements in the North Pacific Ocean Station Papa. *Journal of Geophysical Research* 79, 1771-1780.

Peregrine, D. H., 1983. Breaking waves on beaches. *Annual Review of Fluid Mechanics* 15, 149-178.

Peregrine, D. H., E. D. Cokelet, and P. McIver, 1980. The fluid mechanics of waves approaching breaking. *Amer. Soc. Civ. Eng., Proc. 17th Conf. on Coastal Eng.*, pp. 512-528.

Petrenchuk, O. P., 1980. On the budget of sea salts and sulfur in the atmosphere. *J. Geophys. Res.* 85, 7439-7444.

Phillips, O. M., 1977. *Physics of the Upper Ocean*, 2nd ed. Cambridge University Press, pp. 336.

Plass, G. N., G. W. Kattawar, and J. A. Guinn, Jr., 1975. Radiative transfer in the earth's atmosphere and ocean: Influence of ocean waves. *Applied Optics* 14,

1924-1936.

Podzimek, J., 1980. Advances in marine aerosol research. *Journal de Recherches Atmosphériques* **14**, No. 1, pp. 35-61.

Price, R. K., 1971. The breaking of water waves. *Journal of Geophysical Research* **76**, 1576-1581.

Prodi, F., G. Santachiara, and F. Oliosi, 1983. Characterization of aerosols in marine environments (Mediterranean, Red Sea, and Indian Ocean). *Journal of Geophysical Research* **88**, 10957-10968.

Prospero, J. M., 1979. Mineral and sea salt aerosol concentrations in various ocean regions. *J. Geophys. Res.* **84**, 725-730.

Quinn, J. A., R. A. Steinbrook, and J. L. Anderson, 1975. Breaking bubbles and the water-to-air transport of particulate matter. *Chem. Engr. Sci.* **30**, 1177-1184.

Raiser, V. Yu., E. A. Sharkov, and V. S. Etkin, 1976. Sea foam, physico-chemical properties, radiative and reflective characteristics. Report 306, Institut Kosmicheskich Issledovanii, Moscow. pp. 1-60.

Rayzer, V. Yu., and Ye. V. Sharkov, 1980. On the dispersed structure of sea foam. *(Izvestia) Atmospheric and Oceanic Physics* **16**, (English Edition), 548-550.

Resch, F., and F. Avellan, 1982. Size distribution of oceanic air bubbles entrained in sea water by wave-breaking, in: *Proc. 2nd Intl. Colloquium on Drops and Bubbles*, JPL Pub 82-7, pp. 182-186, Jet Propulsion Lab., Calif. Inst. Tech., Pasadena, CA.

Rizki, M. T. M., 1960. Factors influencing pigment production in a mutant strain of *Serratia marcescens. J. Bact.* **80**, 305-310.

Ross, D. B., V. J. Cardone, and J. W. Conway, Jr., 1970. Laser and microwave observations of sea-surface condition for fetch-limited 17- to 25-m/s winds, *IEEE Transactions on Geoscience Electronics, Vol. GE-8*, No. 4, pp. 326-336.

Rossknecht, G. F., W. P. Elliott, and F. L. Ramsey, 1973. The size distribution and inland penetration of sea-salt particles. *Journal of Applied Meterorology* **12**, 825-830.

Rubin, A. J., 1968. Microflotation: coagulation and foam separation of *Aerobacter aerogenes. Biotechnology and Bioengr.* **10**, 89-98.

Ruskin, R. E., R. K. Jeck, F. K. Lepple, and W. A. Von Wald, 1978. Salt aerosol survey at gas turbine inlet aboard USS *Spruance*. NRL Rpt. 3804, Naval Research Laboratory, Washington D.C. pp. 113.

Salih, A. M. A., 1979. Air bubbles in a convectively accelerated water flow. *Journal of Hydraulic Research* **17**, 315-327.

Salih, A. M. A., 1980. Entrained air in linearly accelerated water flow. *Journal of the Hydraulics Division, American Society of Civil Engineering* **106**, 1595-1605.

Savoie, D. L., and J. M. Prospero, 1977. Aerosol concentration statistics for the northern tropical Atlantic. *J. Geophys. Res.* **82**, 5954-5964.

Schacher, G. E., K. L. Davidson, C. W. Fairall, and D. E. Spiel, 1981. Calculation of optical extinction from aerosol spectral data. *Applied Optics* **20**, 3951-3957.

Schippers, IR. P., 1980. Density of air bubbles below the sea surface, theory and experiments, in: *Cavitation and Inhomogeneities in Underwater Acoustics*, ed. Lauterborn, Springer-Verlag, Berlin, pp. 205-210.

Schnack, E. J., and F. I. Isla, 1982. Sea foam as a sediment transport agent. *Marine Geology* **45**, pp. M9-M14.

Schnell, R. C., 1977. Ice nuclei in seawater, fog water and marine air off the coast of Nova Scotia: summer 1975. *J. Atmos. Sci.* **34**, 1299-1305.

Schnell, R. C., and G. Vali, 1972. Atmospheric ice nuclei from decomposing vegetation. *Nature* **236**, 163-165.

Schnell, R. C., and G. Vali, 1976. Biogenic ice nuclei. Part I: Terrestrial and marine sources. *J. Atmos. Sci.* **33**, 1554-1564.

Schooley, A. H., 1972. Whitecap suppression by cloud shadows on the Potomac River. *Journal of Marine Research* **30**, 315-316.

Scott, J. C., 1975. The preparation of water for surface-clean fluid mechanics. *Journal of Fluid Mechanics* **69**, part 2, 339-351.

Sieburth, J. McN., 1979. *Sea Microbes.* Oxford Uni Press, pp. 491.

Sieburth, J. McN., P. Willis, K. M. Johnson, C. M. Burney, D. M. Lavoie, K. R. Hinga, D. A. Caron, F. W. French III, P. W. Johnson, and P. G. Davis, 1976. Dissolved organic matter and heterotropic microneuston in the surface microlayers of the North Atlantic. *Science* **194**, 1415-1418.

Sigerson, G., 1870. Micro-atmospheric researches. *Proceedings of the Royal Irish Academy, Ser. 2, 1-Sci.*, pp. 13-30.

Smith, S. D., 1981. Comment on 'A new evaluation of the wind stress coefficient over water surfaces'. *J. Geophys. Res.* **86**, C5, p. 4307.

Snyder, R. L., and R. M. Kennedy, 1983. On the formation of whitecaps by a threshold mechanism. Part I: Basic formalism. *Journal of Physical Oceanography* **13**, 1482-1492.

Snyder, R. L., L. Smith, and R. M. Kennedy, 1983. On the formation of whitecaps by a threshold mechanism. Part III: Field experiment and comparison with theory. *Journal of Physical Oceanography* **13**, 1505-1518.

Stewart, R. W., and H. L. Grant, 1962. Determination of the rate of dissipation of turbulent energy near the sea surface in the presence of waves. *Journal of Geophysical Research* **67**, 3177-3180.

Stogryn, A., 1972. The emissivity of sea foam at microwave frequencies. *Journal of Geophysical Research* **77**, No. 9, 1658-1666.

Soulage, G., 1957. Les noyaux de congélation de l'atmosphère. *Ann. Geophys.* **13**, 103-104.

Stuhlman, O., 1932. The mechanics of effervescence. *Physics* **2**, 457-466.

Syzdek, L. D., 1982. Concentrations of *Serratia mar-*

cescens in the surface microlayer. *Limnology and Oceanography* 27, 172-177.

Tanaka, M., 1966. On the transport and distribution of giant sea-salt particles over land (I) Theoretical model. *Special Contributions, Geophysical Institute, Kyoto University, No. 6*, pp. 47-57.

Tang, C. C. H., 1974. The effect of droplets in the air-sea transition zone on the sea brightness temperature. *Journal of Physical Oceanography* 4, 579-593.

Tedesco, R., and D. C. Blanchard, 1979. Dynamics of small bubble motion and bursting in freshwater. *J. Rech. Atmos.* 13, 215-226.

Toba, Y., 1959. Drop production by bursting of air bubbles on the sea surface (II). Theoretical study on the shape of floating bubbles. *Journal of the Oceanographical Society of Japan* 15, 121-130.

Toba, Y., 1962. The air-sea coupling and the sea-salt nuclei. *Journal of the Oceanographical Society of Japan, 20th Anniversary Vol.*, 421-431.

Toba, Y., 1965. On the giant sea-salt particles in the atmosphere I. General features of the distribution. *Tellus* 17, No. 1, 131-145.

Toba, Y., 1966. On the giant sea-salt particles in the atmosphere III. An estimate of the production and distribution over the world ocean. *Tellus* **18**, No. 1, 132-145.

Toba, Y., 1966. Critical examination of the isopiestic method for the measurement of sea-salt nuclei masses. *Special Contributions, Geophysical Institute, Kyoto University, No. 6*, pp. 59-67.

Toba, Y., 1973. Macroscopic principles on the growth of wind waves. *Scientific Reports, Tohoku University, Ser. 5, Geophysics*, 22, No. 2, pp. 61-73.

Toba, Y., 1973. Local balance in the air-sea boundary processes II. Partition of wind stress to waves and current. *Journal of the Oceanographical Society of Japan* 29, 70-75.

Toba, Y., 1973. Local balance in the air-sea boundary processes III. On the spectrum of wind waves. *Journal of the Oceanographical Society of Japan* **29**, 209-220.

Toba, Y., and M. Tanaka, 1963. Study on dry fallout and its distribution of giant sea-salt nuclei in Japan. *Journal of Meteorological Society of Japan, Ser. II*, **41**, 135-144.

Toba, Y., and M. Tanaka, 1965. Dry fallout of sea-salt particles and its seasonal and diurnal variation. *Special Contributions, Geophysical Institute, Kyoto University, No. 5*, pp. 81-92.

Toba, Y., and M. Tanaka, 1967. Simple technique for the measurement of giant sea-salt particles by use of a hand-operated impactor and chloride reagent film. *Special Contributions, Geophysical Institute, Kyoto University, No. 7*, pp. 111-118.

Toba, Y., and M. Tanaka, 1968. A continuous sampler for sea-salt particles especially of a giant class and example of the analysis of data. *Journal de Recherches Atmosphériques* 3, 79-85.

Tokuda, M., and Y. Toba, 1982. Statistical characteristics of individual waves in laboratory wind waves. II. Self-consistents similarity regime. *Journal of the Oceanographical Society of Japan* **38**, 8-14.

Twomey, S., 1953. The identification of individual hygroscopic particles in the atmosphere by a phase-transition method. *J. Appl. Phys.* **24**, 1099-1102.

Twomey, S., 1977. *Atmospheric Aerosols, Developments in atmospheric science*, 7. Elsevier Scientific Pub. Co., Amsterdam, pp. 1-302.

Twomey, S., and K. N. McMaster, 1955. The production of condensation nuclei by crystallizing salt particles. *Tellus* 7, 458-461.

Unna, P. J., 1941. White horses. *Nature* **148**, 226-227.

Valenzuela, G. R., 1982. A note on the generation of capillary waves by steep gravity waves: the effect of wind growth. *Journal of Geophysical Research* **87**, 579-581.

Vali, G., M. Christensen, R. W. Fresh, E. L. Galyan, L. R. Maki, and R. C. Schnell, 1976. Biogenic ice nuclei, Part II: Bacterial sources. *J. Atmos. Sci.* **33**, 1565-1570.

Vanden-Broeck, J.-M., and J. B. Keller, 1980. A new family of capillary waves. *Journal of Fluid Mechanics* **98**, 161-169.

Vonnegut, B., 1974. Electrical potential above ocean waves. *Journal of Geophysical Research* **79**, 3480-3481.

Vorsin, N. N., A. A. Glotov, V. G. Mirovskij, V. Yu. Rajzer, I. A. Troitskij, E. A. Sharkov, and V. S. Etkin, 1982. Microwave emission from sea foam. *Issledovanie Zemli iz Kosmosa* 3, 95-102.

Wallace, G. T., Jr., and R. A. Duce, 1978. Open-ocean transport of particulate trace metals by bubbles. *Deep-Sea Research* 25, 827-835.

Wallace, G. T., Jr., and R. A. Duce, 1978. Transport of particulate organic matter by bubbles in marine waters. *Limnology and Oceanography* **23**, 1155-1167.

Wallace, G. T., Jr., G. I. Loeb, and D. F. Wilson, 1972. On the flotation of particulates in sea water by rising bubbles. *Journal of Geophysical Research* 77, 5293-5301.

Wang, C. S., and R. L. Street, 1978. Transfers across an air-water interface at high wind speeds: the effect of spray. *Journal of Geophysical Research* **83**, C6, 2959-2968.

Weber, M. E., 1981. Collision efficiencies for small particles with a spherical collector at intermediate Reynolds numbers. *J. Separation Process Technology* 2, 29-33.

Weber, M. E., D. C. Blanchard, and L. D. Syzdek, 1983. The mechanism of scavenging of water-borne bacteria by a rising bubble. *Limnol. Oceanogr.* **28**, 101-105.

Webster, W. J., and T. T. Wilheit, 1976. Special characteristics of the microwave emission from a wind-driven foam-covered sea. *Journal of Geophysical Research* 81, 3095-3099.

Wentz, F. J., 1983. A model function for ocean microwave brightness temperatures. *Journal of Geophysical Research* **88**, 1892-1908.

Wenz, G. M., 1962. Acoustic ambient noise in the ocean: spectra and sources. *Journal of the Acoustical Society of America* **34**, 1936-1956.

Wilheit, T. T., Jr., 1977. A review of applications of microwave radiometry to oceanography. *Boundary-Layer Meteorology* **13**, 277-293.

Willett, J. C., 1979. Fair weather electric charge transfer by convection in an unstable planetary boundary layer. *Journal of Geophysical Research* **84**, No. C2, 703-718.

Williams, R. P., 1973. Biosynthesis of prodigiosin, a secondary metabolic of *Serratia marcescens. Appl. Microb.* **25**, 396-402.

Wilson, J. H., 1980. Low frequency wind-generated noise produced by the impact of spray with the ocean's surface. *J. Acoust. Soc. America* **68**, 952-958.

Wilson, J. D., G. W. Thurtell, and G. E. Kidd, 1981. Numerical simulation of particle trajectories in inhomogeneous turbulence: Systems with constant turbulent velocity scale. *Boundary-Layer Meteorology* **21**, 295-313.

Wu, J., 1968. Laboratory studies of wind-wave interactions. *Journal of Fluid Mechanics* **34**, 91-112.

Wu, J., 1974. Evaporation due to spray. *Journal of Geophysical Research* **79**, No. 27, 4107-4109.

Wucknitz, J., 1976. Determination of turbulent fluxes of momentum and sensible heat from fluctuation measurements and structure of wind field above waves at the tropical Atlantic during ATEX. *"Meteor" Forschungsergeb.* **B 11**, 25-50.

Wyman, J., P. F. Scholander, G. A. Edwards, and C. Irvine, 1952. On the stability of gas bubbles in sea water. *J. Mar. Res.* **11**, 47-62.

Zheng, Q. A., V. Klemas, G. S. Hayne, and N. E. Huang, 1983. The effect of oceanic whitecaps and foams on pulse-limited radar altimeters. *Journal of Geophysical Research* **88**, 2571-2578.

INDEX